AF379988

ADVANCED ARTIFICIAL INTELLIGENCE

Third Edition

Series on Intelligence Science

Series Editor: Zhongzhi Shi *(Chinese Academy of Sciences, China)*

ADVANCED ARTIFICIAL INTELLIGENCE

Third Edition

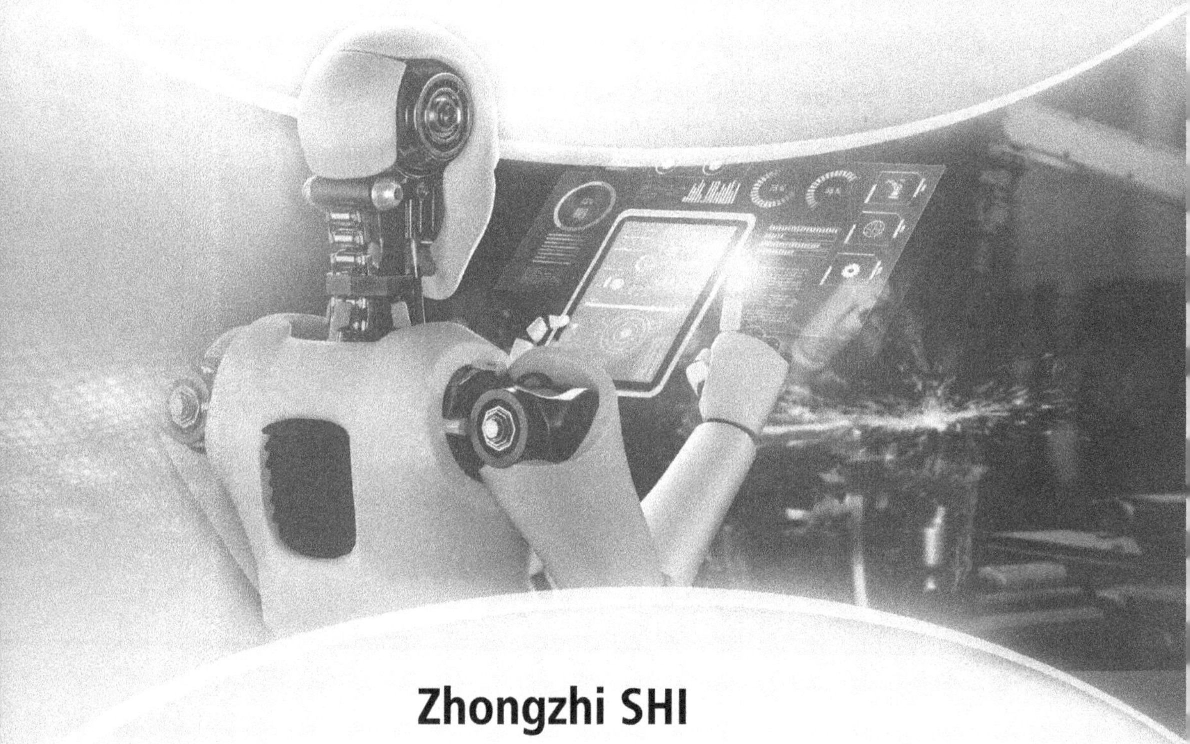

Zhongzhi SHI

Chinese Academy of Sciences, China

World Scientific

NEW JERSEY · LONDON · SINGAPORE · BEIJING · SHANGHAI · HONG KONG · TAIPEI · CHENNAI · TOKYO

Published by

World Scientific Publishing Co. Pte. Ltd.

5 Toh Tuck Link, Singapore 596224

USA office: 27 Warren Street, Suite 401-402, Hackensack, NJ 07601

UK office: 57 Shelton Street, Covent Garden, London WC2H 9HE

Library of Congress Cataloging-in-Publication Data
Names: Shi, Zhongzhi, author.
Title: Advanced Artificial Intelligence / Shi Zhongzhi, Chinese Academy of Sciences, China.
Description: Third edition. | New Jersey : World Scientific, 2025. |
    Series: Series on intelligence science ; vol. 5 | Includes bibliographical references and index.
Identifiers: LCCN 2024028349 | ISBN 9789811293986 (hardcover) |
    ISBN 9789811293993 (ebook) | ISBN 9789811294006 (ebook other)
Subjects: LCSH: Artificial intelligence.
Classification: LCC Q335 .S4668 2025 | DDC 006.3--dc23/eng/20240702
LC record available at https://lccn.loc.gov/2024028349

British Library Cataloguing-in-Publication Data
A catalogue record for this book is available from the British Library.

For any available supplementary material, please visit
https://www.worldscientific.com/worldscibooks/10.1142/13861#t=suppl

Desk Editors: Balasubramanian Shanmugam/Julio Hong

Typeset by Stallion Press
Email: enquiries@stallionpress.com

Preface

Artificial intelligence's long-term goal is to build the human level of artificial intelligence (AI). AI was born 60 years ago and, on the bumpy road, has made encouraging progress, in particular, machine learning, data mining, computer vision, expert systems, natural language processing, planning, robotics, and related applications have brought good economic benefits and social benefits. Widespread use of the Internet is also exploring the application of knowledge representation and reasoning, to build the semantic Web and improve the effectiveness of the rate of Internet information. The inevitable trend of information technology is intelligent. The intelligence revolution with the goal of replacing work performed by human brain work with machine intelligence will open up the history of human post-civilization. If the steam engine created the industrial society, then the intelligent machine must also be able to magically create an intelligent society to realize the social production automation and intelligence which promote the great development of knowledge-intensive economy.

Artificial intelligence is a branch of computer science and is a discipline to study machine intelligence that uses artificial methods and techniques, develops intelligent machines or intelligent systems to emulate, extend, and expand human intelligence, and realizes intelligent behavior. Artificial intelligence in general can be divided into symbolic intelligence and computational intelligence. Symbolic intelligence is the traditional symbolic artificial intelligence; it is the basis of physical symbol system to study the knowledge representation, acquisition and reasoning process. Use of knowledge to solve problems is basic, the most important feature of the current symbol intelligence, so people often put the current stage of artificial intelligence known as knowledge engineering. Knowledge engineering research emphasizes on knowledge information processing methods and technologies, and promotes the development of artificial intelligence.

Computational intelligence includes neural computation, fuzzy systems, genetic algorithms, evolutionary planning, and so on. To achieve intelligence revolution, we must better understand the human brain. Completely revealing the mysteries of the human brain is one of the biggest challenges facing the natural sciences. In the early 21st century, the U.S. National Science Foundation (NSF) and the U.S. Department of Commerce (DOC) jointly funded an ambitious program: Convergent Technology for Improving Human Performance, which viewed the nano-technology, biotechnology, information technology, and cognitive science as the four cutting-edge technology of the 21st century. Cognitive science is a top priority area for development, advocating the development of these four technology integrations, and described the prospects for such a science: cognitive science as the guide of convergent technology because once we are able to on how, why, where, and when to understand the four levels of thinking, we can use nanotechnology to make it with biotechnology and biomedicine to achieve it, and finally the use of information technology to manipulate and control it, make it work. This is a tremendous impact on human society.

On the surface, symbolic intelligence and neural computing are completely different research methods, the former based on knowledge, the latter based on the data; the former uses reasoning, and the latter mapping. In 1996, the invited presentation on "Computers, Emotions and Common Sense" was given by Minsky at the fourth Pacific Region International Conference on Artificial Intelligence in the view that neural computation and symbolic computation can be combined and neural network is the foundation of symbolic system. Hybrid system is committed to the people; it is from this combination, which is consistent with our proposed hierarchical model of the human mind.

New to the Second Edition

This edition captures the progresses in AI that been have made since the last edition in 2011. There has been a great deal of theoretical progress, particularly in areas such as probabilistic graphical inferences, machine learning, and computer vision. There have been algorithmic landmarks, such as deep learning and unsupervised learning. There have been important applications of AI technology, such as practical speech recognition, machine translation, autonomous vehicles, and intelligent robotics. The major changes are as follows:

- Since 2006, deep learning has made great progress and plays an important role in machine learning. Deep learning establishes and simulates the human brain's hierarchical structure to extract the external input data's features from lower to higher, which can explain the external data. In 2016, AlphaGo unquestionably defeated Li Shishi with a score of 3 to 1, thus opening up a wave of artificial intelligence.

- The following chapters from the first edition will be removed:

  Chapter 04: Qualitative Reasoning.
  Chapter 09: Explanation-Based Learning.
  Chapter 11: Rough Set.
  Chapter 15: Artificial Life.

- New material will be added with the following new chapters:

  Chapter 05: Probabilistic Graphical Models.
  Chapter 09: Deep Learning.
  Chapter 11: Unsupervised Learning.
  Chapter 15: Internet Intelligence.

- We estimate that about 30% of the material is quite new. The remaining 70% have been revised.

This edition book is separated into 12 chapters. Chapter 1 is the introduction, which starts with the cognitive questions of artificial intelligence, describing the guiding ideology of writing this book, and an overview of the hot topics for current artificial intelligence research. Chapter 2 discusses logics for artificial intelligence with a more systematic discussion of non-monotonic logic and agent-related logic systems. Chapter 3 provides constraint reasoning and introduces a number of practical constraint reasoning. Chapter 4 describes the graphic model of Bayesian networks for the connection probabilities among variables. Chapter 5 illustrates probabilistic graphical models. Over the years, the author and his colleagues have been engaged in case-based reasoning research, and its main results constitute Chapter 6. Machine learning is the core of the current artificial intelligence research, but also knowledge discovery, data mining, and learnability are important bases for this book with Chapter 7 for discussion. reflecting the latest progress of the study. Chapter 8 discusses support vector machines. Chapter 9 is related to deep learning reflecting the latest progress of the study. Chapter 10 presents reinforcement learning. Chapter 11 describes unsupervised learning for big data. Chapter 12 focuses on association rules. The evolutionary computation is discussed in Chapter 13, focusing on the genetic algorithm. In recent years, significant progress in the distribution of intelligence, combined with the results of our study, Chapter 14 presents the main theories and key technologies for multi-agent systems. The final chapter addresses Internet intelligence and focuses on the Semantic Web, and crowd intelligence.

The author set up the Advanced Artificial Intelligence course for Ph.D. and master students at the University of Chinese Academy of Sciences in 1994. Based on the lecture notes, Science Press published the first, second, and third editions in 1998, 2006, and 2011. This book is defined as a key textbook for regular higher education and is widely used in China. In 2011, World Scientific has published the English edition.

This book can be regarded as a textbook for senior students or graduate students in the information field and related tertiary specialties. It is also suitable as a reference book for relevant scientific and technical personnel.

New to the Third Edition

On July 8, 2017, the Ministry of Science and Technology of China issued a notice on the Development Plan for the New Generation of Artificial Intelligence. The notice points out that artificial intelligence has become a new focus of international competition and a new engine of economic development. It will profoundly change human production, lifestyle, and thinking patterns, achieving an overall leap in social productivity. We must firmly grasp the significant historical opportunities of the development of artificial intelligence and intelligence science, lead the new trend of global artificial intelligence and intelligence science development, and drive the overall leap and leapfrog development of national competitiveness.

On March 27, 2023, the Ministry of Science and Technology of China, together with the Natural Science Foundation of China, launched the special deployment of "AI for Science," closely combining key issues in basic disciplines, such as mathematics, physics, chemistry, and astronomy, focusing on the research needs of key fields, such as drug development, gene research, biological breeding, and new material development, and laying out the cutting-edge technology research and development system of AI-driven scientific research.

The above two measures have greatly promoted research on artificial intelligence. The third edition of advanced artificial intelligence will reflect the research achievements and progress made, and look forward to cutting-edge research topics. We estimate that 50% of the content in this version has been updated and 30% has been modified.

The third edition is separated into 12 chapters. Chapter 1 is the introduction; over the past more than 60 years, the development of artificial intelligence has gone through the formation period, symbolic intelligence, and data intelligence periods. Chapter 2 discusses logic for artificial intelligence with a more systematic discussion of non-monotonic logic and agent-related logic systems. Chapter 3 expounds on reasoning; it is a mental activity that aims to arrive at a conclusion in a rigorous way. Causal reasoning is generally considered a form of inductive reasoning. Chapter 4 describes game theory which is the study of models of strategic interactions among rational agents. Machine learning is the core of the current artificial intelligence research, but also knowledge discovery, data mining, and learnability are important bases for this book with Chapter 5 for statistical learning. Chapter 6 discusses deep learning. Chapter 7 presents reinforcement learning. Chapter 8 describes transfer learning. Chapter 9 focuses on federated learning. Artificial General Intelligence (AGI) could learn to accomplish any intellectual task that human beings or animals can perform; it is a hot topic. Chapter 10 summarizes the most possible approaches. Chapter 11 discusses crowd computing, which is a new collaborative computing paradigm for heterogeneous agents in human–machine–physical integration environments. Chapter 12 discusses AI for science and promotes scientific progress and technical development.

About the Author

Zhongzhi Shi is a Professor at the Institute of Computing Technology, Chinese Academy of Sciences, and an Academician of the International Academy of Information Research Sciences (IAIS), a Fellow of CCF, CAAI, an IEEE senior member, and AAAI and ACM member. His research interests mainly include intelligence science, artificial intelligence, cognitive science, multi-agent systems, machine learning, and neural computing. He has been responsible for 973,863 key projects of NSFC. He has been awarded various honors, such as the National Science and Technology Progress Award (2012), the Beijing Municipal Science and Technology Award (2006), the Achievement Award of Wu Wenjun Artificial Intelligence Science and Technology by CAAI (2013), and the Achievement Award of Multi-Agent Systems by China Multi-Agent Systems Technical Group of AIPR, CCF (2016). He has published more than 20 books, including *Mind Computation, Intelligent Science* (in Chinese), *Advanced Artificial Intelligence, Principles of Machine Learning*, and *Neural Networks*, and more than 500 academic papers. He presently serves as Editor-in-Chief of *International Journal of Intelligence Science*. He served as the Secretary-General of China Computer Federation, the Vice Chair of China Association of Artificial Intelligence, and the Chair of the Machine Learning and Data Mining Group WG12.2, IFIP TC12.

Acknowledgments

I would like to take this opportunity to thank my family, particularly my wife, Zhihua Yu, and my children, Jing Shi and Jun Shi, for their support in the course of this book. I would also like to thank my organization, the Institute of Computing Technology, Chinese Academy of Sciences, for providing good conditions to do research on artificial intelligence.

In Intelligence Science Laboratory, there are 12 post-doctors, 64 Ph.D. candidates, and more than 130 master students who have made contributions to this book. Thanks for your valuable work. In particular, Rui Huang, Liang Chang, Wenjia Niu, Dapeng Zhang, Limin Chen, Zhiwei Shi, Zhixin Li, Qiuge Liu, Huifang Ma, Fen Lin, Zheng Zheng, Hui Peng, Zuqiang Meng, Jiwen Luo, Xi Liu, and Zhihua Cui have given assistance in English version of this book.

This book collects research efforts supported by National Basic Research Priorities Programme (No. 2013CB329502, No. 2007CB311004, No. 2003CB317004), National Science Foundation of China (No. 61035003, No. 60775035, 60933004, 60970088), 863 National High-Tech Program (No. 2007AA01Z132), National Science and Technology Support Plan (No. 2006BAC08B06), Beijing Natural Science Foundation, the Knowledge Innovation Program of Chinese Academy of Sciences, and other funding. I am very grateful for their financial support.

My special thanks to Science Press for publishing this book in Chinese versions 1, 2, and 3 in 1998, 2006, and 2011, respectively. I am most grateful to the editorial staff and artist from World Scientific who provided all the support and help in the course of publishing this book.

Contents

5. Statistical Learning   137

6. Deep Learning   157

10. Artificial General Intelligence   253

11. Crowd Computing   277

Chapter 1

Introduction

Artificial intelligence emphasizes the creation of intelligent machines that work and react like humans. It is usually defined as the science and engineering of emulating, extending, and augmenting human intelligence through artificial means and techniques to make intelligent machines.

1.1 Brief History of AI

Artificial Intelligence (AI) is usually defined as the science and engineering of imitating, extending, and augmenting human intelligence through artificial means and techniques to make intelligent machines. In 2005, John McCarthy pointed out that the long-term goal of AI is human-level AI (McCarthy, 2005).

Humanity's dream and pursuit of intelligent machines can be traced back to more than 3,000 years ago. As early as the Western Zhou Dynasty in China (1066 BC–771 BC), there have been stories about the skillful craftsman Yanshi who presented it to the geisha of King Mu of Zhou. The compass wheel invented by Zhang Heng in the Eastern Han Dynasty (AD 25–220) is the earliest prototype of a robot in the world.

Aristotle (384 BC–322 BC) of ancient Greece laid the foundation for formal logic in his *Treatise on Instruments*. The logical algebra system founded by Boole uses symbolic language to describe the basic rules of reasoning in thinking activities and is called "Boolean algebra" by later generations. These theoretical foundations play an important role in the creation of artificial intelligence.

In 1936, Turing founded the automaton theory of the ideal computer model, proposed the use of discrete recursive functions as the mathematical basis for intelligent description, and gave a standard for testing whether a machine is intelligent based on behaviorism, the Turing test.

In 1943, psychologist McCulloch WS and mathematical logician Pitts W published a mathematical model of neural networks in the *Bulletin of Mathematical Biophysics* (McCulloch & Pitts, 1943). This model is now generally called the M-P neural network model. They summarized some basic physiological characteristics of neurons, proposed formal mathematical descriptions of neurons and structural methods of networks, and thus created the era of neural computing.

In 1956, a two-month academic seminar was held at Dartmouth University in the United States, where the term "artificial intelligence" was proposed, marking the official birth of this discipline.

Over the past more than 60 years, the development of artificial intelligence has gone through the formation period, symbolic intelligence, and data intelligence periods.

1. The formation period of artificial intelligence (1956–1976)

The formative period of artificial intelligence lasted from approximately 1956 to 1976. Major contributions during this period include the following:

(1) In 1956, Newell and Simon's "Logic Theorist" program simulated people's thinking rules when using mathematical logic to prove theorems.

(2) In 1956, Chomsky proposed the theory of formal language. This theory has a profound impact on computer science, especially on the design of programming languages, compilation methods, and computational complexity.

(3) In 1958, McCarthy proposed the table processing language LISP, which could not only process data but also conveniently process symbols, becoming an important milestone in artificial intelligence programming language. At present, LISP language is still an important programming language and development tool for artificial intelligence systems.

(4) Robinson proposed the reduction method in 1965, which was considered a major breakthrough and brought another climax to the research on theorem proof.

(5) In 1968, Feigenbaum of Stanford University and others successfully developed a chemical analysis expert system DENDRAL. From 1972 to 1976, Feigenbaum successfully developed the medical expert system MYCIN.

2. Symbolic intelligence period (1976–2006)

(1) In 1975, Simon and Newell won the Turing Award, the highest award in computer science. In 1976, he proposed the "Physical Symbol System Hypothesis" in his award-winning speech, becoming the founder and representative of the most influential school of symbolism in artificial intelligence.

(2) In 1977, Feigenbaum, a computer scientist at Stanford University in the United States, proposed the new concept of knowledge engineering at the Fifth International Joint Conference on Artificial Intelligence. In the 1980s, the development of expert systems tended to be commercialized, creating huge economic benefits. Knowledge engineering is a discipline that takes knowledge as its research object. It enables the research of artificial intelligence to shift from theory to application, from reasoning-based models to knowledge-based models, and makes the research of artificial intelligence practical.

(3) In 1981, Japan announced its plan to develop a fifth-generation electronic computer. The main features of the computers it develops are intelligent interfaces, knowledge base management, automatic problem-solving capabilities, and other aspects of human intelligent behavior (Shi, 1988).

(4) In 1984, Leslie Valiant proposed the learnable theory after combining his vision and cognitive theory in the fields of computing science and mathematics with other technologies, creating a new era of machine learning and communication (Valiant, 1984).

(5) In 1993, Shoham Y proposed intelligent agent-oriented programming (Shi, 2000).

(6) In 1995, Russell S and Norvig P published the book *Artificial Intelligence* and proposed "defining artificial intelligence as the study of intelligent agents that receive sensory information from the environment and perform actions." (Russell & Norvig, 1995).

(7) In 2011, Judea Pearl won the Turing Award for his fundamental contributions to the field of artificial intelligence. He proposed probabilistic and causal reasoning algorithms, which completely changed the rules and logic of artificial intelligence.

3. Data Intelligence Period (2006–present)

(1) In 2006, Geoffrey Hinton and others published the deep belief network, ushering in a new stage of deep learning.

(2) In 2016, AlphaGo used deep reinforcement learning to defeat Lee Sedol, one of the strongest human Go players, promoting the development and popularization of artificial intelligence.

(3) In 2016, Cambricon launched Cambricon-X, the world's first sparse deep learning processor, which far surpassed CPU and GPU in the two most important indicators of speed (response time) and power consumption.

(4) In 2018, the Turing Award was awarded to Geoffrey Hinton, Yann LeCun, and Yoshua Bengio, who pioneered deep neural networks and contributed to deep learning. They laid the foundation for the development and application of algorithms.

(5) On November 30, 2022, OpenAI in the United States released the chatbot program ChatGPT. On March 16, 2023, Baidu Wenxinyiyan was officially launched. The big language model attracts corporate and institutional customers to use APIs and infrastructure to jointly build AI models and develop applications to achieve inclusive industrial AI.

On July 8, 2017, the State Council of China officially released the "New Generation Artificial Intelligence Development Plan" to deploy relevant research plans. Information technology will continue to develop in the main directions of high performance, low cost, ubiquitous computing, and intelligence. The search for new computing and processing methods and physical implementation is a major challenge facing the future information technology field. The plan points out that artificial intelligence has become a new focus of international competition. Artificial intelligence is a strategic technology that leads the future. Major developed countries in the world regard the development of artificial intelligence as a major strategy to enhance national competitiveness and maintain national security.

1.2 Basic Content of Artificial Intelligence Research

Artificial intelligence is an emerging edge subject and an interdisciplinary subject of natural science and social science. It draws on the latest achievements of natural science and social science, takes intelligence as the core, and forms a new system with its own research characteristics. Research on artificial intelligence covers a wide range of fields, including knowledge representation, search technology, machine learning, and various methods for solving data and knowledge uncertainty problems. The application fields of artificial intelligence include expert systems, games, theorem proving, natural language understanding, image understanding, and robots. Artificial intelligence is also a comprehensive subject. It was born on the basis of cybernetics, information theory, and system theory. It involves philosophy, psychology, cognitive science, computer science, mathematics, and various engineering methods. These disciplines provide rich knowledge and research methods for artificial intelligence research. Figure 1.1 gives a simple diagram of the disciplines related to artificial intelligence and the research and application fields of artificial intelligence.

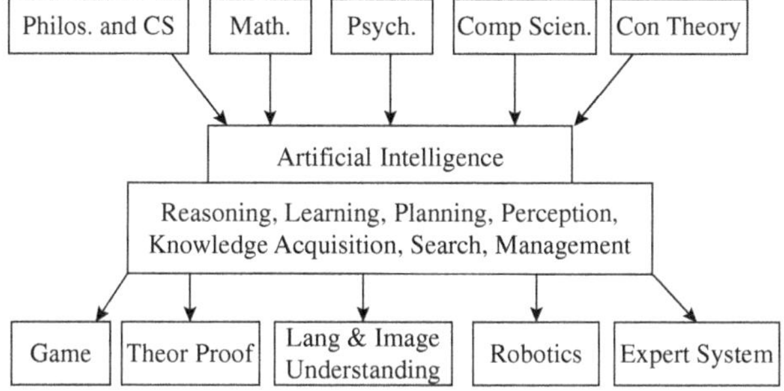

Fig. 1.1. Artificial intelligence research and application.

1.2.1 *Cognitive Modeling*

American psychologist Houston P T and others summarized cognition into the following five types:

(1) Cognition is the processing of information.
(2) Cognition is a mental symbolic operation.
(3) Cognition is problem-solving.
(4) Cognition is thinking.
(5) Cognition is a set of related activities, such as perception, memory, thinking, judgment, reasoning, problem-solving, learning, imagination, concept formation, and language use.

The human cognitive process is very complex. The technology of establishing cognitive models is often called cognitive modeling. The purpose is to explore and study the human thinking mechanism from certain aspects, especially the human information processing mechanism, and also to design corresponding artificial intelligence systems to provide new architectures and technical methods. When cognitive science uses computers to study human information processing mechanisms, it shows that between the input and output of the computer, there is a real information processing process consisting of input classification, symbolic operations, content storage and retrieval, pattern recognition, etc. Although there are substantial differences between the information processing process of computers and the information processing process of humans, we can be inspired by this and realize that humans must also have a corresponding information processing process between stimulation and response. This real process can only be attributed to the process of consciousness. The similarity in symbol processing between computer information processing and human information processing is the origin of the name of artificial intelligence and the basis for its implementation and development. Information processing is also the link between cognitive science and artificial intelligence.

1.2.2 *Knowledge Representation*

The process of human intelligent activities is mainly a process of acquiring and applying knowledge. Knowledge is the basis of intelligence. Through practice, people realize the regularity of the objective world and form knowledge through processing, explanation, selection, and transformation. In order for a computer to be intelligent and enable it to simulate human intelligent behavior, it must have knowledge expressed in an appropriate form. Knowledge representation is a very important research field in artificial intelligence.

The so-called knowledge representation is actually a description of knowledge, or a set of conventions, a data structure acceptable to computers for describing knowledge. Knowledge representation is a feasible, effective, and universal principle and method for studying machine representation of knowledge. The problem of knowledge representation has always been one of the most active parts of artificial intelligence research. Currently, commonly used knowledge representation methods include logical patterns, production systems, frameworks, semantic networks, state spaces, object-oriented, and connectionism.

1.2.3 *Automatic Reasoning*

The form of thinking that logically deduces a new judgment (conclusion) from one or several known judgments (premises) is called reasoning, which is the reflection of the objective connection of things in consciousness. Automatic reasoning is the process of using knowledge. When people solve problems, they use past knowledge and draw conclusions through reasoning. Automatic reasoning is one of the core issues in artificial intelligence research.

According to the way in which new judgments are derived, automatic reasoning can be divided into deductive reasoning, inductive reasoning, and abductive reasoning. Deductive reasoning is a process of reasoning from the general to the specific. Deductive reasoning is an important reasoning method in artificial intelligence. Most of the intelligent systems successfully developed so far are implemented using deductive reasoning.

In contrast to deductive reasoning, inductive reasoning is a process of reasoning from the particular to the general. Inductive reasoning is an important foundation for machine learning and knowledge discovery. It is the most basic and commonly used form of reasoning in human thinking activities.

As the name suggests, abductive reasoning involves inferring causes from conclusions. In abductive reasoning, we are given the rule $p \rightarrow q$ and a reasonable belief in q. Then we want to get the predicate p to be true under some interpretation. Abductive reasoning is unreliable, but because of q, it is also called best explanation reasoning.

According to whether the conclusions derived during the reasoning process increase monotonically, reasoning is divided into monotonic reasoning and non-monotonic reasoning. Its monotonic meaning means that the number of propositions known to be true increases strictly as reasoning proceeds. In monotonic logic, new propositions can be added to the system and new definitions can be proved, and such additions and proofs will never cause previously known propositions or proven propositions to become invalid. In essence, human thinking and reasoning activities are not monotonous. People's understanding, beliefs, and opinions about things in the world around them are always in a state of constant adjustment. For example, a conclusion is derived based on certain premises, but when people obtain other facts, the conclusion is canceled. In this case, the conclusion does not increase as the conditions increase, and this reasoning process is non-monotonic reasoning. Non-monotonic reasoning is one of the results of artificial intelligence automatic reasoning research. In 1978, Reiter R first proposed the non-monotonic reasoning method Closed World Assumption (CWA) (Reiter, 1978) and proposed default reasoning (Reiter, 1980). In 1979, Doyle established the truth maintenance system (TMS) (Doyle, 1979). In 1980, McCarthy proposed circumscription logic (McCarthy, 1980).

There are a lot of uncertainties in the real world. Uncertainty comes from the difference between human subjective understanding and objective reality. The randomness of things happening, the incompleteness, unreliability, imprecision, and inconsistency of human knowledge, and the fuzziness and ambiguity existing in natural language all reflect this difference and will bring uncertainty. Different theories and reasoning methods have been proposed for different causes of uncertainty. In artificial intelligence, representative uncertainty theories and reasoning methods include Bayes theory, Dempster–Shafer evidence theory, and Zadeh fuzzy set theory.

Search is a problem-solving method of artificial intelligence. The search strategy determines the priority relationship of knowledge used in a reasoning step of problem-solving. It can be divided into blind search without information guidance and heuristic search guided by empirical knowledge. Heuristic knowledge is often represented by heuristic functions. The more fully utilized the heuristic knowledge, the smaller the search space for solving problems and the higher the problem-solving efficiency. Typical heuristic search methods include A* and AO* algorithms. In recent years, research on search methods has begun to pay attention to extremely large-scale search problems with millions of nodes.

1.2.4 *Machine Learning*

Machine learning is the study of how computers can simulate or implement human learning behavior to acquire new knowledge or skills, and reorganize existing knowledge structures to continuously improve their own performance. Only by allowing computer systems to have human-like learning capabilities can it be possible to achieve human-level artificial intelligence. Machine learning is one of the core issues in artificial intelligence research and is a very active research field in current artificial intelligence theoretical research and practical applications.

Common machine learning methods include inductive learning, analogy learning, analytical learning, reinforcement learning, genetic algorithm, and connection learning. Deep learning is a new field in machine learning research. Its concept was proposed by Hinton G.E. and others in 2006 (Hinton & Salakhutdinov, 2006). It imitates the analysis and learning mechanism of human brain neural networks to interpret images, sound, and text data. Any progress in machine learning research will promote the improvement of artificial intelligence.

1.3 Main Schools of Artificial Intelligence Research

In the process of more than 60 years of research on artificial intelligence, many different approaches to artificial intelligence research have been formed due to people's different understanding and understanding of the nature of intelligence. Different research approaches have different academic perspectives, adopt different research methods, and form different research schools. At present, the main research schools in the field of artificial intelligence include symbolism, connectionism, and behaviorism. The symbolism method is based on the assumption of physical symbol systems and the principle of limited rationality; the connectionism method is based on the artificial neural network model as the core; the behaviorism method focuses on the study of perception–action response mechanisms.

1.3.1 *Symbolism*

Symbolism school is also known as functional simulation school. The main point of view is that the basis of intelligent activities is the physical symbol system, and the thinking process is the processing of symbol patterns.

In their 1976 American Computer Association (ACM) Turing Award speech, Newell and Simon summarized the physical symbol system hypothesis (Newell & Simon, 1976) and pointed out necessary and sufficient conditions for physical systems that exhibit general intelligent behavior. Yes, it is a physical symbol system. Adequacy states that intelligence can be obtained through any rationally organized system of physical symbols. Necessity states that a generally intelligent agent must be an instance of a physical symbol system. The postulated necessity of physical symbolic systems requires that an intelligent agent, whether it be a human, an alien, or a computer, acquires intelligence through the physical implementation of operations on symbolic structures.

General intelligent behavior represents the same actions and behaviors in human activities. Within physical limits, a system will exhibit behavior that is suitable for its purpose and adapted to the requirements of its environment.

In the years that followed, both artificial intelligence and cognitive science conducted extensive research in the field described by this hypothesis. The assumption of physical symbol systems leads to three important methodological guarantees: (a) the use of symbols and symbol systems as mediators to describe the world; (b) the design of search mechanisms, especially heuristic searches, to explore the possibilities that these symbol systems can support the space of reasoning; (c) the separation of cognitive architecture, which means assuming that a reasonably designed symbol system can provide a complete causal reason for intelligence, regardless of the method of its implementation. Based on this view, artificial intelligence eventually became an empirical and constructive discipline, which attempts to understand intelligence by establishing working models of intelligence.

From a symbolic point of view, knowledge representation is the core of artificial intelligence. Cognition is the processing of symbols. Reasoning is the process of solving problems using heuristic knowledge and heuristic search. The reasoning process can be expressed in some formal language. Symbolism advocates using logical methods to establish a unified theoretical system of artificial intelligence, but there are problems with "common sense" and the representation and processing of uncertain things. Therefore, it has been criticized by other schools of thought.

What is often referred to as "classical artificial intelligence" is research guided by a symbolic perspective. Classic artificial intelligence research can be divided into cognitive school and logic school. The cognitive school, represented by Simon, Minsky, and Newell, starts from human thinking activities and uses computers to simulate macroscopic functions. The logic school, represented by McCarthy and Nelson, advocates using logic to study artificial intelligence, that is, using formal methods to describe the objective world.

1.3.2 Connectionism

The school of artificial intelligence based on the connection mechanism and learning algorithm of neurons and neural networks is connectionism, also known as the structural simulation school. This method studies the nature and ability of information processing methods that are non-programmed, adaptable to environmental changes, and similar to the human brain style. The main view of this school of thought is that the brain is the basis of all intelligent activities and therefore starts from the brain neurons. Starting from the study and its connection mechanism, understanding the structure of the brain and its information processing process and mechanism is expected to reveal the mystery of human intelligence, thereby truly realizing the simulation of human intelligence on machines.

The main characteristics of this method are storing information in a distributed manner, processing information in a parallel manner, having self-organization and self-learning capabilities, being suitable for simulating human image thinking, and obtaining an approximate solution relatively quickly. It is these characteristics that enable neural networks to provide a new method and approach for people to use machines to process information. However, this method is not suitable for simulating people's logical thinking process, and people have found that existing models and algorithms also have certain problems, and theoretical research also has certain difficulties. Therefore, the connection mechanism alone can solve all problems of artificial intelligence. It's also unrealistic.

The representative achievement of connectionism is a mathematical model of neurons proposed by McCulloch and Pitts in 1943, namely the M-P model, from which a feed-forward network is formed. It can be said that M-P is the original model of artificial neural network, ushering in the era of neural computing and creating a new way for artificial intelligence to use electronic devices to simulate the structure and function of the human brain. Since then, the continuous development of neural network theory and technical research, as well as important breakthroughs in image processing, pattern recognition, and other fields, has created conditions for the realization of connectionist intelligent simulation.

1.3.3 *Behaviorism*

The behavioral school, also known as the behavioral simulation school, believes that the basis of intelligent behavior is the "perception–action" response mechanism. Based on the theories, methods, and technologies of intelligent control systems, we study anthropomorphic intelligent control behaviors.

In 1991, Brooks R A proposed intelligence without knowledge representation and intelligence without reasoning (Brooks, 1991). He believed that intelligence is only manifested in interaction with the environment and should not adopt a centralized model but needs to have different characteristics. The behavior module interacts with the environment to generate complex behaviors. He believes that no expression can perfectly represent the real concepts in the objective world, so it is inappropriate to use symbol strings to represent intelligent processes. This is in many ways a reflection of behavioral psychology in artificial intelligence. The basic behavior-based perspective can be summarized as follows:

(1) The formal expression and modeling methods of knowledge are one of the important obstacles to artificial intelligence.
(2) Intelligence depends on perception and action. It should directly use the machine to act on the environment and use the environment's response to the action as a prototype.
(3) Intelligent behavior can only be reflected in the world and manifested through interaction with the surrounding environment.
(4) Artificial intelligence can evolve gradually, developing and enhancing in stages just like human intelligence.

Brooks's behavioral (evolutionary) perspective opens up new avenues for artificial intelligence research. Based on these views, Brooks developed a robot bug that used some relatively independent functional units to realize basic functions such as avoidance, advancement, and balance respectively to form a hierarchical asynchronous distributed network, which achieved certain success, especially in creating a new method for robot research.

After the idea of behaviorism was proposed, it attracted widespread attention. Some people believed that the behavioral success of Brooks's robot bug could not lead to advanced control behavior. It was just an illusion to expect machines to evolve from insect intelligence to human intelligence. Nonetheless, the rise of the behaviorist school shows that the ideas of cybernetics and systems engineering will further influence the development of artificial intelligence.

The above three research methods study human natural intelligence from different aspects and have a corresponding relationship with the thinking model of the human brain. Roughly divided, it can be considered that symbolism studies abstract thinking, connectionism studies image thinking, and behaviorism studies perceptual thinking (Shi, 1992). The three major schools and three approaches to artificial intelligence research each have their own strengths, and they must learn from each other's strengths and integrate them comprehensively.

1.4 New Generation Artificial Intelligence

Layout the basic theory of frontier research. On the basis of the direction of artificial intelligence paradigm change, we shall discuss forward-looking layout advanced machine learning, brain-like intelligence calculation, quantum intelligence computing, and other cross-domain basic theory research. Advanced machine learning theory focuses on breakthroughs in adaptive learning, autonomous learning, and other theoretical methods to achieve a strong ability of artificial intelligence high interpretability. The theory and method of brain-like intelligence focus on the breakthrough of the theory and method of information coding, processing, memory, learning, and reasoning, form brain-like complex system and brain-like control, and establish a new model and brain-inspired cognitive computing model. Quantum intelligent computing theory focuses on the quantum acceleration of machine learning methods and the establishment of high-performance computing and quantum algorithm hybrid model and forms highly efficient and accurate quantum artificial intelligence system architecture.

To carry out interdisciplinary exploratory research, we shall promote the artificial fusion of artificial intelligence and neuroscience, cognitive science, quantum science, psychology, mathematics, economics, sociology, and other related basic disciplines, strengthen the introduction of artificial intelligence algorithm, model development of mathematical basic theory research, pay attention to artificial intelligence law and ethical theory of basic theory to support the original strong, non-consensus exploratory research, encourage scientists to explore freely and to capture the forefront of artificial intelligence problems, and put forward more original theory to make more original discoveries.

The basic theory contains the following directions:

(1) **Large data intelligence theory:** This chapter studies the new methods of artificial intelligence combined with data-driven and knowledge-guided approaches, the theory and method of cognitive computing with natural language understanding and image graphics as the core, the theory and method of comprehensive depth reasoning and creative artificial intelligence, the basic theory of intelligent decision-making under incomplete information, and the framework, data-driven general artificial intelligence mathematical model and theory.

(2) **Cross-media perceptual computing theory:** We shall study transcendental perception of human visual perception, active visual perception and computation for the real world, auditory perception and calculation of natural acoustic scenes, verbal perception and computation of natural interactive environment, human perception and computation for asynchronous sequences, and intelligent perception of self-learning, urban full-scale intelligent perception reasoning engine.

(3) **Hybrid enhanced intelligence theory:** We shall do research on "people in the loop" of the hybrid enhanced intelligence, man–machine intelligent symbiosis

behavior and brain-computer collaboration, machine intuitionistic reasoning and causal models, associative memory model and knowledge evolution methods, complex data and tasks to enhance intelligent learning methods, cloud Robot collaborative computing methods, contextual understanding in the real world environment, and man–machine group synergy.

(4) **Group intelligence theory:** We shall do research on group intelligence structure theory and organization method, group intelligence incentive mechanism and emergence mechanism, group intelligence learning theory and method, and group intelligence general computing paradigm and model.

(5) **Autonomous collaborative control and optimization decision theory:** We shall do research on cooperative awareness and interaction for autonomous unmanned systems, collaborative control and optimization decision-making for autonomous unmanned systems, knowledge-driven anthropogenic ternary coordination, and interoperability theory.

(6) **Advanced machine learning theory:** We shall study statistical learning basic theory, uncertainty reasoning and decision-making, distributed learning and interaction, privacy protection learning, small sample learning, deep reinforcement learning, unsupervised learning, semi-supervised learning, active learning, and other learning theories and efficient models.

(7) **Brain intelligent computing theory:** We shall do research on brain perception, brain learning, brain memory mechanism and computational fusion, brain-like complex system, brain control, and other theories and methods.

(8) **Quantum intelligence computing theory:** We shall explore the quantum mode and intrinsic mechanism of brain cognition, study efficient quantum intelligence models and algorithms, high-performance high-bit quantum artificial intelligence processors, and real-time quantum artificial intelligence systems that can interact with external environment.

The new generation of artificial intelligence key common technology of the research and deployment should focus on the algorithm, the data, and hardware to enhance the perceptions, knowledge, cognitive reasoning, and human–computer interaction capacity as the focus and form open compatible, stable, and mature technology system. The key common technology contains the following directions:

(1) **Knowledge computing engine and knowledge service technology:** We shall do research on knowledge computing and visual interaction engine, innovation design, digital creativity, and visual media as the core business intelligence and other knowledge service technologies and carry out large-scale biological data knowledge discoveries.

(2) **Cross-media analysis reasoning technology:** We shall do research on cross-media unified characterization, association understanding and knowledge mining, knowledge map construction and learning, knowledge evolution and reasoning, intelligent description and generation technologies and develop cross-media analysis reasoning engine and verification system.

(3) **Group key technology of intelligence:** We shall carry out the active knowledge and discovery of group intelligence, knowledge acquisition and acquisition, collaboration and sharing, evaluation and evolution, human–computer integration and enhancement, self-maintenance and security interaction, and other key technology research and build group intelligence space service architecture, intelligent collaborative decision, and control technology.

(4) **Hybrid enhanced intelligent new architecture and new technology:** We shall do research on hybrid enhanced intelligent core technology, cognitive computing framework, new hybrid computing architecture, man–machine sharing and online intelligent learning technology, and parallel management and control of hybrid enhanced intelligent framework.

(5) **Independent unmanned system of intelligent technology:** We shall do research on intelligent robot, marine robot, polar robot technology, unmanned workshop/intelligent factory intelligent technology, high-end intelligent control technology, autonomous unmanned operation, and so on. This chapter studies the autonomous control technology of robot and robot arm based on computer vision localization, navigation, and recognition in complex environment.

(6) **Virtual reality intelligent modeling technology:** This chapter studies the mathematical expression and modeling method of intelligent behavior of virtual object, the problem of natural, continuous, and deep interaction between virtual object and virtual environment and users, and the technology and method system of intelligent object modeling.

(7) **Intelligent computing chip and system:** We shall do research and develop neural network processor and energy-efficient, reconfigurable brain computing chip, a new type of sensor chip and system, intelligent computing architecture and systems, artificial intelligence operating system, and mixed computing architecture suitable for artificial intelligence.

(8) **Natural language processing technology:** We shall study the short text of the calculation and analysis technology, cross-language text mining technology, and machine-aware intelligence for semantic understanding of technology and the man–machine dialogue system of multimedia information understanding.

We shall construct an innovative platform of artificial intelligence and strengthen the support for the application, research, and development of artificial intelligence.

Artificial intelligence open-source hardware and software infrastructure platform focuses on building unified computing framework platform supporting knowledge reasoning, probability statistics, depth learning, and other artificial intelligence paradigms and forms a promotion of artificial intelligence software, hardware, and intelligent clouds between the ecological chain. The group intelligent service platform focuses on the construction of the knowledge resource management and the open sharing tool based on the large-scale cooperation of the Internet, and forms the platform and the service environment for the innovation of the industry and university. The hybrid enhanced intelligent support platform focuses on the construction of a heterogeneous real-time computing engine supporting large-scale training and a new computing cluster, providing a service-oriented, systematic platform and solution for complex intelligent computing. Autonomous unmanned system support platform focuses on the construction of autonomous systems for autonomous systems in the environment of environmental awareness, autonomous collaborative control, intelligent decision-making, and other artificial intelligence common core technology support systems and forms development and test environment of open, modular, reconfigurable autonomous unmanned system. Artificial intelligence basic data and security detection platforms focus on the construction of artificial intelligence for the public data resource library, the standard test dataset, and cloud service platform and form artificial intelligence algorithms and platform security test evaluation methods, techniques, norms, and tools. We shall promote all kinds of open source of common software and technology platforms. All kinds of platforms shall be in accordance with the requirements of the depth of integration of military and civilian and related provisions and promote the sharing of military and civilian sharing.

1.5 Artificial Intelligence Applications

Currently, almost all branches of science and technology share the theories and technologies provided by the field of artificial intelligence. Here are some classic, representative, and important application fields of artificial intelligence.

1.5.1 *Expert System*

Expert system is a type of computer intelligent program system with specialized knowledge and experience. By modeling the problem-solving ability of human experts, it uses knowledge representation and knowledge reasoning technology in artificial intelligence to simulate complex problems that are usually solved by experts, reaching a level of problem-solving ability equivalent to that of experts.

This knowledge-based system design method is centered on the knowledge base and inference engine, that is,

Expert system = knowledge base + inference engine

It separates knowledge from other parts of the system. Expert systems emphasize knowledge rather than methods. Many problems do not have algorithm-based solutions, or the algorithm solutions are too complex. Using expert systems can make use of the rich knowledge of human experts, so expert systems are also called knowledge-based systems. Generally speaking, an expert system should have the following three elements:

(1) have expert-level knowledge in a certain application field,
(2) able to simulate the thinking of experts,
(3) able to reach expert-level problem-solving skills.

Since the 1980s, driven by knowledge engineering, many expert system development tools have emerged, such as EMYCIN, CLIPS (OPS5, OPS83), G2, KEE, and OKPS.

1.5.2 *Data Mining*

Data mining is an exciting and successful application that can meet people's requirements to mine implicit, unknown, and potentially valuable information and knowledge from large amounts of data. For the data owner, in his specific work or living environment, the information and knowledge hidden in the data that can be exploited are automatically discovered. To achieve these goals, we need to have a large amount of raw data, clear mining goals, corresponding domain knowledge, and a friendly human–machine interface and find appropriate development methods. Mining results are used for decision-making by the data owner and must receive the owner's support, recognition, and participation.

Currently, data mining has many successful cases in marketing, banking, manufacturing, insurance, computer security, medicine, transportation, telecommunications, and other fields. Currently, representative data mining tools or platforms include SAS Enterprise Miner of the American SAS Company, Intelligent Miner of the IBM Company, Clementine of the Solution Company, Scenario of the Canadian Cognos Company, Palantir, the American big data company, and Intelligent Information of the Institute of Computing Technology of the Chinese Academy of Sciences. Process the big data mining cloud engine CBDME developed by the Key Laboratory, etc.

1.5.3 *Natural Language Processing*

Natural language processing studies computers' communication technologies such as listening, speaking, reading, and writing with users through natural languages familiar to humans. It is an interdisciplinary subject connected with linguistics, computer science, mathematics, psychology, acoustics, and other disciplines. The research content of natural language processing mainly includes language computing (computing at various levels, such as phonetics and phonemes, morphology, syntax, semantics, and pragmatics), language resource construction (computational vocabulary, terminology, electronic dictionaries, corpora, and knowledge ontology), machine translation or machine-assisted translation, Chinese and minority language text input and output and intelligent processing, Chinese handwriting and print recognition, Chinese speech recognition and text-to-text conversion, information retrieval, information extraction and filtering, text classification, Chinese search engine, and using natural language as the basis Hub multimedia retrieval.

Chinese information processing (including information processing of Chinese and minority languages) occupies a special position in the scientific and technological progress and industrial development of my country's information field and promotes the development of my country's information technology and industry: Wang Xuan's Chinese character laser phototypesetting (won the first prize of the National Science and Technology Progress Award twice), Lenovo Hanka (won the first prize of the National Science and Technology Progress Award), Liu Yingjian's Hanwang Chinese character input system (won the first prize of the National Science and Technology Progress Award), Chen Zhaoxiong's machine translation system (won the first prize of the National Science and Technology Progress Award), and Ding Xiaoqing's Tsinghua Wentong Chinese Character OCR system (won the second prize of the National Science and Technology Progress Award), etc. These achievements, which embody the distinct spirit of independent innovation, are not only witnesses of the development of my country's Chinese information processing industry but also provide valuable spiritual wealth for continued vigorous development in the future.

We have entered the era of massive information with the Internet as its main symbol. A related grim fact is that the effective use of digital information has become a global bottleneck restricting the development of information technology. Natural language processing will inevitably become a new strategic high ground for the medium and long-term development of information science and technology. The "National Medium and Long-term Science and Technology Development Plan" states that our country will promote "the development of 'human-centered' information technology based on image and natural language understanding and promote innovation in multiple fields."

1.5.4 *Intelligent Robot*

An intelligent robot is an automated machine with a highly developed "brain" and some intelligent abilities similar to those of humans or living things, such as perception, planning, action, and collaboration. It is a highly flexible automation machine. As people's understanding of the intelligent nature of robot technology deepens, robot technology begins to penetrate into various fields of human activities. Combining the application characteristics of these fields, people have developed a variety of special robots and various intelligent machines with sensing, decision-making, action, and interaction capabilities, such as mobile robots, micro-robots, underwater robots, medical robots, military robots, Aerial space robots, and entertainment robots.

Intelligent robots have broad development prospects. Although domestic and foreign research on intelligent robots has achieved many results, the level of intelligence is not yet very high. Therefore, the development of intelligent robots must also be accelerated. The operating environment of intelligent robots is quite complex. If we want robots to develop better and solve many of the problems they are facing now, then we humans should learn more from nature, through the process of learning, imitating, copying, and recreating natural creatures, discover relevant theories and technical methods, and apply them to robots, making robots continue to make breakthroughs in function and technical level, thus producing more advanced and smarter robots.

The Robot World Cup (RoboCup) is currently a high-profile competition that promotes the research and development of robots. Some people predict that a football team composed of robots will defeat professional football teams in the 2050 World Cup. Of course, these robots currently developed still only have partial intelligence, and they are still far from the true meaning of life intelligence. Robot vision and natural language communication are two of the main difficulties.

1.5.5 *Pattern Recognition*

Pattern recognition refers to the process of processing and analyzing various forms of information that characterize things or phenomena in order to describe, identify, classify, and explain things or phenomena. Pattern is the form in which information exists and is transmitted, such as spectrum signals, graphics, text, the shape of objects, the way of behavior, and the state of the process, all belong to the category of pattern.

People perceive various things or phenomena in the external world through patterns, which is the basis for acquiring knowledge, forming concepts, and responding.

Early pattern recognition research emphasized simulating the psychological and physiological processes of the human brain forming concepts and recognizing patterns. In the late 1950s, the perceptron proposed by Rosenblatt was both a pattern recognition system and a mathematical model of the human brain. However, with the need for practical applications and the development of computing technology, pattern recognition research often adopts mathematical technical methods that are different from biological cybernetics, physiology, and psychology. In 1957, Zhou Shaokang first proposed using decision theory methods to identify patterns. Narasimhan R proposed the syntactic method of pattern recognition in 1962. Since then, the Chinese-American scholar Jingsun Fu has conducted in-depth research in this area and published his first monograph "Syntactic Pattern Recognition and Its Application" in 1974. (Fu, 1974). Various pattern recognition methods developed in modern times can basically be classified into two categories: decision theory methods and structural methods.

With the popularization of information technology applications, pattern recognition has shown a trend of diversity and diversification and can be carried out at different conceptual granularities. Among them, biometric recognition has become an active field of pattern recognition research, including speech recognition, text recognition, image recognition, character recognition, and scene recognition. Biometric identification technology, such as fingerprint (palmprint) identification, face identification, signature identification, iris identification, and behavioral posture identification, has also become a hot research topic. Fuzzy clustering, genetic algorithm, Bayeux, etc. Bayesian theory, support vector machine, and other methods are used for image segmentation, feature extraction, classification, clustering, and pattern matching, making identity recognition an important tool to ensure economic security and social security.

1.5.6 *Distributed Artificial Intelligence*

Distributed artificial intelligence studies how a group of distributed, loosely coupled agents use their knowledge, skills, and information to work together to achieve their own or global goals. Since the 1990s, the rapid development of the Internet has provided excellent conditions for the development of new information systems, decision-making systems, and knowledge systems. They have increased rapidly in scale, scope, and complexity. The development of distributed artificial intelligence technology has applications that are increasingly critical to the success of these systems.

Research on distributed artificial intelligence can be traced back to the late 1970s. Early research on distributed artificial intelligence was mainly about distributed problem solving, and its goal was to create large-grained collaborative groups that work together to solve a certain problem. In 1983, Hewitt C and his colleagues developed a concurrent programming system based on the ACTOR model (Hewitt & DeJong, 1983). The ACTOR model provides the theory of parallel computing in distributed systems and the ability of a group of experts, or ACTORs, to obtain intelligent behavior. In 1991, Hewitt proposed the semantics of open information systems (Hewitt, 1991), pointing out that the properties of competition, commitment, collaboration, negotiation, etc. should be used as the scientific basis of distributed artificial intelligence, trying to provide a new basis for theoretical research on distributed artificial intelligence. In 1983, Lesser V R and others from the University of Massachusetts developed the distributed vehicle monitoring and testing system DVMT (Lesser & Corkill, 1983). In 1987, Gasser L and others developed the MACE system, which is an experimental distributed artificial intelligence system development environment (Gasser *et al.*, 1987). Each computing unit in MACE is called an agent. They have knowledge representation and reasoning capabilities. Agents communicate through message transmission.

Since the 1990s, agents and multi-agent systems have become the mainstream of distributed artificial intelligence research. An intelligent agent can be regarded as an automatically executing entity that senses the environment through sensors and acts on the environment through effectors. The BDI model of an agent is a formal model based on the thinking attributes of the agent, where B represents Belief, D represents Desire, and I represents Intention. A multi-agent system is a system composed of multiple agents. The core of research is how to coordinate behaviors among a group of autonomous agents. Multi-agent systems can form a society of agents, in the form of groups, teams, organizations, alliances, etc., with greater flexibility and adaptability, more suitable for open and dynamic world environments, and have become a hot topic in artificial intelligence research today.

1.5.7 *Internet Intelligence*

If the emergence of computers provides a material foundation for the realization of artificial intelligence, then the emergence and development of the Internet provide a broader space for artificial intelligence and have become a symbol of the informatization of today's human society. The Internet has become a "digital library" for more and more people. People generally use search engines such as Google and Baidu to serve their daily work and life.

The goal pursued by the Semantic Web is to make the information on the Web understandable by machines, thereby realizing the automatic processing of Web information to adapt to the rapid growth of Web information resources and better serve humans (Berners-Lee *et al.*, 2001). The Semantic Web provides a common framework that allows data to be shared and reused across the boundaries of different applications, enterprises, and groups. The Semantic Web is a collaborative project under the leadership of W3C, involving a large number of researchers and industry partners. The Semantic Web is based on the Resource Description Framework (RDF). RDF uses XML as syntax and URI as naming mechanism to integrate various applications.

The Semantic Web successfully applies artificial intelligence research results to the Internet, including knowledge representation and reasoning mechanisms. People expect the future Internet to be an on-demand encyclopedia that can customize search results, search hidden Web pages, consider the user's location, search for multimedia information, and even provide personalized services to users.

1.5.8 *Game*

Game playing is a ubiquitous phenomenon in human society and nature, such as chess, cards, and war. The two parties in the game can be individuals, groups, or biological groups or intelligent machines. Each party strives to use its own wisdom to succeed or defeat the other party. The game process may generate a surprisingly large search space. Searching these large and complex spaces requires the use of powerful techniques for judging alternative states and exploring the problem space. These techniques are called heuristic search. Games provide a good experimental place for artificial intelligence to test artificial intelligence technologies to promote the development of these technologies.

In the history of the development of artificial intelligence, Samuel developed a checkers program in 1956 and defeated himself. On May 11, 1997, IBM's "Deep Blue" computer defeated the chess master Kasparov G. On August 9, 2006, in the first Chinese Chess Human-Computer Competition held in Beijing, the computer defeated the human chess master by a narrow margin of 3 wins, 5 losses, and 2 losses.

Playing chess is a typical intellectual problem, and the solution process is usually a heuristic search process. In the chess game, the entire pattern of the chessboard is used as the state, legal moves are used as operations, and heuristic knowledge is used as navigation to find a path to victory in a state space. The positions in a game are easy to represent on a computer and do not require the complex formats necessary to represent more complex problems. The simplicity of the game makes testing the game program free of any financial or ethical burden. State space search is the basis of most game research.

1.5.9 *AI for Science*

AI is the magic weapon to dispel the dark clouds in various fields of science, AI for Science (AI4S) will be the next main battlefield of AI, and it will greatly expand. They discovered that AI4S will empower technology and all aspects of industry, helping us speed up scientific research, and the last mile between technological innovations will also help scientists overcome confusion uncovering the chaotic natural and social characteristics and discovering the behind-the-scenes of things the key rules at work.

1.6 Summary

Artificial intelligence is a discipline that studies computational models that can think and perform actions rationally. It is a simulation of human intelligence on a computer. As a subject, artificial intelligence has gone through several stages of conception, formation, and development, and is still developing. Although artificial intelligence has also created some practical systems, we have to admit that these are far from reaching the level of human intelligence.

Knowledge representation, reasoning, learning, intelligent search, and uncertainty processing of data and knowledge are the basic research fields of artificial intelligence. Typical application fields of artificial intelligence include expert systems, data mining, natural language processing, intelligent robots, pattern recognition, distributed artificial intelligence, Internet intelligence, and gaming.

The research approaches of artificial intelligence mainly include methods with symbol processing, connection mechanism methods with network connection, and behaviorism methods with perception and action. The integration and synthesis of these methods have become the basis of today's artificial intelligence research trend. This chapter pointed out the basic theories and key technologies of new generation artificial intelligence.

Entering the 21st century, the popularity of the Internet and the rise of big data have once again pushed artificial intelligence to a new peak. Knowledge automation based on big data and cyberspace will open up mankind's march into the artificial world, deeply develop big data and intellectual resources, and deepen the intelligent revolution in agriculture and industry. Intelligence science, an interdisciplinary study of brain science, cognitive science, artificial intelligence, and other disciplines, will guide the development of brain-like computing and realize human-level artificial intelligence (Shi, 2021).

Exercises

1.1 What is Artificial Intelligence (AI)? What is the research objective of AI?

1.2 Please briefly introduce the main stages of development in the history of AI.

1.3 What are the five fundamental questions for AI research?

1.4 What is the physical symbol system? What is the physical symbol system assumption?

1.5 What is symbolic intelligence? What is computational intelligence?

1.6 Please describe the simple model of machine learning and its basic elements.

1.7 What is Distributed Artificial Intelligence (DAI)? What are the main research domains of DAI?

1.8 Please refer to relevant literature and discuss whether the following tasks can be solved by current computers:

(a) Defeat an international grandmaster in the world's chess competition.
(b) Defeat a 9 duan professional in a game of Go.
(c) Discover and prove a new mathematical theorem.
(d) Find bugs in the programs automatically.

1.9 How to classify knowledge-based systems? How to achieve collective intelligence behaviors?

Chapter 2

Logic Foundation

Logic is a primary tool in the study of computer science as well as in the study of artificial intelligence. McCarthy claimed human-level artificial intelligence (AI) is logical AI, based on the formalizing of commonsense knowledge and reasoning in mathematical logic.

2.1 Introduction

Logic as a formal science was founded by Aristotle. Leibniz reaffirmed Aristotle's logical developing direction of mathematics form and founded mathematical logic. From the thirties of the last century, various mathematical methods were extensively introduced and used in mathematical logic, with the result that mathematical logic becomes one branch of mathematics and is as important as algebra and geometry. Mathematical logic has spread out many branches, such as model theory, set theory, recursion theory, and proof theory.

Logic is a primary tool in the study of computer science as well as in the study of artificial intelligence. It is widely used in many domains, such as semasiology, logic programming language, the theory of software specification and validation, the theory of database, the theory of knowledge base, intelligent systems, and the study of robots. The objective of computer science is essentially coincident with the goal of logic. On the one hand, the objective of computer science is to simulate with the computer the function and behavior of the human brain and bring the computer to be an extension of the brain. Here the simulation of the function and behavior of the human brain is in fact to simulate the thinking process of persons. On the other hand, logic is a subject focused on the discipline and law of human thinking. Therefore, the methods and results obtained in logic are naturally selected and put to use during the research of computer science. Furthermore, the intelligent behavior of human beings is largely expressed by language and character; therefore, the simulation of human natural language is the point of departure for the simulation of human thinking process. In 2005, McCarthy wrote, "I think the best hope for human-level AI is logical AI, based on the formalizing of commonsense knowledge and reasoning in mathematical logic" (McCarthy, 2005).

Language is the starting point for the study of human thinking in logic, as well as for the simulation of human thinking in computer science. Topics related to language are important issues that run through the domain of computer science. Many subjects of computer science, such as programming languages and their formal semantics, knowledge representation and reasoning, and natural language processing, are all related to language. Generally speaking, representation and reasoning are two basic topics in computer science and artificial intelligence. Majority of the intelligent behavior relies on a direct representation of knowledge, for which formal logic provides an important approach.

Knowledge, especially the so-called common knowledge, is the foundation of intelligent behavior. Intelligent behavior such as analyzing, conjecturing, forecasting, and deciding are all based on the utilization of knowledge. Accordingly, in order to simulate with computer the intelligent behavior, one should first make knowledge represented in the computer and then enable the computer to utilize and reason about the knowledge. Representation and reasoning are two basic topics of knowledge in the study of artificial intelligence. They are entirely coincident with the two topics focused on the study of natural language, i.e., the accurate structure and reasoning of natural languages. Therefore, the methods and results obtained in logic are also useful for the study of knowledge in artificial intelligence. The ability of representation and the performance of reasoning are a pair of contradictions for any logic system applied to intelligent systems. A trade-off between such a pair is often necessary.

The logic applied in majority of logic-based intelligent systems is first-order logic or its extensions. The representation ability of first-order logic is so strong that many experts believe that all the knowledge representation problems arising in the research of artificial intelligence can be carried out within the framework of first-order logic. First-order logic is suitable for representing knowledge with uncertainty. For example, the expression $\exists x\, P(x)$ states that there exists an object for which the property P holds, while it is not pointed out which one is such an object. For another example, the expression $P \vee Q$ states that at least one of P and Q holds, but it is not determined whether P (or Q) really holds. Furthermore, first-order logic is equipped with a complete axiom system, which can be treated as a standard of reference in the designing of strategies and algorithms on reasoning. Although first-order logic is capable of representing majority of knowledge, it is not convenient and concise for many applications. Driven by various requirements, lots of logic systems have been proposed and studied; in the following, we enumerate some typical examples:

(1) In order to represent knowledge on epistemic, such as believe, know, desire, intention, goal, and commitment, various modal logics were proposed.

(2) In order to represent knowledge which is related to time, various temporal logics were proposed.

(3) In order to represent knowledge with uncertainty, the so-called fuzzy logic was proposed. As a system built upon the natural language directly, fuzzy logic adopts many elements from the natural language. According to Zadeh, the founder of fuzzy logic, fuzzy logic can be regarded as a computing system on words; in other words, fuzzy logic can be defined by the formula "fuzzy logic = computing with words".

(4) Knowledge of humans is closely interrelated to human activities. Accordingly, knowledge on behavior or action is important for intelligent systems. Compared with various static elements of logic, action is distinguished by the fact that the execution of actions will affect the properties of intelligent systems. Representation and reasoning about actions are classical topics in the study of artificial intelligence; many problems, such as the frame problem and the qualification problem, were put forward and well studied. Many logic systems, such as the dynamic logic and the dynamic description logic, were also proposed.

(5) Computer-aided decision-making has become one of the important applications of computers. Persons always hold their predilections as while as they are making a decision. In order to represent the rule and simulate the behavior of people's decision-making process, it is inevitable to deal with the predilection. As a result, based on management science, a family of so-called partial logics was proposed and studied.

(6) Time is one of the most important terms present in intelligent systems. Some adverbs, such as occasionally, frequently, and often, are used in the natural language to represent time. Knowledge about the time which is described by these adverbs cannot be represented with classical temporal logic. Therefore, an approach similar to the integral of mathematics was introduced into logic. With the resulting logic, time that is described by various adverbs can be formally represented and operated.

2.2 Logic Programming

In this section, we give a brief introduction to the logic programming language Prolog. Prolog was first developed by a group around Alain Colmerauer at the University of Marseilles, France, in the early 1970s. Prolog was one of the first logic programming languages, and it is now the major artificial intelligence and expert systems programming language.

Prolog is declarative in style rather than procedural. Users just need to represent the facts and rules, over which the execution is triggered by running queries; the execution is then carried out according to find a resolution refutation of the negated query. In other words, users just need to tell the Prolog engine what to do but not how to do it. Furthermore, Prolog holds the following features:

(1) Prolog is a unification of data and program. Prolog provides a basic data structure named terms. Both data and programs of prolog can be constructed over terms. This property is fit for the intelligent program since the outputs of certain programs can be executed as newly generated programs.
(2) Prolog supports automatic backtracking and pattern-matching, which are two of the most useful and basic mechanisms used in intelligent systems.
(3) Prolog uses recursion. Recursion is extensively used in the Prolog program and data structure so that a data structure with a big size can be manipulated by a short program. In general, the length of a program represented with Prolog is only 10% of which is written with the C++ language.

All of these features make Prolog suitable for encoding intelligent programs and suitable for applications, such as natural language processing, theorem proving, and expert systems.

2.2.1 *Definitions of Logic Programming*

First, we introduce the Horn clause which is the constituent of logic programs. A clause consists of two parts: the head and the body. As an IF-THEN rule, the condition portion of a clause is called the head and the conclusion portion of it is called the body.

Definition 2.1. A Horn clause is a clause that contains at most one literal (proposition/predicate) at the head.

Horn clauses in Prolog can be separated into three groups:

(1) Clauses without conditions (facts): A.
(2) Clauses with conditions (rules): $A: - B_1, \ldots, B_n$.
(3) Goal clauses (queries): $? :- B_1, \ldots, B_n$.

Semantics of the above Horn clauses is informally described as follows:

(1) The clause A states that A is true for any assignments on variables.
(2) The clause $A :- B_1, \ldots, B_n$ states that for any assignments on variables if $B_1, \ldots,$ and B_n are evaluated to be true, then A must also be true.

(3) The goal clause? $:- B_1, \ldots, B_n$ represents a query that will be executed. Execution of a Prolog program is initiated by the user's posting of a query; the Prolog engine tries to find a resolution refutation of the negated query.

For example, here are three Horn clauses:

(a) $W(X, Y) :- P(X), Q(Y)$.
(b) $?-R(X, Y), Q(Y)$.
(c) $?-R(X, Y) \wedge Q(Y)$.

The Horn clause indexed by (a) is a rule, with $P(X)$, $Q(Y)$ the body and $W(X, Y)$ the head. The Horn clause indexed by (b) is a query with $R(X, Y)$, $Q(Y)$ the body. The intuition of the query indexed by (c) is whether $R(X, Y)$ and $Q(Y)$ hold and what are the values of X and Y in the case that $R(X, Y) \wedge Q(Y)$ holds.

We are now to formally define Logic Programs.

Definition 2.2. A logic program is a collection of Horn clauses. In logic, program clauses with the same predicate symbol are called the definition of the predicate.

For example, the following two rules form a logic program:

$$\text{Father}(X, Y) :- \text{Child}(Y, X), \text{Male}(X).$$

$$\text{Son}(Y, X) :- \text{Child}(Y, X), \text{Male}(Y).$$

This program can also be extended with the following facts:

$$\text{Child}(\text{xiao-li, lao-li}).$$

$$\text{Male}(\text{xiao-li}).$$

$$\text{Male}(\text{lao-li}).$$

Taking these rules and facts as inputs of the Prolog engine, we can compile and execute it. Then the following queries can be carried out:

(1) query: $?-$ Father(X, Y), we will get the result Father(lao-li, xiao-li).
(2) query: $?-$ Son(Y, X), we will get the result Son(xiao-li, lao-li).

2.2.2 Data Structure and Recursion in Prolog

An important and powerful tool in problem-solving and programming, recursion is extensively used in data structures and programs of Prolog.

Term is a basic data structure in Prolog. Everything including program and data is expressed in the form of term. Terms of Prolog are defined recursively by the

Table 2.1. Prolog list structure.

[ ] or nil	empty table
[*a*]	cons(*a*,nil)
[*a*, *b*]	cons(*a*,cons(*b*,nil))
[*a*, *b*, *c*]	cons(*a*,cons(*b*,cons(*c*,nil))
[*X* | *Y*]	cons(*X*,*Y*)
[*a*, *b* | *c*]	cons(*a*,cons(*b*,*c*))

following BNF rule:

$$\text{<term>} ::= \text{<constant>} \mid \text{<variable>} \mid \text{<structure>} \mid (\text{<term>})$$

where structures are also called compound terms and are generated by the BNF rule:

$$\text{<structure>} ::= \text{<function>}(\text{<term>}\{, \text{<term>}\})$$

$$\text{<function>} ::= \text{<atom>}$$

List is an important data structure supported by Prolog. A list can be represented as a binary function cons(X, Y), with X the head of the list and Y the tail. The tail Y of a list cons(X, Y) is also a list which can be generated by deleting the element X from cons(X, Y). Elements of a list can be atoms, structures, terms, and lists. Table 2.1 shows some notations on lists that are used in Prolog.

Finally, we present an example on which recursion is used in programs of Prolog. Consider a simple predicate that checks if an element is a member of a list. It has the two clauses listed as follows:

member(X, [X |_]).

member(X,[_| Y]) :- member(X, Y).

In this example, the predicate member is recursively defined, with the first Horn clause being the boundary condition and the second the recursive case.

2.2.3 *SLD Resolution*

SLD resolution is the basic inference rule used in logic programming. It is also the primary computation procedure used in PROLOG. Here the name SLD is an abbreviation of "Linear resolution with Selection function for Definite clauses". First, we introduce definitions of definite clause.

Definition 2.3. A Definite clause is a clause of the form

$$A\text{:- } B_1, B_2, \ldots, B_n$$

where the head is a positive literal and the body is composed of zero, one, or more literals.

Definition 2.4. A definite program is a collection of definite clauses.

Definition 2.5. A definite goal is a clause of the form

$$?\text{:-}B_1, B_2, \ldots, B_n$$

where the head is empty.

Let P and G be a program and a goal, respectively, then the solving process for the corresponding logic program is to seek an SLD resolution for $P \cup \{G\}$. Two rules should be decided for the resolution process: one is the computation rule on how to select the sub-goal; the other is the search strategy on how to go through the program. Theoretically, any search strategy used in artificial intelligence can be adopted. However, in practice, strategies should be selected according to their efficiency. Following is the standard SLD resolution process:

(1) Sub-goals are selected with a "left then right" strategy.
(2) The program has gone through with a strategy based on the depth-first search and the backtracking method.
(3) Clauses of the program P are selected with the same priority of their appearance in the program.
(4) The occur-check is omitted from the unification algorithm.

There are some characteristics for such a resolution process.

1. There exists simple and efficient method for the realization of depth-first search strategy
The depth-first search strategy can be realized with just a goal stack. A goal stack for the SLD tree consists of branches which are going through. Correspondingly, the searching process is composed of the pop and push operators on the stack. In the case that the sub-goal on the top of the stack is unified with the head of some clause of the program P, the corresponding resolvent will be put into stack. While in the case that no clause could be unified, a backtracking operator will be triggered with the result that an element was popped from the stack; in this case, the resulted stack should be inspected for unification.

Example 2.1. Consider the following program:

$$p(X, Z)\text{:-} q(X, Y), p(Y, Z).$$

$$p(X, X).$$

$$q(a, b).$$

Let "$?\text{-}p(X, b)$" be the goal. Then the evolvement of the goal stack is shown in Table 2.2.

Table 2.2. Prolog goal stack.

$?\text{-}p(X, b)$.				G is put into stack
$?\text{-}p(X, b)$.	$?\text{-} q(X, Y), p(Y, b)$.			A resolvent is put into stack
$?\text{-}p(X, b)$.	$?\text{-} q(X, Y), p(Y, b)$.	$?\text{-} p(b, b)$.		A resolvent is put into stack
$?\text{-}p(X, b)$.	$?\text{-} q(X, Y), p(Y, b)$.	$?\text{-} p(b, b)$.	$?\text{-}q(b, W), p(W, b)$	A resolvent is put into stack
$?\text{-}p(X, b)$.	$?\text{-} q(X, Y), p(Y, b)$.	$?\text{-} p(b, b)$.	$\square$	An element is popped, then the resolvent $\square$ is put into stack
$?\text{-}p(X, b)$.				The pop operation is triggered for three times
$\square$				the resolvent $\square$ is put into stack $\square$ is popped

2. Completeness of SLD resolution process is destroyed by the depth-first search strategy

This problem can be partially solved by changing the order of sub-goals and the order of clauses of the program. For example, consider the following program:

(1) $p(f(X)){:\text{-}}\, p(X)$.

(2) $p(a)$.

Let "$?\text{-}p(Y)$" be the goal. Then it is obvious that the SLD resolution process will fall into an endless loop. However, if we exchange the order of clause (1) and clause (2), then we will get the result $Y = a, Y = f(a), \ldots$.

Consider another program:

(1) $q(f(X)) :\text{-} q(X)$.

(2) $q(a)$.

(3) $r(a)$.

Let "G: $?\text{-}q(Y), r(Y)$" be the goal. Then the SLD resolution process will fall into an endless loop. However, if we exchange the order of sub-goals contained in G and get the goal "G: $?\text{-} r(Y), q(Y)$", then we will get the result $Y = a$ after the SLD resolution process.

In order to guarantee the completeness of the SLD resolution process, the width-first search strategy must be embodied in the search rules of Prolog. However, as a result, both the time and space efficiency of the process will be decreased, and the complexity of the process will be increased. A trade-off is to maintain the depth-first search strategy that is used in Prolog and supplement it with some programs which embody other search strategies and are written with the Prolog language.

3. Soundness of SLD resolution is not guaranteed without occur-check

Occur-check is a time-consuming operation in the unification algorithm. In the case that occur-check is called, the time needed for every unification process is linear to the size of the table, consequently, the time needed for the append operation on predications is $O(n_2)$; here n is the length of the table. Since little unification process in the Prolog program uses occur-check, the occur-check operator is omitted from the unification algorithms of most Prolog systems.

In fact, without the occur-check, we no longer have soundness of SLD resolution. A sub-goal might not be unified with a clause in the case that some variables occur in the term. However, since the occur-check is omitted, the unification will still be executed and reach a wrong result. For example, let "$p(Y, f(Y))$" and "$?\text{-}p(X, X)$" be the program and the goal, respectively. Then the unification algorithm will generate a replacement $\theta = \{Y/X, f(Y)/Y\}$ for the pair $\{p(X, X), p(Y, f(Y))\}$. Such a mistake will be covered if the variable Y is not used in the SLD resolution anymore. However, once the variable Y is used again, the resolution process will fall into an endless loop.

2.2.4 *Non-Logic Components: CUT*

Program is the embodiment of algorithm. Algorithm in the logic programming is characterized by the following formula:

$$\text{algorithm} = \text{logic} + \text{control}$$

where the logic component determines the function of the algorithm and the control component determines the strategy which will be used to realize the function. Theoretically, a programmer just needs to specify the logic component, and then the corresponding control component can be automatically determined by the logic programming system. However, most Prolog systems in practice cannot reach such automation. As set forth, in order to guarantee a valid execution of the program, a programmer has to take the order of clauses into consideration. Another problem is the fact that an endless branch might be generated during the SLD resolution, according to the depth-first search strategy adopted by Prolog. In such a situation, the goal stack used in the resolution algorithm will be overflowed and bring the resolution process into an error state. The "CUT" component is introduced to solve this problem.

From the point of declarative semantics, CUT is a non-logical control component. Represented as the character "!", CUT can be treated as an atomic component and be inserted into clauses of the program or the order. Declarative semantics of a program is not affected by any "!" which appeared in the program or in the order.

From the point of operational semantics, some control information is carried by the CUT component. For example, let G be the following goal:

$$?\text{-}A_1, \ldots, A_{m-1}, A_m, A_{m+1}, \ldots, A_k$$

Let the following, which is denoted by C, be one of the clauses of the program:

$$A\text{:-}B_1, \ldots, B_i, !, B_{i+1}, \ldots, B_q$$

Consider the state that sub-goals $A_1, \ldots, A_{m-1}$ have been solved, and let G' be the current goal. Suppose Am can be unified with A. After the unification operation, the body of the clause C is added into the goal G'. Now a cut "!" is contained in the current goal G'. We call Am a cut point and call the current goal G' as the father-goal of "!". Now it is the turn to solve sub-goals $B_1, \ldots, B_i, !, B_{i+1}, \ldots, Bq$ one by one. As a typical sub-goal, "!" is valid and can be jumped over. Suppose a backtracking is triggered by the fact that some sub-goals behind "!" cannot be unified, then the goal stack will be tracked back to $Am\text{-}1$, the sub-goal prior to the cut point Am. From the point of SLD tree, all the nodes which are rooted by the father-goal of "!" and are accessed still will be cut out.

For example, let P be the following program:

(1) $p(a)$.
(2) $p(b)$.
(3) $q(b)$.
(4) $r(X)\text{:-}\ p(X), q(X)$.
(5) $r(c)$.

Let G be the sub-goal "$?\text{-}r(X)$". Then the SLD tree generated during the process of SLD resolution is presented as Figure 2.1. However, if a cut is inserted into the clause (4) of program P, i.e., the clause (4) of program P is changed as follows:

$$(4)'\, r(X)\text{:-}\ p(X),\ !,\ q(X).$$

Then the corresponding SLD tree should become that presented in Figure 2.2. In the latter case, no solution will be generated since a critical part is cut out from this SLD tree.

According to the example, soundness of SLD resolution might be destroyed by the CUT mechanism. Furthermore, incorporation of the CUT mechanism will cause the inconsistency between the declarative semantics and the operational semantics of

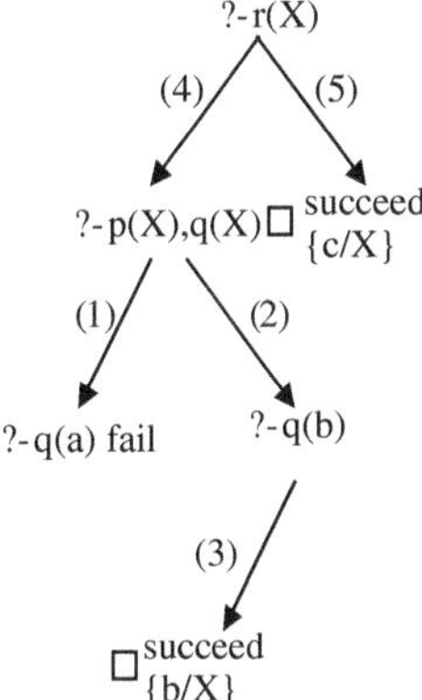

Fig. 2.1.   A SLD tree without !.

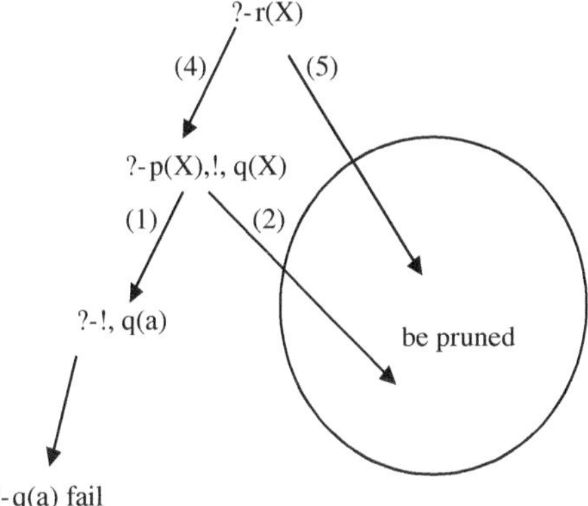

Fig. 2.2.   A SLD tree with !

logic programming. For example, let P be the following program which is designed to calculate the maximum value of two data:

$$\max(X, Y, Y):\text{-} X = <Y$$

$$\max(X, Y, X):\text{-} X > Y$$

We can check that declarative semantics and operational semantics of P are consistent. Now, we insert a CUT predication into P and get the following program P_1:

$$\max(X, Y, Y):\text{-}X = <Y, !$$

$$\max(X, Y, X):\text{-}X > Y$$

Efficiency of the program is obviously increased, although both the declarative semantics and the operational semantics of $P1$ are not changed with respect to those of P. Efficiency of the program can be further increased if we replace P_1 with the following program P_2:

$$\max(X, Y, Y)\text{:-}X = <Y, !$$

$$\max(X, Y, X)$$

However, although the operational semantics of P_2 is still the same as that of $P1$, the declarative semantics of $P2$ is changed as follows: the maximum value of X and Y is always X; it can also be Y in the case of $X \leq Y$. Obviously, the semantics of P_2 is different from our original intention.

The "fail" is another predication used by Prolog. Acting as a sub-goal, the predication "fail" cannot be solved at all and therefore will give rise to a backtracking. The predication CUT can be placed prior to the predication "fail" and forms the so-called cut-fail composition. During the SLD resolution process, in the case that a clause which contains a cut-fail composition is examined for resolution, resolution of the father-goal of the CUT predication will be finished directly. Therefore, efficiency of the search will be increased.

For example, consider the following program P:

(1) strong(X):- heart_disease(X), fail.
(2) strong(X):- tuberculosis(X), fail.
(3) strong(X):- nearsight(X), fail.
(4) strong(X).
(5) heart_disease(Zhang).

Let "?- strong(Zhang)" be the goal. According to the first clause of program P, the goal "strong(Zhang)" can be first unified with a resolvent "heart_disease(Zhang), fail", then a backtracking will be triggered by the "fail" after the unification of "heart_disease(Zhang)". In the following steps, sub-goals "tuberculosis(Zhang)" (or "nearsight(Zhang)") which are generated according to the second clause (resp., the third clause) cannot be unified. Finally, the goal "strong(Zhang)" will be unified according to the fourth clause of P and produce a positive result. Backtracking is triggered three times in this example. We can reduce the backtracking by placing a CUT prior to the "fail" occurring in the P. For example, the first three clauses of P can be changed as follows:

(1) strong(X):- heart_disease(X), !, fail.
(2) strong(X):- tuberculosis(X),!, fail.
(3) strong(X):- nearsight(X), !, fail.

Then, according to the first clause of the program, a backtracking will be triggered by the "fail" after the unification of the goal "strong(Zhang)" and the unification of the newly generated sub-goal "heart_disease(Zhang)". Since "strong(Zhang)" is the father-goal of the CUT that contained in the first clause of P, it will be popped from the goal stack also. Therefore, the SLD resolution will be finished right away and return a negative result.

Since first-order logic is undecidable, there is no terminable algorithm to decide whether G is a logic inference of P, for G and P any program and any goal, respectively. Certainly, SLD resolution algorithm will return a corresponding result if G is a logic inference of P. However, in the case that G is not a logic inference of P, the SLD resolution process (or any other algorithm) will fall into an endless loop. In order to solve this problem, the rule "negation as failure" is introduced into logic programming. For any clause G, if it cannot be proved, then the rule will enable the result that $\neg G$ is reasonable.

Based on the "negation as failure" rule, the predication "not" is defined in Prolog as follows:

$$\text{not}(A):\text{- call}(A), !, \text{fail.}$$

$$\text{not}(A).$$

Here "call" is a system predication and "call(A)" will trigger the system to solve the sub-goal "A". If the answer to the sub-goal "A" is positive, then a backtracking will be triggered by the "fail" occurring in the first clause, and the SLD resolution will be finished right away; in this case, the result for the clause "not(A)" is negative. However, if the answer to the sub-goal "A" is negative, then a backtracking will be triggered right away so that the "fail" occurring in the first clause will not be visited; in this case, the result for the clause "not(A)" is positive according to the second clause.

2.3 Closed World Assumption

With respect to any base set KB of beliefs, the closed world assumption (CWA) provides an approach to complete the theory $T(KB)$ which is defined by KB. Here, a theory $T(KB)$ is complete if either every ground atom in the language or its negation is in the theory. The basic idea of the CWA is that everything about the world is known (i.e., the world is closed); therefore, if a ground atom P cannot be proved according to the theory, then P will be considered to be negative. The CWA completes the theory by including the negation of a ground atom in the completed theory whenever that ground atom does not logically follow from KB.

One of the important applications of the CWA is to complete the database system. For example, let *KB* be the following database which contains information about the contiguities of countries:

Neighbor(China, Russia).

Neighbor(China, Mongolia).

$\forall x \forall y$(Neighbor$(x, y) \leftrightarrow$ Neighbor(y, x)).

Then, it is obvious that T(*KB*) is incomplete since neither Neighbor(Russia, Mongolia) nor $\neg$ Neighbor(Russia, Mongolia) can be logically inferred from *KB*. According to the CWA, the database *KB* can be completed by adding the assertion $\neg$ Neighbor(Russia, Mongolia) into it. It is obvious that the CWA is non-monotonic because the set of augmented beliefs would shrink if we added a new positive ground literal to *KB*.

Let *KB*asm be the set of all of the assertions added into *KB* during the completing process. According to the CWA, it is obvious that for any ground atom P,

$$\neg P \in KB_{asm} \text{ if and only if } P \notin T(KB)$$

For example, with respect to the database *KB* presented in the previous example, we have *KB*asm $= \{\neg$neighbor(Russia, Mongolia)$\}$.

Let CWA(*KB*) be the CWA-augmented theory, i.e., CWA(*KB*) $=$ T(*KB*$\cup$ *KB*asm). It is obvious that CWA(*KB*) is more powerful compared with T(*KB*) since many results that cannot be deduced from *KB* can now be derived from *KB*$\cup$ *KB*asm.

The augmented theory CWA(*KB*) might be inconsistent. For example, let *KB* $= \{P(A) \lor P(B)\}$, then it is *KB*asm $= \{\neg P(A), \neg P(B)\}$ since neither $P(A)$ nor $P(B)$ can be derived from *KB*, therefore the set *KB*$\cup$*KB*asm is inconsistent. Inconsistency of the CWA-augmented theory is an important problem that needs to be solved.

Theorem 2.1. *CWA(KB) is consistent if and only if, for every positive-ground-literal clause $P_1 \lor P_2 \lor \ldots \lor P_n$ that follows from KB, there is at least one ground literal P_i which is entailed by KB.*

In other words, CWA(KB) is inconsistent if and only if there are positive ground literals $P_1, P_2, \ldots, P_n$ such that $KB \mathrel{|} = P_1 \lor P_2 \lor \ldots \lor P_n$ and $KB \mathrel{|} \neq P_i$ for each $1 \leq i \leq n$.

Example 2.2. Let *KB* $= \{P(A) \lor P(B)\}$. It is obvious that CWA(*KB*) is inconsistent.

Example 2.3. Let *KB* $= \{\forall x(P(x) \lor Q(x)), P(A), Q(B)\}$. With respect to the atoms *A* and *B*, *KB* will be augmented with $\neg P(B)$ and $\neg Q(A)$, and will result in a consistent theory. However, if there is an atom C, then the resulting theory is inconsistent since it is both $(P(x) \vee Q(x)) \,\big|\! \neq P(C)$ and $(P(x) \vee Q(x)) \,\big|\! \neq Q(C)$.

Generally speaking, theory augmented by the CWA might be inconsistent. However, if the knowledge base KB is composed of Horn clauses and is consistent, then the augmented theory CWA(KB) is also consistent, i.e., we have the following theorem:

Theorem 2.2. *If the clause form of KB is Horn and consistent, then the CWA augmentation CWA(KB) is consistent.*

The condition that KB be Horn is too strong for many applications. In fact, according to Theorem 2.1, such a condition is not absolutely necessary for the CWA augmentation of KB to be consistent. An attempt to weaken this condition leads to the idea of the CWA with respect to a predicate P. Under that convention, if KB is Horn in some predicate P and P is not provable from KB, then we can just add the negation of P into the set KBasm. Here, we say that a set of clauses is Horn in a predict P if there is at most one positive occurrence of P in each clause.

For example, suppose KB is $\{P(A) \vee Q(A), P(A) \vee R(A)\}$. It is obvious that KB is Horn in the predicate P, even though both $P(A) \vee Q(A)$ and $P(A) \vee R(A)$ are not Horn clauses. Set KBasm $= \{\neg P(A)\}$, then we have $KB \cup KB$asm $\big| = Q(A)$ and $KB \cup KB$asm $\big| = R(A)$, and thereby get a consistent augmented theory CWA$'$ (KB) with respect to the predicate P.

But in fact, with respect to some predicate, consistency of the augmented theory cannot be guaranteed still. For example, let $KB = \{P(A) \vee Q, P(B) \vee \neg Q\}$, and let P be the particular predicate; then we have KBasm $= \{\neg P(A), \neg P(B)\}$. Since $KB \big| = P(A) \vee P(B)$, the augmented theory CWA$'$(KB) with respect to the predicate P is inconsistent.

2.4 Non-monotonic Logic

Driven by the development of the intelligence science, various non-classical logics were preposed and studied since the eighties of the last century. Non-monotonic is one of these logics (McDermott, 1980).

The human understanding of the world is a dialectical developing process which obeys the negation-of-negation law. During the cognitive process, man's understanding of the objective world is always uncertain and incomplete; it will be negative or completed as while as some new knowledge is acquired. As pointed by Karl Popper, the process of scientific discovery is a process of falsification. Under certain condition and environment, every theory always has its historical limitations. Along with the increase of human understands of the world and along with the development of scientific research, old theories will not meet the new needs and will be overthrew by the new discovery; upon that, old theories are negated and new theories are born. In this sense, the growth of human knowledge is in fact a non-monotonic development process.

Classical logics such as the formal logic and the deductive logic are all monotonic in their dealing with the human cognitive process. With these logics, new knowledge acquired according to rigorous logic inference must be consistent with the old knowledge. In another word, if there is a knowledge base A and it is known that A implies the knowledge B, i.e. $A \to B$, then the knowledge B can be inferenced by these logics. However, as stated above, human cognitive process is in fact non-monotonic and is not consistent with such a process at all.

Non-monotonic reasoning is characterized by the fact that the theorem set of an inference system is not monotonic increased along with the progress of inference. Formally, let F be the set of knowledge holded by humans at some stage of the cognitive process, and let $F(t)$ be the corresponding function on time t. Then the set $F(t)$ is not monotonic increased along with the progress of time. In another word, $F(t_1) \subseteq F(t_2)$ is not always holds for any $t_1 < t_2$. At the same time, human understanding of the world is in fact enhanced. A basic reason for such a phenomenon is the incomplete knowledge base used in the reasoning process. Non-monotonic logic is a family of tools for the processing of incomplete knowledge.

Inference rules used in monotonic logics are monotonic. Let Γ be the set of inference rules of a monotonic logic, then the language $\text{Th}(\Gamma) = \{A | \Gamma \to A\}$ determined by these rules holds the following monotonicity:

(1) $\in \text{Th}(\Gamma)$
(2) if $\Gamma_1 \subseteq \Gamma_2$, then $\text{Th}(\Gamma_1) \subseteq \text{Th}(\Gamma_2)$
(3) $\text{Th}(\text{Th}(\Gamma)) = \text{Th}(\Gamma)$ (idempotence)
   where (3) is also called as fixed point. A marked feature of monotonic inference rules is that the language determined by them is a bounded least fixed point, i.e., $\text{Th}(\Gamma_1) = \cap\{s | \Gamma_1 \to S \text{ and } \text{Th}(S) = \Gamma_2\}$.

In order to deal with the property of nonmonotonic, the following inference rule is introduced:

(4) if $\Gamma \neg + \neg P$, then $\Gamma | \sim MP$
   Here M is a modal operator. The rule states that if $\neg P$ cannot be deduced from Γ, then P is in default treated as true.
   It is obvious that a fixed point $\text{Th}(\Gamma) = \Gamma$ cannot be guaranteed any more as while as the inference rule (4) is incorporated into monotonic inference systems.

In order to solve this problem, we can first introduce an operator NM as follows: for any first-order theory Γ and any formula set $S \subseteq L$, set

(5) $NM\Gamma(S) = Th(\Gamma \cup AS\Gamma(S))$

where $AS\Gamma(S)$ is a default set of S and is defined as follows:

(6) $AS\Gamma(S) = \{MP|P \in L \vee P \in S\} - Th(\Gamma)$

Then, $Th(\Gamma)$ can be defined as the set of theorems that can be deduced from Γ nonmonotonically, i.e.,

(7) $Th(\Gamma) =$ the least fixed point of $NM\Gamma$

Rule (7) is designed to blend the inference rule (4) into the first-order theory Γ so that reasoning can be carried out with a closed style. However, since the definition of $Th(\Gamma)$ is too strong, not only the calculation but also the existence of $Th(\Gamma)$ cannot be guaranteed. Therefore, definition of $Th(\Gamma)$ is revised as follows:

(8) $Th(\Gamma) = \cap(\{L\} \cup \{S|NM\Gamma(S) = S\})$

Now, let L be the language determined by these rules, then L must be a fixed point according to $NM\Gamma(L) = L$.

Furthermore, according to these rules, Γ is inconsistent if $Th(\Gamma)$ does not exist. The definition of $Th(\Gamma)$ presented in (8) can also be rewritten as follows:

(9) (9) $Th(\Gamma) = \{P|\Gamma| \sim P\}$

where $\Gamma| \sim P$ represent $P \in Th(\Gamma)$. We also use $FP(\Gamma)$ to denote the set $\{S|NM\Gamma(S) = S\}$ and call each element of this set as a fixed point of the theory Γ.

There are three major schools on non-monotonic reasoning: the circumscription theory proposed by McCarthy, the default logic proposed by Reiter, and the autoepistemic logic proposed by Moore. In the circumscription theory, a formula S is true with respect to a limited range if and only if S cannot be proved to be true w.r.t. a bigger range. In the default logic, "a formula S is true in default" means that "S is true if there is no evidence to prove the false of S". In the autoepistemic logic, S is true if S is not believed and there are no facts which are inconsistent with S.

Various non-monotonic logic systems have been proposed by embracing the non-monotonic reasoning into formal logics. These non-monotonic logics can be roughly divided into two categories: non-monotonic logics based on minimization, and non-monotonic logics based on fixed point. Non-monotonic logics based on minimization can again be divided into two groups: one is these based on the minimization of model, such as the logic with the closed world assumption and the circumscription proposed by McCarthy, and the other is these based on the minimization of knowledge model, such as the ignorance proposed by Konolige. Non-monotonic logics based on fixed point can be divided into default logics and autoepistemic logics. The non-monotonic logic (NML) proposed by McDermott and Doyle is a general default logic and was used for study the general foundation of non-monotonic logics, and the default logic proposed by Reiter is a first-order formalization of default rules. Autoepistemic logic was firstly proposed by Moore to solve the so-called Hanks-McDermott problem on nonmonotinic logics.

2.5 Default Logic

Default reasoning is a family of plausible reasoning. The intuition of various forms of default reasoning is to derive conclusions based upon patterns of inference of the following form:

> In the ordinary situation, A holds.
> In the typical situation, A holds.
> Then it is a default assumption that A holds.

A typical example of default reasoning is the statement "birds fly". As we know, the statement "birds fly" is different from the statement "All birds will fly", since there are many exceptions, such as the penguins, ostriches, and Maltese falcon. Given a particular bird, we will conclude that it flies according to the following plausible proposition:

> In the ordinary situation, birds can fly, or
> In the typical situation, birds can fly, or
> if x is a bird, then it is a default assumption that x can flies.

However, if we know that this bird is an ostrich according to the subsequent discovery, we will revise our conclusion with a new result that this bird cannot fly. Therefore, it is obvious that what is reflected in this example is a process of plausible reasoning instead of deductive reasoning.

Based on the study of reasoning about incompletely specified worlds, a logic system named default logic was proposed by Reiter (1980).

"By default" is an ordinary technology used in computer program designing. For example, let P be a program, let Q be a procedure specified in P, and let x be a variable that occurs in both P and Q. Then, the type of x which occurs in P will by default be the type of x which occurs in Q, unless the type of x is redeclared in Q. In other words, with the "by default" technology, operations of the system will be carried out according to predetermined rules, unless other requirements are explicitly specified by the programmer.

The idea of "by default" is introduced into logic by Reiter and forms the so-called default logic. In classical logics, new facts about a world are deduced from the known facts; all the facts that can be deduced are determined by facts contained in the knowledge base. In the default logic, knowledge base can be expanded with default knowledge so that more facts can be deduced; in spite that this default knowledge may be unreliable.

Default rules used in default logic are of the following form:

$$\frac{\alpha(\overline{x}) : M\beta_1(\overline{x}), \ldots, M\beta_m(\overline{x})}{W(\overline{x})} \tag{2.1}$$

It can also be represented as follows:

$$\alpha(\overline{x}) : M\beta_1(\overline{x}), \ldots, M\beta_m(\overline{x}) \rightarrow W(\overline{x}) \tag{2.2}$$

Here $\overline{x}$ is a parameter vector, $\alpha(\overline{x})$ is called the prerequisite of the default rule, $W(\overline{x})$ is the consequent, $\beta_i(\overline{x})$ is the default condition, and M is the default operator. The default rule is to be read as "If the prerequisite $\alpha(\overline{x})$ holds and it is consistent to assume $\beta_1(\overline{x}), \ldots, \beta_m(\overline{x})$, then infer that the consequent holds." For example, consider the following default rule:

$$\frac{bird(x) : M\ flies(x)}{flies(x)}$$

It states that if x is a bird and it is consistent to assume that x can fly, then infer that x can fly.

A default rule is closed if and only if none of $\alpha, \beta_1, \ldots, \beta_m, W$ contains a free variable.

Definition 2.6. A default theory is a pair (D,W), where D is a set of default rules and W a set of closed formulas.

A default theory (D, W) is closed iff every default rule contained in D is closed. Default theory is non-monotonic. For example, suppose $T = \langle W, D \rangle$ is a default theory with $D = \left\{ \frac{:M A}{B} \right\}$ and $W = \emptyset$, then the formula B can be derived from T. However, if we add the knowledge $\neg A$ into W and get the default theory $T' = \langle D, W' \rangle$, where $W' = \{\neg A\}$, then the formula B cannot be derived from T' anymore, despite that T' is an extension of T together with $W' \supseteq W$.

Example 2.4. Suppose $W = \{bird(tweety), \forall x(ostrich(x) \rightarrow \neg flies(x))\}$ and

$$D = \left\{ \frac{bird(x) : M\ flies(x)}{flies(x)} \right\}$$

Then the formula flies(tweety) can be deduced from the default theory. However, if we add the knowledge ostrich(tweety) into W, then flies(tweety) cannot be deduced any more.

Example 2.5. Suppose $W = \{feathers(tweety)\}$ and

$$D = \left\{ \frac{bird(x) : M\ flies(x)}{flies(x)}, \frac{feathers(x) : M\ bird(x)}{bird(x)} \right\}$$

Then the formula flies(tweety) can be deduced from the default theory. However, it cannot be deduced any more if we add the following knowledge into W:

$$ostrich(tweety),$$
$$\forall x(ostrich(x) \rightarrow \neg flies(x))$$
$$\forall x(ostrich(x) \rightarrow feathers(x)).$$

Definition 2.7. Let $\Delta = <D, W>$ be a closed default theory. Γ is an operator defined w.r.t. Δ such that, for any set S of closed formulas, $\Gamma(S)$ is the smallest set satisfying the following three properties:

(1) $W \subseteq \Gamma(S)$.
(2) $\Gamma(S)$ is deductively closed, i.e., $\text{Th}(\Gamma(S)) = \Gamma(S)$.
(3) For any default rule $\alpha : M\beta_1, \ldots, M\beta_m \rightarrow w$ contained in D, if $\alpha \in \Gamma(S)$ and $\neg\beta_1, \ldots, \neg\beta_m \notin S$, then it must be $w \in \Gamma(S)$.

Definition 2.8. A set E of closed formulas is an extension for $\Delta = <D, W>$ iff E is a fixed point of the operator Γ w.r.t. Δ, i.e., iff $\Gamma(E) = E$.

Definition 2.9. A formula F can be deduced from a default theory $\Delta = <D, W>$, in symbols $\Delta| \sim F$, iff F is contained in the extension of Δ.

Example 2.6. Suppose $D = \left\{ \frac{:M A}{\neg A} \right\}$ and $W = \emptyset$. Then the default theory $\Delta = <D, W>$ has no extension.

The result of this example can be demonstrated as follows. Suppose there is a fixed point E of the operator Γ w.r.t. Δ, then we have the following: (a) If $\neg A \notin E$, we will get $\neg A \in E$ according to the third property of Definition 5.2 and arrive at a contradiction. (b) If $\neg A \in E$, then the default rule of D must have been applied in such a way that $\neg A$ was added into E, therefore it must be $\neg A \notin E$, otherwise the rule cannot be applied. So, we arrive at a contradiction again. As a result, there is no fixed point of the operator Γ w.r.t. Δ, i.e., the default theory $\Delta = <D, W>$ has no extension.

Example 2.7. Suppose $D = \{\frac{:MA}{\neg B}, \frac{:MB}{\neg C}, \frac{:MC}{\neg F}\}$, $W = \emptyset$. Then the default theory $\Delta = <D, W>$ has a unique extension $E = Th(\{\neg B, \neg F\})$.

For this example, it is easy to demonstrate that E is a fixed point of the operator Γ w.r.t. Δ. However, for any set $S \subseteq \{\neg B, \neg C, \neg F\}$ except $\{\neg B, \neg F\}$, we can demonstrate that Th(S) is not a fixed point of Γ w.r.t. Δ.

Example 2.8. Suppose $D = \{\frac{:MA}{A}, \frac{B:MC}{C}, \frac{F\vee A:ME}{E}, \frac{C\wedge E:M\neg A, M(F\vee A)}{G}\}$, $W = \{B, C \rightarrow F \vee A, A \wedge C \rightarrow \neg E\}$. Then there are three extensions for the default theory $\Delta = <D, W>$:

$$E1 = Th(W \cup \{A, C\})$$

$$E2 = Th(W \cup \{A, E\})$$

$$E3 = Th(W \cup \{C, E, G\})$$

According to the above example, we can see that not all default theories have their extensions; at the same time, the number of extensions for a default theory is not limited to one. Effective default reasoning on a default theory is based on the existence of extensions. Therefore, it is important to study and discuss the conditions for the existence of extension.

Theorem 2.3. *Let E be a set of closed formulas, and let* $\Delta = <D, W>$ *be a closed default theory. Define E0* $= W$ *and for* $i > 0$ *it is*

$$E_{i+1} = \text{Th}(E_i) \cup \{w | (\alpha{:}M\beta_1, \ldots, M\beta_m \rightarrow w) \in D, \alpha \in E_i, \neg\beta_1, \ldots, \neg\beta_m \notin E\}$$

Then E is an extension for Δ iff $E = \cup_{i=0}^{\infty} E_i$.

With this theorem, the three extensions of Example 2.8. can be examined to be right.

There is a special default rule $\frac{:M\neg A}{\neg A}$. A natural question about it is whether the extension of a default theory determined by this default rule is the same as the corresponding CWA-augmented theory. The answer to this question is negative. For example, suppose $W = \{P \vee Q\}$ and $D = \{\frac{:M\neg P}{\neg P}, \frac{:M\neg Q}{\neg Q}\}$. Then it is obvious that CWA(Δ) is inconsistent, but the sets $\{P \vee Q, \neg P\}$ and $\{P \vee Q, \neg Q\}$ are all consistent extensions for Δ.

Example 2.9. Suppose $D = \{\frac{:MA}{\neg A}\}$, $W = \{A, \neg A\}$. Then the extension for $\Delta = <D, W>$ is $E = \text{Th}(W)$.

This example is surprising since the extension for Δ is inconsistent. In fact, some conclusions on the inconsistency of extensions have been summed up as follows:

(1) A closed default theory $<D, W>$ has an inconsistent extension if and only if the formula set W is inconsistent.

Let E be an extension for $<D, W>$. The result can be demonstrated as follows. On the one hand, if W is inconsistent, then the extension E is also inconsistent since $W \subseteq E$. On the other hand, if E is inconsistent, then any default rule of D can be applied since any formula can be deduced from E; therefore, according to Theorem 2.3, we will get the result that $E = T_h(W)$. So, W is also inconsistent.

(2) If a closed default theory has an inconsistent extension, then this is the unique extension for this default theory.

In the case that there is more than one extension for a default theory, some conclusions on the relationship between these extensions have been summed up also as follows:

(3) If E and F are extensions for a closed normal default theory and if $E \subseteq F$, then $E = F$.
(4) Suppose $\Delta_1 = <D_1, W_1>$ and $\Delta_2 = <D_2, W_2>$ are two different default theories, and that $W_1 \subseteq W_2$. Suppose further that extensions of Δ_2 are consistent. Then extensions of Δ_1 are also consistent.

Definition 2.10. A default rule is normal iff it has the following form:

$$\frac{A : MB}{B} \tag{2.3}$$

where A and B are any formulas. A default theory $\Delta = <D, W>$ is normal iff every default rule of D is normal.

Normal default theories hold the following properties:

(1) Every closed normal default theory has an extension.
(2) Suppose E and F are distinct extensions for a closed normal default theory, then $E \cup F$ must be inconsistent.
(3) Suppose $\Delta = <D, W>$ is a closed normal default theory, and that $D' \subseteq D$. Suppose further that E'_1 and E'_2 are distinct extensions of $<D', W>$. Then Δ has distinct extensions E_1 and E_2 such that $E'_1 \subseteq E_1$ and $E'_2 \subseteq E_2$.

2.6 Circumscription Logic

Circumscription logic (CIRC) is proposed by McCarthy for non-monotonic reasoning. The basic idea of circumscription logic is that "the objects that can be shown to have a certain property P by reasoning from certain facts A are all the objects that satisfy P" (McCarthy, 1980). During the process of human informal reasoning, the objects that have been shown to have a certain property P are often treated as all the objects that satisfy P; such a treatment will be used in further reasoning and will not be revised until other objects are discovered to have the property P. For example, it is ever guessed by the famous mathematician Erdos that the mathematical equation $xxyy = zz$ has only two trivial solutions: $x = 1$, $y = z$ and $y = 1$, $x = z$. But later it was proved by Chinese mathematical Zhao He that this mathematical equation has infinite number of trivial solutions and therefore overthrew Erdos's guess.

Circumscription logic is based on minimization. In the following, starting with a propositional circumscription which is based on minimal model, we first introduce basic definitions of circumscription. Then we introduce some basic results on predicate circumscription.

Definition 2.11. Let p_1, p_2 be two satisfying truth assignments for a propositional language L_0. Then p_1 is called smaller than p_2, written as $p_1 \succeq p_2$, if and only if $p_2(x) = l$ for any proposition x which holds $p_1(x) = l$.

Definition 2.12. Let p be a satisfying truth assignment of a formula A. We say that p is a minimal satisfying assignment of A if and only if there is no other satisfying truth assignment p' of A such that $p' \succeq p$.

Definition 2.13. A formula B is called a minimal entailment of a formula A, written as $A \models_M B$, if and only if B is true with respect to any minimal model of A.

Minimal model is non-monotonic. The following example reflects the property of minimal model:

$$p \models_M \neg q$$

$$p \vee q \models_M \neg p \vee \neg q \tag{2.4}$$

$$p, q, p \vee q \models_M p \wedge q$$

Definition 2.14. Let $Z = \{z_1, z_2, \ldots, z_n\}$ be all the propositions occurring in a formula A. Then, a satisfying truth assignment P is called a $\succeq^Z$-minimal satisfying assignment of A if and only if there is no other satisfying truth assignment P' of A such that $P \succeq^Z P'$. Where, $P \succeq^Z P'$ if and only if $P'(z) = l$ for any proposition z which holds $z \in Z$ and $P(z) = l$.

Definition 2.15. Let $P = \{p_1, p_2, \ldots, p_n\}$ be all the propositions occurring in a formula A. Then, a formula φ is entailed by the propositional circumscription of P in A, written as $A \models_P \varphi$, if and only if φ is true with respect to any $\succeq^Z$-minimal satisfying assignment of A.

The propositional circumscription CIRC(A, P) is defined as the following formula:

$$A(P) \wedge \forall P'(A(P') \wedge P' \to P)) \to (P \to P') \tag{2.5}$$

where $A(P')$ is the result of replacing all the occurrences of P in A with P'. If we use $P' \succ P$ to replace $P' \to P$, then CIRC(A, P) can also be rewritten as follows:

$$A(P) \wedge \neq \exists P'(A(P') \wedge P' \succ P) \tag{2.6}$$

Therefore, logical inferences in the propositional circumscription can be represented as schemas of the form $A \models_P \varphi$ or CIRC(A, P) $\models \varphi$. The following theorem on the soundness and completeness has been proved:

Theorem 2.4. $A \models_p \varphi$ *if and only if* $A \models_p \varphi$.

In the following, we advance the idea of propositional circumscription into predicate circumscription.

Definition 2.16. Let T be a formula of a first-order language L, and let ρ be a set of predicates contained in T. Let $M[T]$ and $M^*[T]$ be two models of T. Then, $M^*[T]$ is called smaller than $M[T]$, written as $M^*[T] \succeq M[T]$, if and only if

(1) M and M^* have the same domain,
(2) all the relations and functions occurring in T, except these contained in ρ, have the same interpretation in M and M^*,
(3) the extension of ρ in the M^* is a subset of ρ in the M.

A model M of T is called $\succeq$ P-minimal if and only if there is no other model M' of T such that $M \succeq_P M'$.

Definition 2.17. M_m is a minimal model of ρ if and only if $M = M_m$ for any model M such that $M \succeq_\rho M_m$.

For example, let the domain be $D = \{1, 2\}$,

$$T = \forall x \exists y (P(y) \wedge Q(x, y))$$

$$= [(P(1) \wedge Q(1, 1)) \vee (P(2) \wedge Q(1, 2))]$$

$$\wedge [(P(1) \wedge Q(2, 1)) \vee (P(2) \wedge Q(2, 2))]$$

Let M and M* be the following models:

M:	$P(1)$	$P(2)$	$Q(1, 1)$	$Q(1, 2)$	$Q(2, 1)$	$Q(2, 2)$
	True	True	False	True	False	True
M^*:	$P(1)$	$P(2)$	$Q(1, 1)$	$Q(1, 2)$	$Q(2, 1)$	$Q(2, 2)$
	False	True	False	True	False	True

Then, model M and model M^* have the same true assignments on Q. At the same time, P is true in both (1) and (2) of model M; however, for model M^*, P is true in just (2). Therefore, we have $M^* \succeq_P M$. Furthermore, since $M^* \neq M$, we have $M^* \succ_P M$.

Let T be a set of beliefs, and let P be a predicate that occurs in T. During the extension process, we should seek formula φ_p such that for any model M of $T \wedge \varphi_P$ there is no model M^* of T which satisfies

$$M^* \succ_P M$$

The formula $T \wedge \varphi_P$ which satisfies such a principle of minimization is called the circumscription of P on T.

Let P^* be a predicate constant which has the same number of variables as that of P. Then, it can be demonstrated that any model of the following formula is a minimal model of P on T:

$$(\forall x\, P^*(x) \to P(x)) \wedge \neg(\forall x\, P(x) \to P^*(x)) \wedge T(P^*)$$

Therefore, any model of the following formula is a minimal model of P on T:

$$\neg((\forall x\, P^*(x)) \to P(x)) \wedge \neg(\forall x\, P(x) \to P^*(x)) \wedge T(P^*))$$

As a result, the following is a circumscription formula of P on T:

$$\phi_P = \forall P^* \neg((\forall x\, P^*(x) \to P(x)) \wedge \neg(\forall x\, P(x) \to P^*(x)) \wedge T(P^*))$$

Definition 2.18. A formula ϕ is entailed by the predicate circumscription of P in A, written as $T \models_P \phi$ or $\mathrm{CIRC}(T, P) \models \phi$, if and only if ϕ is true with respect to all the $\succeq^P$-minimal model of P.

The predicate circumscription $\mathrm{CIRC}(T, P)$ of P in T is defined as

$$\mathrm{CIRC}(T, P) = T \wedge \forall P^* \neg((\forall x)(P^*(x) \to P(x)) \wedge \neg(\forall x)(P(x)$$
$$\to P^*(x)) \wedge T(P^*)) \tag{2.7}$$

It can also be rewritten as

$$\mathrm{CIRC}(T, P) = T \wedge \forall P^*((T(P^*) \wedge (\forall x)(P^*(x) \to P(x)))$$
$$\to (\forall x)(P(x) \to P^*(x))) \tag{2.8}$$

Since it is a formula of high-order logic, we can rewrite it as

$$\phi_P = \forall P^*((T(P^*) \wedge (\forall x)(P^*(x) \to P(x))) \to (\forall x)(P(x) \to P^*(x))) \tag{2.9}$$

It states that if there is a P^* such that $T(P^*)$ and $\forall x\, (P^*(x) \to P(X))$, then $\forall x\, (P(x) \to P^*(x))$ can be deduced as a conclusion.

If we use $P \wedge P'$ to replace P^* (here P' is a predicate constant with the same number of variables as that of P), then $\text{CIRC}(T, P)$ can be written as

$$\phi_P = T(P \wedge P')\forall x(P(x) \wedge P'(x) \to P(x)) \to \forall x)(P(x) \to P(x) \wedge P'(x)) \tag{2.10}$$

And therefore we get the following formula:

$$T(P \wedge P') \to (\forall x)(P(x) \to P'(x)) \tag{2.11}$$

If we replace $(\forall x)(P^*x) \to P(x))$ by $P^* \succeq P$, then

$$P^* \succ P \text{ represent}(P^* \succeq P) \wedge \neg(P \succeq P^*) \text{ and}$$

$$P^* = P \text{ represent}(P^* \succeq P) \wedge (P \succeq P^*)$$

And therefore we get

$$\phi_P = \forall P^*(T(P^*) \wedge (P^* \succeq P) \to (P \succeq P^*)) \tag{2.12}$$

$$\text{i.e., } \phi_P = \forall P^*(T(P^*) \to \neg(P^* \succ P))$$

$$= \neg(\exists P^*)(T(P^*) \wedge (P^* \succ P)) \tag{2.13}$$

Theorem 2.5. *Let T be a formula of a first-order language, and let P be a predicate contained in T. Then, for any P' such that $T(P) \vdash T(P') \wedge (P' \succeq P)$, it must be*

$$\text{CIRC}(T, P) = T(P) \wedge (P = P') \tag{2.14}$$

According to this theorem, if $T(P') \wedge (P' \succeq P)$ can be deduced from $T(P)$, then $P = P'$ is the circumscription formula of P in T.

2.7 Non-Monotonic Logic NML

The non-monotonic logic NML proposed by McDermott and Doyle is a general default logic for the study of general foundation of non-monotonic logics (McDermott & Doyle, 1980). McDermott and Doyle modify a standard first-order logic by introducing a modal operator $\Diamond$, which is called compatibility operator. For example, the following is a formula for NML:

$$\forall x(\text{Bird}(x) \wedge \Diamond \text{Fly}(x) \to \text{Fly}(x))$$

It states that if x is a bird and it is consistent to assert that x can fly, then x can fly.

According to the example, it is obvious that default assumptions of default theory can be represented in NML, and therefore default theory can be treated as a special case of NML. However, in non-monotonic logic, $\Diamond A$ is treated as a proposition in the formation of formulas, but in default theory, $\Diamond A$ can only appear in default rules. Therefore, there are many fundamental differences between NML and default theory.

In the following, starting with the compatibility operator $\Diamond$, we give an introduction to the non-monotonic reasoning mechanisms.

First, according to the intuitive sense of $\Diamond$, we might introduce the following rule from the point of syntax:

$$\text{if} | \text{-}/\neg A, \text{ then} | \text{-}\Diamond A$$

It states that if the negation of A is not derivable, then A is compatible. We can see that rules like this are in fact unsuitable since the negation of each formula which is not a theorem will be accepted as a formula, and consequently the non-monotonic is eliminated.

Therefore, McDermott and Doyle adopted a different form as follows:

$$\text{if} | \text{-}/\neg A, \text{ then} | \sim \Diamond A$$

Here the notation $| \sim$ is introduced to represent non-monotonic inference, just like that used in default theory.

We can also distinguish $| \sim$ from the inference relation $|$- of first-order logic according to the following discussion. We know that in the monotonic first-order logic, it is

$$T \subseteq S \rightarrow Th(T) \subseteq Th(S)$$

Suppose

$$T | \text{- fly(tweety)} \tag{2.15}$$

and

$$S = T \cup \{\neg \text{fly(tweety)}\} \tag{2.16}$$

Then, since $T |$- fly(tweety) and $T \subseteq S$, we will get

$$S | \text{- fly(tweety)} \tag{2.17}$$

At the same time, since $\neg \text{fly(tweety)} \in S$, we have

$$S | \text{- } \neg \text{fly(tweety)} \tag{2.18}$$

Therefore, it is obvious that $Th(T) \subseteq Th(S)$ does not hold. So, the notation $| \sim$ is different from $|$-.

Let FC be a first-order predicate calculus system with the compatibility operator $\Diamond$ embraced in, and let LFC be the set of all the formulas of FC. Then, for any set $\Gamma \subseteq$ LFC, Th(Γ) is defined as

$$\text{Th}(\Gamma) = \{A|\Gamma \vdash_{FC} A\} \tag{2.19}$$

Th(Γ) can also be defined according to another approach. For any set $S \subseteq$ LFC, a non-monotonic operator NMΓ is first defined as

$$NM_\Gamma(S) = \text{Th}(\Gamma \cup ASM_\Gamma(S))$$

where ASM$_\Gamma$ (S) is the assumption set of S and is defined as

$$ASM_\Gamma(S) = \{\Diamond Q|Q \in L_{FC} \wedge \neg Q \notin S\}$$

then Th(Γ) can be defined as

$$\text{Th}(\Gamma) = \cap(\{L_{FC}\} \cup \{S|NM_\Gamma(S) = S\}) \tag{2.20}$$

According to this definition, we can see that Th(Γ) is the intersection of all fixed points of NMΓ, or the entire language if there are no fixed points.

Now, the non-monotonic inference $|\sim$ can be defined as $\Gamma|\sim P$ if and only if $P \in$ Th(Γ).

It should be noted that $\Gamma|\sim P$ requires that P is contained in each fixed point of NMΓ in the case that there are fixed points. However, in default theory, what is needed for P to be provable in Δ is just that P is contained in one of Δ's extensions, i.e., P is contained in one of the fixed points.

Example 2.10. Suppose Γ is an axiom theory which contains $\Diamond P \rightarrow \neg Q$ and $\Diamond Q \rightarrow \neg P$, i.e.,

$$\Gamma = FC \cup \{\Diamond P \rightarrow \neg Q, \Diamond Q \rightarrow \neg P\}$$

Then there are two fixed points for this theory: $(P, \neg Q)$ and $(\neg P, Q)$.

However, for another theory $\Gamma = FC \cup \{\Diamond P \rightarrow \neg P\}$, we can demonstrate that it has no fixed points. The demonstration is as follows. Suppose $NM\Gamma(S) = S'$. If $\neg P \notin S$, then we will have $\Diamond P \in ASM\Gamma$ (S) and consequently $\neg P \in S'$; on the contrary, if $\neg P \in S$, then we will have $\Diamond P \notin ASM\Gamma(S)$, and consequently $\neg P \notin S'$. Therefore, S will never be equal to S', i.e., there is no fixed point for NMΓ.

The above phenomenon can be further explained according to the following results:

$$\{\Diamond P \to \neg Q, \Diamond Q \to \neg P\}| \sim (\neg P \vee \neg Q)$$

$$\{\Diamond P \to \neg P\}| \sim \text{contradiction}$$

McDermott and Doyle pointed out the following two problems on the reasoning process of *NML*:

(1) $\Diamond A$ cannot be deduced from $\Diamond(A \wedge B)$.
(2) What can be deduced from $\{\Diamond P \to Q, \neg Q\}$ is surprising.

In order to overcome these problems, McDermott and Doyle introduced another modal operator $\Box$ called necessity. The relationship between $\Diamond$ and $\Box$ is as follows:

$$\Box P \equiv \neg \Diamond \neg P$$

$$\Diamond P \equiv \neg \Box \neg P$$

Here the first definition states that P is necessary if and only if its negation is incompatible; the second definition states that P is compatible if and only if its negation is not necessary.

2.8 Autoepistemic Logic

Autoepistemic logic was proposed by Moore as an approach to represent and reason about the knowledge and beliefs of agents (Moore, 1985). It can be treated as a modal logic with a modal operator B which is informally interpreted as "believe" or "know". Once the beliefs of agents are represented as logical formulas, then a basic task of autoepistemic logic is to describe the conditions which should be satisfied by these formulas. Intuitively, an agent should believe these facts that can be deduced from its current beliefs. Furthermore, if an agent believes or does not believe some fact, then the agent should believe that it believes or does not believe this fact.

An autoepistemic theory T is sound with respect to an initial set of premises A if and only if every autoepistemic interpretation of T in which all the formulas of A are true is an autoepistemic model of T. The beliefs of an ideally rational agent should satisfy the following conditions:

(1) If $P_1, \ldots, P_n \in T$ and $P_1, \ldots, P_n \vdash Q$, then $Q \in T$ (where $\vdash$ means ordinary tautological consequence).
(2) If $P \in T$, then $BP \in T$.
(3) If $P \notin T$, then $\neg BP \in T$.

No further conditions could be drawn by an ideally rational agent in such a state; therefore, the state of belief characterized by such a theory is also described by Moore as a stable autoepistemic theory. If a stable autoepistemic theory T is consistent, it will satisfy the following two conditions:

(4) If $BP \in T$, then $P \in T$.
(5) If $\neg BP \in T$, then $P \notin T$.

An autoepistemic logic named $\mathcal{L}_B$ was proposed and studied by Moore. This logic is built up of a countable set of propositional letters, the logical connectives $\neg$ and $\wedge$, and a modal connective B.

2.9 Truth Maintenance System

Truth Maintenance System (TMS) is a problem-solver subsystem for recording and maintaining beliefs in knowledge base (Doyle, 1979). The relationship between TMS and default inference is similar to the relationship between production system and first-order logic. A truth maintenance system is composed of two basic operations: (a) Make assumptions according to incomplete and finite information, and take these assumptions as a part of beliefs. (b) Revise the current set of beliefs when discoveries contradict these assumptions.

There are two basic data structures in TMS: nodes, which represent beliefs, and justifications, which represent reasons for beliefs. Some fundamental actions are supported by the TMS. First, it can create a new node, to which some statements of a belief will be attached. Second, it can add (or retract) a new justification for a node, to represent a step of an argument for the belief represented by the node. Finally, the TMS can mark a node as a contradiction, to represent the inconsistency of any set of beliefs which enter into an argument for the node. In this case, the TMS invokes the truth maintenance procedure to make any necessary revisions in the set of beliefs. The TMS locates the set of nodes to update by finding those nodes whose well-founded arguments depend on changed nodes. When this happens, another process of the TMS, dependency-directed backtracking, is also carried out to analyze the well-founded argument of the contradiction node; then the contradiction can be eliminated accordingly to locate and delete the assumptions occurring in the argument.

The TMS provides two services: truth maintenance and dependency-directed backtracking. Both of these services are carried out on the basis of the representation of reasons for beliefs.

1. Representation of Reasons for Beliefs

A node may have several justifications, each justification representing a different reason for believing the node. A node is believed if and only if at least one of its justifications is valid, i.e., at least one of its justifications can be deduced from the current knowledge base (where these beliefs generated according to assumptions are also included in this knowledge base).

In the TMS, each proposition or each rule can be represented as a node. Each node is of the following two types:

(1) the IN-node which has at least one valid justification and
(2) the OUT-node which has no valid justifications.

Therefore, there are four states for the knowledge of each proposition p: an IN-node for p, an OUT-node for p, an IN-node for $\neg p$, and an OUT-node for $\neg p$.

Each node has its justifications. The TMS employs two forms for justifications, called support-list (SL) and conditional-proof (CP) justifications. The former is used to represent reasons for believing the node, while the latter is used to record the reasons for contradiction.

Each SL justification is of the following form:

$$(SL(\text{<IN-list>})(\text{<OUT-list>})) \tag{2.21}$$

An SL justification is valid if and only if each node in its IN-list is IN-node, and each node in its OUT-list is OUT-node.

For example, consider the following SL justifications:

(1) It is now summer. $(SL\,(\,)\,(\,))$
(2) The weather is very humid. $(SL\,(1)\,(\,))$

In this example, IN-list and OUT-list of the SL justification of node (1) are all empty, which means that the justification of node (1) is always valid and therefore node (1) will always be an IN-node. We call nodes of this type as premise. IN-list of the SL justification of node (2) is composed of node (1), which means that node (2) is believed if node (1) is an IN-node. According to this example, we can see that the inference of TMS is in fact similar to the inference of predicate logic. The difference between them is that premises in the TMS can be retracted and correspondingly the knowledge base can be revised.

Based on the above example, we add an item to the OUT-list of node (2) and get the following SL justifications:

(1) It is now summer. $(SL\,(\,)\,(\,))$
(2) The weather is very humid. $(SL\,(1)\,(3))$
(3) The weather is very dry.

In this case, the condition for node (2) to be believed is that node (1) is an IN-node and node (3) is an OUT-node. All of these SL justifications state that "if it is now summer and there is no evidence to prove that the weather is very dry, then it can be derived that the weather is very humid". We call nodes whose SL justification has a non-empty OUT-list as assumptions.

Each CP justification is of the following form:

$$(\text{CP} <\text{consequent}> <\text{IN-hypotheses}> <\text{OUT-hypotheses}>) \tag{2.22}$$

A CP justification is valid if (1) the consequent node is an IN-node, (2) each node of the IN-hypotheses is an IN-node, and (3) each node of the OUT-hypotheses is an OUT-node.

The set of hypotheses must be divided into two disjoint subsets, since nodes may be derived both from some IN-nodes and some OUT-nodes.

2. Default Assumptions

Let $\{F_1, \ldots, F_n\}$ be the set of alternative default nodes, and let G be a node which represents the reason for making an assumption to choose the default. To make Fi the default, justify it with the following SL justification:

$$(SL\,(G)\,(F_1, \ldots, F_{i-1}, F_{i+1}, \ldots, F_n)) \tag{2.23}$$

If no additional information about the value exists, none of the alternative nodes except F_i will have a valid justification, so F_i will be an IN-node and each F_j with $j \neq i$ will be an OUT-node. However, if a valid justification is added to some other alternative node and causes that alternative to become an IN-node, then the above SL justification will be invalid and make F_i an OUT-node. Consider the case that F_i has been selected as the default assumption and a contradiction is derived from F_i, then the dependency-directed backtracking mechanism will recognize F_i as an assumption because it depends on the other alternative nodes being OUT. The backtracker may then justify one of the other alternative nodes, say F_j, and make F_i an OUT-node where the backtracker-produced justification for F_j will have the following form:

$$(SL <\text{various nodes}> <\text{remainder nodes}>) \tag{2.24}$$

where <remainder nodes> represent the set of nodes except F_i and F_j.

The above approach will not work in the case that the complete set of alternatives cannot be known in advance but must be discovered piecemeal. To solve this problem, we can use a slightly different set of justifications with which the set of alternatives can be gradually extended.

Retain the above notation and let $\neg F_i$ be a node which represents the negation of F_i. Then, arrange F_i to be believed if $\neg F_i$ is an OUT-node, and set up justifications so that if F_j is distinct from F_i, then F_j supports $\neg F_i$, i.e., F_i is justified with

$$(SL\ (G)\ (\neg F_i)) \qquad (2.25)$$

and $\neg F_i$ is justified with

$$(SL\ (F_j)\ (j \neq i)) \qquad (2.26)$$

where F_j is an alternative distinct from F_i.

According to these justifications, F_i will be assumed if no reasons exist for using any other alternative. However, if some contradiction is derived from F_i, then $\neg F_i$ will become an IN-node and correspondingly F_i will become an OUT-node. The dependency-directed backtracking mechanism will be used to recognize the cause of the contradiction and construct a new default assumption.

3. Dependency-Directed Backtracking

When the TMS makes a contradiction node as an IN-node, it will invoke the dependency-directed backtracking to find and remove at least one of the current assumptions in order to make the contradiction node an OUT-node. Let C be the contradiction node. The dependency-directed backtracking is composed of the following three steps:

Step 1. Trace through the foundations of the contradiction node C to find the set $S = \{A_1, \ldots, A_n\}$ which is composed of maximal assumptions underlying C. Where A_i is called a maximal assumption underlying C if and only if A_i is in C's foundations and there is no other assumption B in the foundations of C such that A_i is in the foundations of B.

Step 2. Create a new node NG to represent the inconsistency of S. NG is also called a nogood node for representing the following formula:

$$A_1 \wedge \ldots \wedge A_n \rightarrow \text{false}$$

which is equivalent with

$$\neg(A_1 \wedge \ldots \wedge A_n) \qquad (1)$$

Node NG has the following CP justification:

$$(CP\ C\ S\ ()) \qquad (2)$$

Step 3. Select some maximal assumption A_i from S. Let $D_1, \ldots, D_k$ be the OUT-nodes in the OUT-list of A_i's supporting justification. Select D_j from this set and justify it with

$$(SL \ (NG \ A_1 \ldots A_{i-1} A_{i+1} \ldots A_n) \ (D_1 \ldots D_{j-1} D_{j+1} \ldots D_k)) \tag{3}$$

If the TMS finds other arguments so that the contradiction node C is still an IN-node after the addition of the new justification for Dj, repeat this backtracking procedure.

As an example, consider a program scheduling a meeting. First, suppose the date for the meeting is Wednesday. The corresponding knowledge base is as follows:

(1) The date for the meeting is Wednesday (*SL* ( ) (2)).
(2) The date for the meeting is not Wednesday.

Here, node (1) is an IN-node since there are no arguments for the statement "the date for the meeting is not Wednesday".

Next, suppose it can be deduced from beliefs represented in other nodes, node (32), node (40), and node (61), such that the time for the meeting is 14:00. Then the corresponding knowledge base of the TMS is as follows:

(1) The date for the meeting is Wednesday (*SL* ( ) (2)).
(2) The date for the meeting is not Wednesday.
(3) The time for the meeting is 14:00 (*SL* (32, 40, 61) ( )).
   Now suppose a previously scheduled meeting rules out the combination of the data of Wednesday and the time of 14:00, by supporting a new node with node (1) and node (3) and then declaring this new node to be a contradiction:
(4) Contradiction (*SL* (1, 3) ( ))
   Then the dependency-directed backtracking system will trace the foundations of node (4) to find two assumptions, (1) and (3), both maximal. Correspondingly, the following nogood node is constructed to record the result.
(5) nogood (*CP* 4 (1, 3) ( ))

The TMS arbitrarily selects node (1) and justifies (1)'s only OUT antecedent (2) and correspondingly changes node (2) as follows:

(2) The date for the meeting is not Wednesday (*SL* (5) ( ))

Now, node (2) and node (5) are IN-nodes, and consequently node (1) and node (4) are OUT-nodes. Therefore, the contradiction is eliminated.

De Kleer pointed out some limitations of TMS and correspondingly proposed an assumption-based TMS (ATMS) (de Kleer, 1986). A typical characteristic of ATMS is the capability of working with multiple contradictory assumptions at once.

The ATMS consists of two components: a problem solver and a TMS. The problem solver includes all domain knowledge and inference procedures. Every inference made is communicated to the TMS. The TMS's job is to determine what data are believed and disbelieved given the justification records thus far.

An ATMS justification describes how a node is derivable from other nodes and is of the following form:

$$A_1, A_2, \ldots, A_n \Rightarrow D$$

where D is the node being justified and is called the consequent and $A_1, A_2, \ldots, A_n$ is a list of nodes and is called the antecedents. The non-logical notation "$\Rightarrow$" is used here because the ATMS does not allow negated literals and treats implication unconventionally.

Limited to the space, a detailed discussion of ATMS is omitted here. Readers may refer to the relevant literature.

2.10 Situation Calculus

Action is a basic concept in many branches of computer science. For example, in the branch of database theory, delete, insert, and update of data are frequently used operations (or actions). These operations play an important role in the database. Another example is the multi-agent system of distributed artificial intelligence, where various behaviors (or actions) of agents are the basis of the cooperation of agents. The knowledge and beliefs of agents are an important research topic for multi-agent systems, where the update and revision of knowledge and beliefs are also based on the study of action theory.

Situation calculus is the most commonly used formalism for the study and process of actions. With respect to the progress of a database, Fangzhen Lin and Reiter embed situation calculus into a many-sorted first-order logic framework **LR** and established a formal foundation for action (Lin & Reiter, 1994). In the **LR** framework, individuals are divided into three sorts: state, action, and object. Based on these three sorts, to characterize an action, they described the precondition (under which the action could be performed) and the effect (the change of the world after the execution of the action) by sentences of **LR**. In **LR**, a system with actions was treated uniformly as a logic theory which was called basic action theory. The **LR** framework and the corresponding basic action theory provide a theoretical foundation for the study of actions.

Based on the **LR** framework, we present a many-sorted logic for the representation and reasoning about actions (Tian *et al.*, 1997). In this logic, actions are treated as functions rather than individual sorts. Such a treatment is more coincident with the intuitive understanding of actions and also has a clear semantics in model theory. In the logic, the so-called minimal action theory is introduced to characterize systems with actions; a model theory is correspondingly established to analyze the progression in minimal action theory; finally, some results on the definability of progression in minimal action theory are presented.

2.10.1 *Many-Sorted Logic for Situation Calculus*

LR is defined to be a many-sorted first-order logic (Lin & Reiter, 1994). In its signature $\mathcal{L}$, there are three sorts: sort s for state, sort o for object, and sort a for action. The constant S_0 of sort s is used to denote the initial state. There is a distinguished binary function **do** of type $<a, s; s>$; do(a, s) denotes the successor state to s resulting from performing the action a. The binary relation **Poss** of type $<a, s>$ is also introduced; Poss(a, s) means that it is possible to perform the action a in the state s. Finally, a binary relation $<$ of type $<s, s>$ is introduced to denote the sequential relation between states.

A relation symbol is called to be state independent if all its parameters are of sort o. A function symbol is called to be state independent if all of its parameters as well as its function value are of sort o. A relation symbol is called to be a fluent if there is a parameter of sort s while the other parameters are of sort o. The number of state-independent relations, state-independent functions, and fluents are all supposed to be limited in $\mathcal{L}$.

Term, atom formula, and formula of **LR** are defined in the usual way.

For any state term st, $\mathcal{L}_{st}$ is defined to be the subset of $\mathcal{L}$ that does not mention any other state terms except st, does not quantify over state variables, and does not mention Poss and $<$. Formally, $\mathcal{L}_{st}$ is the smallest set satisfying the following conditions:

(1) If $\psi \in \mathcal{L}$ and it does not mention any state term, then $\psi \in \mathcal{L}_{st}$.
(2) For every fluent $F(x_1, \ldots, x_n, st) \in \mathcal{L}_s$, $F(x_1, \ldots, x_n, st) \in \mathcal{L}_{st}$.
(3) If $\psi, \varphi \in \mathcal{L}_{st}$, then $\neg\psi$, $\psi \wedge \varphi$, $\psi \vee \varphi$, $\varphi \to \psi$, $\psi \leftrightarrow \varphi$, $(\forall x)\psi$, $(\exists x)\psi$, $(\forall a)$ ψ and $(\exists a)\psi$ are all in $\mathcal{L}_{st}$, where x and a are variables of sort o and sort a, respectively.

$\mathcal{L}_{st}^2$ is defined to denote the second-order extension of $\mathcal{L}_{st}$ by n-ary predicate variables on the domain of sort o, n $\geq$ 0. Formally, $\mathcal{L}_{st}^2$ is the smallest set satisfying the following conditions:

(1) $\mathcal{L}_{\text{st}} \subseteq \mathcal{L}_{\text{st}}^2$.

(2) If p is an n-ary predicate variable on the domain of sort o, and $x_1, \ldots, x_n$ are terms of sort o, then $p(x_1, \ldots, x_n) \in \mathcal{L}_{\text{st}}^2$.

(3) If $\psi, \varphi \in \mathcal{L}_{\text{st}}^2$, then $\neg\psi$, $\psi \wedge \varphi$, $\psi \vee \varphi$, $\varphi \to \psi$, $\psi \leftrightarrow \varphi$, $(\forall p)\psi$, $(\exists p)\psi$, $(\forall x)\psi$, $(\exists x)\psi$, $(\forall a)\psi$, and $(\exists a)\psi$ are all in $\mathcal{L}_{\text{st}}^2$, where x and a are variables of sort o and sort a, respectively, and p is an n-ary predicate variable on domain of sort o.

2.10.2 Basic Action Theory in LR

A basic action theory D is of the following form:

$$\mathbf{D} = \sum \cup \mathrm{D_{ss}} \cup \mathrm{D_{ap}} \cup \mathrm{D_{una}} \cup \mathrm{D_{s0}} \tag{2.27}$$

where

(1) $\sum$ is a set of logic formulas stating that the domain of sort s is a branching temporal structure, with S_0 the root and ***do*** the successive function. Formally, it contains the following axioms:

$$S_0 \neq \mathrm{do}(a, s)$$

$$\mathrm{do}(a_1, s_1) = \mathrm{do}(a_2, s_2) \to (a_1 = a_2 \wedge s_1 = s_2)$$

$$\forall P[(P(S_0) \wedge \forall a, s(P(s) \to P(\mathrm{do}(a, s)))) \to \forall s P(s)]$$

$$\neg(s < S_0);$$

$$s < \mathrm{do}(a, s') \leftrightarrow (\mathrm{Poss}(a, s') \wedge s \leq s')\circ$$

(2) D_{ss} is a set of logic sentences expressing the effect after an action is performed. Generally, each sentence is of the form

$$\mathrm{Poss}(a, s) \to (F(\vec{x}, \mathrm{do}(a, s)) \to \psi_F(\vec{x}, a, s)) \tag{2.28}$$

where F is a fluent and ψ_F is a logic formula in $\mathcal{L}_s.\circ$

(3) D_{ap} is a set of logic sentences expressing the precondition under which an action could be performed. Each sentence has the following general form:

$$\mathrm{Poss}(A(\vec{x}, s)) \to \psi_A(\vec{x}, s) \tag{2.29}$$

where A is an action and ψ_A is a logic formula in $\mathcal{L}_s$.

(4) D_{una} is a set of logic formulas expressing that two actions performed on two groups of objects will not have the same effect unless these two actions as well

as these two groups of objects are the same, respectively. Generally, each pair of formulas is of the following forms:

$$A(\vec{x}) \neq A'(\vec{y})$$

$$A(\vec{x}) = A(\vec{y}) \rightarrow (\vec{x} = \vec{y})$$

(5) D_{S0} is a finite set of logic formulas in $\mathscr{L}_{S0}$ and is called the initial condition of the action theory.

2.11 Dynamic Description Logic

2.11.1 *Description Logic*

Description logic is a kind of formalization of knowledge representation based on an object, and it is also called concept representation language or terminological logic. Description logic is a decidable subclass of first-order logic. It has well-defined semantics and possesses strong expression capability. One description logic system consists of four parts: constructors which represent concept and role, *TBox* subsumption assertion, *ABox* instance assertion, and reasoning mechanism of *TBox* and *ABox*. The representation capability and reasoning capability of the description logic system lie in the aforementioned four elements and different hypotheses (Baader *et al.*, 2003).

There are two essential elements, i.e., concept and role, in description logic. Concept is interpreted as a subclass of the domain. Role represents interrelation between individuals, and it is a kind of binary relation of domain set.

In certain domains, a knowledge base K $= <T, A>$ consists of two parts: *TBox* T and *ABox A*. *TBox* is a finite set of subsumption assertions, and it is also called a terminological axiom set. The general format of subsumption assertion is $C \sqsubseteq D$, where C and D are concepts. *ABox* is a finite set of instance assertions. Its format is C(a), where a is the individual name; or its format is P(a,b), where P is a primitive role and a and b are two individual names.

In general, *TBox* is an axiom set which describes domain structure, and it has two functions: one is to introduce concept name, and the other is to declare the subsumption relationship of concepts. The process of introducing concept name is expressed by $A \doteq C$ or $A \sqsubseteq C$, where A is the concept which is introduced. The format of subsumption assertion of concepts is $C \sqsubseteq D$. As to concept definition and subsumption relation definition, the following conclusion comes into existence:

$$C \doteq D \Leftrightarrow C \sqsubseteq D \quad \text{and} \quad C \sqsubseteq D$$

ABox is an instance assertions set, and its function is to declare attributes of individuals or relationships of individuals. There are two kinds of formats: one is to declare the relationship between the individual and concept, and the other is to declare the relationship between two individuals. In *ABox*, as to arbitrary individual a and concept C, the assertion which decides whether individual a is a member of concept C is called concept instance assertion, i.e., concept assertion. $a \in C$ is denoted as $C(a)$; $a \notin C$ is denoted as $\neg C(a)$.

Given two individuals a, b and a role R, if individual a and individual b satisfy role R, then aRb is role instance assertion, and it is denoted as $R(a, b)$.

In general, according to the constructor provided, description logic may construct complex concept and role based on simple concept and role. Description logic includes the following constructors at least: intersection ($\cap$), union ($\cup$), negation ($\neg$), existential quantification ($\exists$), and value restriction ($\forall$). The description logic which possesses these constructors is called ALC. Based on ALC, different constructors may be added to it so that different description logics may be formed. For example, if number restrictions "$\leq$" and "$\geq$" are added to the description logic ALC, then a new kind of description logic ALCN is formed. Table 2.3 shows the syntax and semantics of description logic ALC.

An interpretation $I = (\Delta^I, \cdot^I)$ consists of a domain Δ^I and an interpretation function $\cdot I$, where interpretation function $\cdot I$ maps each primitive concept to a subset of domain Δ^I and maps each primitive role to the subset of domain $\Delta^I \times \Delta^I$. With respect to an interpretation, the concept of ALC is interpreted as a domain subset, and the role is interpreted as a binary relation:

(1) An interpretation I is a model of subsumption assertion $C \subseteq D$, if and only if $C^I \subseteq D^I$.
(2) An interpretation I is a model of $C(a)$, if and only if $a \in C^I$; an interpretation I is a model of $P(a, b)$, if and only if $(a, b) \in P^I$.
(3) An interpretation I is a model of knowledge base K, if and only if I is a model of each subsumption assertion and instance assertion of knowledge base K.
(4) If knowledge base K has mode I, then K is satisfiable.
(5) As to each model of knowledge base K, if assertion δ is satisfiable, then we say that knowledge base K logically implicates δ, and it is denoted as $K| = \delta$.
(6) As to concept C, if knowledge base K has a model I, and $C^I \neq \phi$, then concept C is satisfiable. The concept C of knowledge base K is satisfiable if and only if $K| \neq C \subseteq \perp$.

The basic reasoning problems of description logic include concept satisfiability, concept subsumption relation, instance checking, consistency checking, and so on,

Table 2.3.   Syntax and semantics of ALC.

Constructor	Syntax	Semantics	Example
Primitive concept	A	$A^I \forall \Delta^I$	Human
Primitive concept	P	$P^I \subseteq \Delta^I \times \Delta^I$	has-child
top	$\perp$	Δ^I	True
Bottom	$\perp$	Φ	False
intersection	$C \cap D$	$C^I \cap D^I$	Human $\cap$ Male
Union	$C \cup D$	$C^I \cup D^I$	Doctor $\cup$ Lawyer
Negation	$\neg C$	$\Delta^I - C^I$	$\neg$Male
Existential quantification	$\exists R.C$	$\left\{ x \mid \exists y, (x, y) \in R^I \wedge y \in C^I \right\}$	$\exists$has-child.Male
Value restriction	$\forall R.C$	$\left\{ x \mid \forall y, (x, y) \in R^I \Rightarrow y \in C^I \right\}$	$\forall$has-child.Male

where the concept satisfiability is the most basic reasoning problem, and other reasoning problems may be reduced to concept satisfiability problems.

In description logic, reasoning problem may be reduced to concept satisfiability problem through the following properties. As to concepts C, D, there exists the following proposition:

(1) $C \subseteq D \Leftrightarrow C \cap \neg D$ is unsatisfiable.
(2) $C \doteq D$ (concept C and D are equivalent) $\Leftrightarrow$ both $(C \cap \neg D)$ and $(D \cap \neg C)$ are unsatisfiable.
(3) C and D are disjoint $\Leftrightarrow C \cap D$ is unsatisfiable.

2.11.2 *Syntax of Dynamic Description Logic (DDL)*

Dynamic description logic DDL is formed by extending traditional description logic (Shi *et al.*, 2005), while traditional description logic has many species. The dynamic description logic DDL studied here is based on description logic ALC.

Definition 2.19. The primitive symbols in DDL are as follows:

· concept names: $C_1, C_2, \ldots$
· role names: $R_1, R_2, \ldots$
· individual constant: $a, b, c, \ldots$
· individual variable: x, y, z, $\ldots$
· concept operator: $\neg$, $\sqcap$, $\sqcup$ and quantifier $\exists$, $\forall$
· formula operator: $\neg$, $\wedge$, $\rightarrow$ and quantifier $\forall$
· action names: $A_1, A_2, \ldots$
· action constructs: ;(sequence), $\cup$(choice), *(iteration), ?(test)

· action variable: $\alpha, \beta, \ldots$
· formula variable: $\phi, \psi, \pi, \ldots$
· state variable: $u, v, w, \ldots$.

Definition 2.20. Concepts in DDL are defined as follows:

(1) Primitive concept P, top $\top$ and bottom $\bot$ are concepts.
(2) If C and D are concepts, then $\neg C$, $C \cap D$, and $C \cup D$ are concepts.
(3) If C is concept and R is role, then $\exists R.C$ and $\forall R.C$ are concepts.
(4) If C is concept and α is action, then $[\alpha]C$ is action too.

Definition 2.21. Formulas in DDL are defined as follows, where C is concept, R is role, a, b are individual constants, and x, y are individual variables:

(1) $C(a)$ and $R(a, b)$ are called assertion formulas.
(2) $C(x)$ and $R(x, y)$ are called general formulas.
(3) Both assertion formulas and general formulas are all formulas.
(4) If φ and ψ are formulas, then $\neg\varphi$, $\varphi \wedge \psi$, $\varphi \rightarrow \psi$, and $\forall x\varphi$ are all formulas.
(5) If φ is a formula, then $[\alpha]\varphi$ is also a formula.

Definition 2.22. A finite set of $\{a_1/x_1, \ldots, a_n/x_n\}$ is an instance substitution, where $a_1, \ldots, a_n$ are instance constants which are called substitution items and $x_1, \ldots, x_n$ are variables which are called substitution bases, $x_i \neq x_j$ for each pair i, $j i_n\{1, \ldots, n\}$ such that $i \neq j$.

Definition 2.23. Let φ be a formula, let $x_1, \ldots, x_n$ be all the variables occurring in φ, and let $a_1, \ldots, a_n$ be instance constants. If $\varphi\prime$ is a substitution result of φ with $\{a_1/x_1, \ldots, a_n/x_n\}$, then $\varphi\prime$ is called an instance formula of φ.

Definition 2.24. A condition in DDL is an expression of the form: $\forall C$, $C(p)$, $R(p, q)$, $p = q$ or $p \neq q$, where N_C is a set of individual constants, N_X is a set of individual variables, N_I is the union N_C and N_X, $p, q \in N_I$, C is concept of DDL, and R is role of DDL.

Definition 2.25. *An action description is of the form* $A(x_1, \ldots, x_n) \equiv (P_A, E_A)$, where we have the following:

(1) A is the action name.
(2) $x_1, \ldots, x_n$ are individual variables, which denote the objects on which the action operates.
(3) P_A is the set of *pre-conditions*, which must be satisfied before the action is executed, i.e., $P_A = \{\text{con} \mid \text{con} \in \text{condition}\}$.

(4) E_A is the set of post-conditions, which denote the effects of the action; E_A is a set of pair *head/body*, where head $= \{$con $\mid$ con $\in$ condition$\}$ and body is a condition.

Remark:

(1) Action defines the transition relation of state, i.e., an action A transits a state u to a state v, if action A can produce state v under state u. The transition relation depends on whether states u, v satisfy the pre-conditions and post-conditions of action A. The transition relation is denoted as u *TA* v.
(2) Since some states that happened before action A may influence the post-condition of action A, there is some difference between pre-conditions and post-conditions. As to post-conditions head/body, if each condition of head can be satisfied in state u, then each condition of body can also be satisfied in state v.

Definition 2.26. Let $A(x_1, \ldots, x_n) \equiv (P_A, E_A)$ be an action description and let $A(a_1, \ldots, a_n)$ be the substitution of $A(x_1, \ldots, x_n)$ by $\{a_1/x_1, \ldots, a_n/x_n\}$. Then $A(a_1, \ldots, a_n)$ is called an action instance of $A(x_1, \ldots, x_n)$. $A(a_1, \ldots, a_n)$ is called atom action, $P_A(a_1, \ldots, a_n)$ is the precondition of $A(a_1, \ldots, a_n)$, and $E_A(a_1, \ldots, a_n)$ is the result set of $A(a_1, \ldots, a_n)$.

Definition 2.27. Actions in DDL are defined as follows:

(1) Atomic action $A(a_1, \ldots, a_n)$ is action.
(2) If α and β are actions, then $\alpha; \beta$, $\alpha \cup \beta$, and α^* are all actions.
(3) If φ is an assertion formula, then $\varphi?$ is action.

2.11.3 *Semantics of Dynamic Description Logic*

The semantics of DDL can be illustrated by a structure composed of the following components:

(1) non-empty set Δ, which is the set of all individuals discussed in a specified domain,
(2) the set of state W, which is the set of all states of the world in a specified domain,
(3) an interpretation I which explains each individual, concept, and role in DDL as follows:

① Each individual constant is interpreted as an element of Δ.
② Each concept is interpreted as a subset of Δ.
③ Each role is interpreted as a binary relation on Δ.

(4) Each action is mapped into a binary relation on W.

Next, we explain the semantics of DDL in detail. First, for a state u in DDL, an explanation $I(u) = (\Delta, \bullet^{I(u)})$ in u is composed of two components, written as $I(u) = (\Delta, \bullet^{I(u)})$, where explanation function $\bullet^{I(u)}$ maps each concept into a subset of Δ and maps each role into a binary relation on $\Delta \times \Delta$:

- $\top^{I(u)} = \Delta$
- $\bot^{I(u)} = \varnothing$
- $C^{I(u)} \subseteq \Delta$
- $R^{I(u)} \subseteq \Delta \times \Delta$
- $(\neg C)^{I(u)} = \Delta - C^{I(u)}$
- $(\neg R)^{I(u)} = \Delta \times \Delta - R^{I(u)}$
- $(C \sqcap D)^{I(u)} = C^{I(u)} \cap D^{I(u)}$
- $(C \sqcup D)^{I(u)} = C^{I(u)} \cup D^{I(u)}$
- $(\exists R.C)^{I(u)} = \{x \mid \exists y . ((x, y) \in R^{I(u)} \wedge y \in C^{I(u)})\}$
- $(\forall R.C)^{I(u)} = \{x \mid \forall y . ((x, y) \in R^{I(u)} \Rightarrow y \in C^{I(u)})\}$
- $([\alpha]C)^{I(u)} = \{x \mid uT_\alpha v \wedge x \in (C)^{I(v)}\}$

Since the interpretation of the object's name does not depend on the particular world, we use the rigid designator and assume that each name of individual is uniform and does not change with state changes. Usually, we write aI(u) as a for short.

Given action $A(x_1, \ldots, x_n) \equiv (P_A, E_A)$, N_X^A is all variable set which happened in action A, $I = (\Delta^I, \cdot I)$ is an interpretation, and map $\gamma : N_X^A \to \Delta^I$ is a variable evaluation which happened in action A. As to individual constant $a \in N^C$ of ABox of DDL, $\cdot^I$ interprets a as an element of Δ^I, i.e., $aI \in \Delta^I$. As to individual variable or individual constant $p \in N^I$, their interpretation is given as follows:

$$p^{I,\gamma} = \begin{cases} \gamma(p), & \text{if } p \in N_X \\ p^I, & \text{if } p \in N_C \end{cases}$$

then the condition interpretations of DDL are as follows:

- If $C^I = \Delta^I$, then I and γ satisfy condition $\forall C$.
- If $a^{I,\gamma} \in C^I$, then I and γ satisfy condition $C(a)$.
- If $a^{I,\gamma} = b^{I,\gamma}$, then I and γ satisfy condition $a = b$.
- If $a^{I,\gamma} \neq b^{I,\gamma}$, then I and γ satisfy condition $a \# b$.
- If $<a^{I,\gamma}, b^{I,\gamma}> \in R^I$, then I and γ satisfy condition $R(a,b)$.

In each state u, assertion formulas connect individual constants to concepts and roles. So there are two kinds of assertion formulas: concept assertions with the form of $C(a)$ and role assertions with the form of $R(a_1, a_2)$. As to concept assertions, they declare the relation of individual constant and concept, i.e., the relation of element and set. Semantics of concept assertion can be interpreted as follows:

- $u \models C(a)$ iff $a \in C^{I(u)}$.
- $u \models \neg C(a)$ iff $a \notin C^{I(u)}$.

For example, under certain states, individual constant a denotes a block, and this means individual constant a belongs to the concept of block, and it is denoted as Block(a); individual constant b is a button, and it is denoted as Button(b).

Role assertions declare the relation of two individual objects or the attribute of an individual object. It is a binary relation, and its semantics can be explained as follows:

- $u \models R(a_1, a_2)$ iff $(a_1, a_2) \in R^{I(u)}$.
- $u \models \neg R(a_1, a_2)$ iff $(a_1, a_2) \notin R^{I(u)}$.

For example, under certain states, agent a1 and agent a_2 are acquaintances, and it is denoted as $hasAquaintance(a_1, a_2)$. Object a presses on object b, and it is denoted as $On(a, b)$. Role assertions may also declare some attributes of an individual object. For example, Button b_1 is in open state, and it is denoted as $hasState(b_1, O_N)$; the length of object a is 10, and it is denoted as $hasLength(a, 10)$.

Analogously, for a particular state u, those formulas that are composed of assertion formulas can be interpreted as follows, where φ and ψ are assertion formulas:

- $u \models \neg\phi$ iff $u\phi$.
- $u \models \phi \wedge \psi$ iff $u \models \phi$ and $u \models \psi$.
- $u \models \phi \rightarrow \psi$ iff $u \models \phi \Rightarrow u \models \psi$.

Action execution results in the change of world state, so action also may be defined as a state transition relationship. But the process of action change is the process of individual attribute change or individual relation change in fact, so under certain states, all individual attributes, relation descriptions, and so on consist of the world state description. These individual attribute and relation descriptions may be defined based on action description (Definition 2.33). State in DDL corresponds to all condition interpretation of action description under the corresponding state in fact, so action may be interpreted based on aforementioned condition interpretation and description logic interpretation. Before the action of DDL is defined, state transition (one state transits to another state under action) is defined first.

Definition 2.28. Given two interpretations $I(u) = (\Delta, \bullet^{I(u)})$ and $I(v) = (\Delta, \bullet^{I(v)})$, under states u and v, an action $\alpha = (P\alpha, E\alpha)$ can produce state v when applied to state u (written $u \rightarrow_\alpha v$) if there exists an assignment map $\gamma . N_X^\alpha \rightarrow \Delta$ such that $\gamma, I(u)$, and $I(v)$ satisfy the following conditions:

(1) I(u) and γ satisfy each condition of pre-conditions P_α.
(2) As to each pair head/body of post-conditions E_α, if $I(u)$ and γ satisfy head, then $I(v)$ and γ satisfy body.

In this case, we can say that action α can produce state v when applied to state u and assignment map γ, and it is denoted as $u \rightarrow^\gamma_\alpha v$.

The semantics of atomic and complex actions are defined as follows:

- $\alpha = \{<u, v> \mid u, v \in W, u \rightarrow^\gamma_\alpha v\}$
- $\alpha; \beta = \{<u, v> \mid u, v, w \in W, u \rightarrow^\gamma_\alpha w \wedge w \rightarrow^\gamma_\beta v\}$
- $\alpha \cup \beta = \{<u, v> \mid u, v \in W, u \rightarrow^\gamma_\alpha v \vee u \rightarrow^\gamma_\beta v\}$
- $\alpha* = \{<u, v> \mid u, v \in W, u \rightarrow^\gamma_\alpha v \vee u \rightarrow^\gamma_{\alpha;\alpha} v \vee u \rightarrow^\gamma_{\alpha;\alpha;\alpha} v \vee \ldots\}$
- $\phi? = \{<u, u> \mid u \in W, u \models \phi\}$.

Since the aforementioned action interpretation is based on description logic, some new concepts about action may be given, and these new concepts supplement description logic.

Definition 2.29. Given action α (primitive action or complex action), if there exist two interpretations $I(u) = (\Delta, \bullet^{I(u)})$ and $I(v) = (\Delta, \bullet^{I(v)})$ that satisfy $u \rightarrow_\alpha v$, then action α is realizable.

Definition 2.30. Given action α (primitive action or complex action), if there exists an interpretation $I(u) = (\Delta, \bullet^{I(u)})$ corresponding to *ABox A*, and an interpretation $I(v) = (\Delta, \bullet^{I(v)})$ corresponding to *ABox A* and *TBox T*, and $u \rightarrow_\alpha v$ is satisfied, then action a is realizable corresponding to *ABox A*.

Definition 2.31. Given two actions α and β (primitive action or complex action), as to arbitrary interpretations $I(u) = (\Delta, \bullet^{I(u)})$ and $I(v) = (\Delta, \bullet^{I(v)})$, the following condition is satisfied: if the existence of $u \rightarrow_\alpha v$ can make $u \rightarrow_\beta v$ come true, then action β subsumes action α, denoted as $\alpha \sqsubseteq \beta$.

Definition 2.32. Given two actions α and β (primitive action or complex action), an arbitrary interpretation $I(u) = (\Delta, \bullet^{I(u)})$ corresponding to *ABox A* of DDL, and an arbitrary interpretation $I(v) = (\Delta, \bullet^{I(v)})$ corresponding to *ABox A* and *TBox T*, the following condition is satisfied: if the existence of $u \rightarrow_\alpha v$ can make $u \rightarrow_\beta v$ come true, then action β subsumes action α corresponding to *ABox A*, denoted as $\alpha \sqsubseteq A\beta$.

Remark. Similar to Definition 2.31 and Definition 2.32, action realization or subsumption corresponding to *TBox* T may also be defined. In order to understand the action subsumption, the following is a simple example.

Given four action descriptions: $\alpha_1 = (\{A(a), \neg A(b)\}, \{\Phi/\neg A(a), \Phi/A(b)\})$, $\alpha_2 = (\{A(c), \neg A(d)\}, \{\Phi/\neg A(c), \Phi/A(d)\})$, $\alpha_3 = (\{A(x), \neg A(y)\}, \{\Phi/\neg A(x),$

$\Phi/A(y)\}$), and $\alpha_4 = (\{A(y), \neg A(x)\}, \{\Phi/\neg A(y), \Phi/A(x)\})$, where a, b, c, d are individual constants and x, y are individual variables, then we have $a1 \subseteq a3$, $a2 \subseteq a3$, $a1 \subseteq a4$, $a2 \subseteq a4$, $a3 \subseteq a4$, and $a4 \subseteq a3$, but there is no subsumption relation between α_1 and α_2.

Exercises

2.1 What is monotonic reasoning? What is non-monotonic reasoning?

2.2 How are default rules represented in default theory? What representation forms does there exist?

2.3 A default theory is a pair $T = \langle W, D \rangle$, with D a set of default rules and W a set of closed formulas. Please represent the following sentences with such a default theory.

    (1) Some mollusks have shells.
    (2) Cephalopods are mollusks.
    (3) Not all cephalopods have shells.

2.4 Both the closed world assumption and the circumscription are formalisms for non-monotonic reasoning. Please exposit these two formalisms and compare the differences between them.

2.5 Represent the following situations with the truth maintenance system:

    (1) It is now summer.
    (2) The weather is very humid.
    (3) The weather is very dry.

2.6 How to maintain the consistency of a knowledge base by the truth maintenance system? Please illustrate it with an example.

2.7 Please represent the following monkey-and-bananas problem with situation calculus.

    The world is composed of a monkey in a room, a bunch of bananas hanging from the ceiling, and a box that can be moved by the monkey. The bananas are out of the reach of the monkey; however, the box will enable the monkey to reach the bananas if the monkey climbs on it. The actions available to the monkey include *Go* from one place to another, *Push* an object from one place to another, *ClimbUp* onto or *ClimbDown* from an object, and *Grasp* or *Ungrasp* an object. Initially, the monkey is at A, the bananas at B, and the box at C. The question is how to get the bananas.

2.8 What are the basic elements of the description logic?

2.9 How are actions represented in the dynamic description logic?

Chapter 3

Causal Reasoning

Reasoning is a mental activity that aims to arrive at a conclusion in a rigorous way. Causal reasoning is generally considered a form of inductive reasoning. More concretely, causal reasoning aims at an epistemological problem of establishing precise causal relationships between causes and effects, focusing on detecting genuine, real causes for some effects, and genuine, real effects of some causes.

3.1 Reasoning

Human thought mainly involves perceptual thought, imagery thought, abstract thought, and inspirational thought. Perceptual thought is the primary level of thought. When people begin to understand the world, perceptual materials are simply organized to form self-consistent information, thus only phenomena are understood. The form of thought based on this process is perceptual thought. Perceptual thought about the surface phenomena of all kinds of things can be obtained in practice via direct contact with the objective environment through sensories, such as eyes, ears, noses, tongues, and bodies, thus its sources and contents are objective and substantial.

Imagery thought mainly relies on generalization through methods of typification and the introduction of imagery materials in thinking. It is common to all higher organisms. Imagery thought corresponds to the connection theories of neural mechanisms. AI topics related to imagery thought, include pattern recognition, image processing, and visual information processing.

Abstract thought is a form of thought based on abstract concepts, through thinking with symbol information processing. Only with the emergence of language is abstract thought possible: language and thought boost each other and promote each other. Thus, physical symbol system can be viewed as the basis of abstract thought.

Research has been done on inspirational thought. Some researchers hold that inspirational thought is the extension of imagery thought to sub-consciousness, during which a person does not realize that part of his brain is processing information. While some others argue that inspirational thought is sudden enlightenment. Despite all these disagreements, inspirational thought is particularly important to creative thinking and needs further research.

In the process of human thinking, attention plays an important role. Attention sets certain orientation and concentration for noetic activities to ensure that one can promptly respond to the changes of the objective realities and be better accustomed to the environment. Attention limits the number of parallel thinking. Thus for most conscious activities, the brain works serially, with an exception of parallel looking and listening.

Based on the above analysis, we propose a hierarchical model of human thought, as shown in Figure 3.1 (Shi, 1990a). In the figure, perceptual thought is the simplest form of thought, which is constructed from the surface phenomena through sensories, such as eyes, ears, noses, tongues, and bodies. Imagery thought is based on the connection theories of neural networks for highly parallel processing. Abstract thought is based on the theory of physical symbol system in which abstract concepts are represented with languages. With the effect of attention, different forms of thought are processed serially most of the time.

Reasoning is an important form of thinking that derives new knowledge from existing knowledge. The creativity of human thinking can be clearly seen in reasoning. When people use reasoning methods to understand those realistic processes that are directly observed by instinct, as long as they can verify the necessary important links in the complex chain of reasoning in practice and make logical inferences,

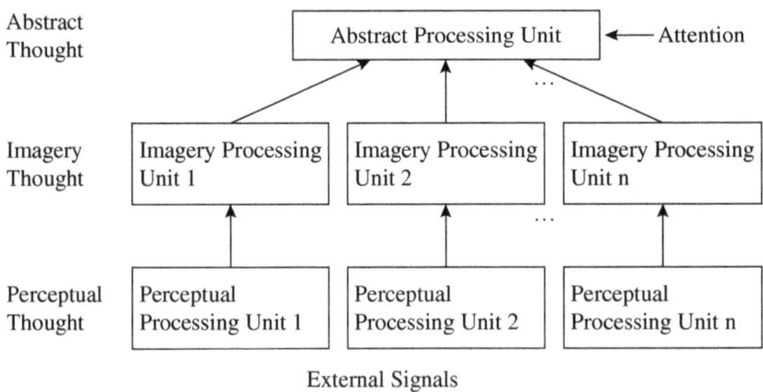

Fig. 3.1.   Hierarchical model of thought.

then the new judgments (conclusions) they make, the new concepts proposed are scientific.

Reasoning is an important means for us to understand the world and solve problems. The eight common forms of reasoning are inductive reasoning, deductive reasoning, abductive reasoning, analogical reasoning, probabilistic reasoning, causal reasoning, common sense reasoning, and non-monotonic reasoning. By learning and understanding these forms of reasoning, we can analyze problems more accurately, infer conclusions, improve the logic and accuracy of thinking, and obtain better applications in various fields. Therefore, mastering these forms of reasoning is of great significance to improving our thinking and problem-solving abilities.

3.1.1 *Deductive Reasoning*

Deductive reasoning is reasoning in which there is an implicit relationship between the premise and the conclusion, or in other words, reasoning in which the premise and the conclusion are inevitably connected, and the general is deduced from the particular.

A is the logical inference of Γ (that is, the formula in Γ), denoted as $\Gamma \models A$, if and only if any assignment φ in any non-empty domain, if $\varphi(\Gamma) = 1$, then $\varphi(A) = 1$.

Given a non-empty domain S. When any assignment φ in S makes

$$\varphi(\Gamma) = 1 \Rightarrow \varphi(A) = 1$$

When established, we say that A in S is the logical inference of Γ, which is written as $\Gamma \models A$ in S.

One form of reasoning that is often used is the syllogism. It consists of only three property judgments: two of which are premises and the other one is the conclusion. In terms of subject and predicate, it contains only three different concepts, each of which appears once in each of the two judgments. These three different concepts are called major term, minor term, and middle term, respectively. The major term is the concept that serves as the predicate of the conclusion, represented by P. The minor term is the concept that is the subject of the conclusion, represented by S. The middle term is the concept that appears in both premises, represented by M. The various syllogism forms formed due to the different positions of the major term, the middle term, and the minor term in the premise are called syllogism figures.

Rule-based deduction systems are generally divided into forward systems, reverse systems, and two-way comprehensive systems. In a rule-based forward deduction system, the implication used as an F rule operates on the total database of facts until a termination condition of the target formula is obtained. In the rule-based inverse deduction system, the implication used as the *B* rule operates on the total database of targets until the termination condition containing these facts is obtained. In the rule-based forward and reverse synthesis system, different rules (F rules or B rules) are applied from two directions to perform operations. This system is a direct proof system rather than a reduction inversion system.

3.1.2 *Inductive Reasoning*

Inductive reasoning is reasoning that deduces the universal laws of this type of thing or phenomenon from individual things or phenomena. This kind of reasoning reflects the probability connection between the premises and the conclusion. This is a general thought process derived from the particular.

In recent decades, research on inductive logic abroad has been carried out in two directions. One is to continue to find logical ways to derive corresponding universal principles from empirical facts in the classical sense of Bacon's inductive logic; the other direction is to use probability theory and formal principles. It uses the method of rationalization and rationalization to explore the degree of "support" or "confirmation" of limited empirical facts for universal propositions that are suitable for a certain range. This logic is actually a logic of theoretical evaluation.

Bacon's main ideas about induction are as follows:

(1) The senses must be helped and guided to overcome the one-sidedness and super-ficiality of perceptual knowledge.
(2) Appropriate inductive procedures must be followed when constructing concepts.
(3) The axioms should be constructed in a step-by-step manner.
(4) We must pay attention to the role of the inversion method and the exclusion method in the induction process.

These ideas formed the basis of the four rules of induction later proposed by John S. Mill. In his book *Logical System*, Mill proposed the methods of agreement, difference, covariation, and remainder of induction. He believed that "induction may be defined as the operation of discovering and proving general propositions."

Probabilistic logic emerged in the 1930s. Based on relative frequency, Reichenbach used the mathematical tools of probability theory to find the frequency limit of a proposition and use it to predict future events. In the 1940s and 1950s, Carnap established probabilistic logic based on reasonable beliefs. He adopts a Bayesian stance, depicting reasonable beliefs directly as probability functions, and viewing probability as representing the logical relationship between one statement and another evidence statement.

3.1.3 *Abductive Reasoning*

In abductive reasoning, we are given the rule $p \Rightarrow q$ and a reasonable belief in q. Then hope that the predicate p is true under some interpretation.

It should be noted that in general, there may be many sets of hypotheses, that is, there may be many sets of potential explanations for a given phenomenon. The definition of logic-based abductive explanation implies that there are corresponding mechanisms for discovering explanations of content in knowledge base systems. If an explanatory hypothesis must be able to derive an explanation of phenomenon O, then the way to build a complete explanation is to reason backward from O.

3.1.4 *Reasoning by Analogy*

Based on the existence of some similarity or similar relationship between two objects, the reasoning process of inferring that another object has a certain corresponding property from one known object has a certain property is called analogical reasoning.

Analogical reasoning is based on previously acquired knowledge about a system as a means of inferring information about another similar system. The objective basis of this kind of analogical reasoning lies in the universal connection between various elements among things, processes, and systems, as well as the comparable objective basis that exists between such connections.

When we use the experience gained in solving a certain problem to solve similar problems, the key is to be good at finding similarities between different problems. Polish mathematician Banach once said, "A person is a mathematician because he is good at discovering similarities between judgments; if he can identify similarities between arguments, he is an excellent mathematician; if he can see through If he sees the similarities between theories, then he becomes an outstanding mathematician. However, I think there should be such mathematicians who can see the similarities between similar theories." This is true not only in mathematics but also in physics, chemistry, biology, astronomy, genetics, and earth science as well. Not only are there similarities within disciplines, but there are also similarities between disciplines. Cognitive science is founded on the similarity between information processing in the human brain and computer processing. Similarities are found in social sciences, literature, art, music, etc. Similarity is the unity of similarity and variation in the existence of objective things. Variation is the difference in the development process of things. According to similar factors, similar phenomena can be divided into functional similarity, structural similarity, dynamic similarity, geometric similarity, etc.

3.1.5 *Non-Monotonic Reasoning*

Classical logic, such as formal logic and deductive logic, is monotonous in its treatment of human understanding of the world. Assuming A represents a set of inference rules, the language $\text{Th}(S) = \{P|S \to P\}$ of monotonic logic has the following monotonicity:

(1) $A \subseteq \text{Th}(A)$
(2) If $A \subseteq S$, then $\text{Th}(A) \subseteq \text{Th}(S)$
(3) $\text{Th}(\text{Th}(A)) = \text{Th}(A)$

One of the distinctive features of a monotonic inference rule is that its language is a closed minimum fixed point, that is,

$$\text{Th}(A) = \cap\{s|A \to S \quad \text{and} \quad \text{Th}(A) = S\}$$

Specifically, there is a knowledge system A. If it is known that A contains knowledge S, that is, $A \to S$, then the knowledge S can be inferred. However, the reasoning system established on this basis is often inconsistent with the human understanding process of the objective world. Reasoning based on classical logic is an idealized model of people's reasoning. In daily life or in some artificial intelligence application systems, people often have to reason based on some rules that are generally correct but not absolutely correct, or when reasoning is carried out with incomplete information, the conclusion obtained by this kind of reasoning is temporary and may be modified, so it is not monotonic, so it is called non-monotonic reasoning.

In first-order logic, we use $\forall x P(x) = 1$ to represent the fact that "all x have property P". However, in real life, such sentences are close to the truth but not absolutely correct, that is, most x has the property P, but some exceptions may occasionally be encountered. For example, all birds can fly, with the exception of penguins and ostriches. All oranges are yellow, except for unripe and mutant varieties. Since such comprehensive general statements are not absolutely correct, the reasoning using these statements will inevitably produce errors. One way to solve this problem is to abandon such statements completely. This will not only produce errors, but it will also lose something close to the truth and many conclusions that could have been drawn. Another way is to modify this type of statement and then use it when it is completely correct. However, this modification is very difficult. Even if it is modified, the structure of the sentence has become quite complicated and cannot be used flexibly. A simple and appropriate way to deal with it is to first assume that such statements are correct and make inferences based on them. If after acquiring new facts, it is found that the original conclusion is problematic, the inference will be non-monotonic.

In order to obtain strong logical support for non-monotonic reasoning, various non-monotonic logics have been proposed. Among the more famous works are R. Reiter's default logic, J. McCarthy's restriction logic, Doyle's truth maintenance system, and R.C. Moore's self-cognitive logic.

1. Default logic

The basic idea is that in the process of reasoning, some propositions whose truth or falsity cannot be determined but must be determined. If it is assumed to be true and no contradiction occurs, these propositions will be established by default.

A default theory is defined as a binary (D, W), where D is the default set and W is the closed set of suitable formulas. The general form of the default rule is expressed as

$$\frac{\alpha(X)\colon M\beta_1(X), \ldots, M\beta_m(X)}{W(X)} \tag{3.1}$$

where $\alpha(X), \beta_1(X), \ldots, \beta_{\mathrm{m}}(X)$ are all suitable formulas, and the free variables are among $X = x_1, \ldots, x_n$. $\alpha(X)$ is called the antecedent, and $W(X)$ is called the conclusion. If $\alpha, \beta_1, \ldots, \beta_{\mathrm{m}}, W$ does not contain free variables, it is said to be closed by default.

2. Restricted logic

Restricted logic and its reasoning system were first proposed by McCarthy. The original qualification reasoning mainly dealt with "predicate qualification", and its basic method was to minimize the expansion of a specific predicate. In finite logic, only when it is proved that something satisfies property P, it is considered to have property P.

3. Truth maintenance system

Doyle's truth maintenance system is an implemented non-monotonic reasoning system. It is used to assist other reasoning programs in maintaining the correctness of the system. Its role is not to generate new inferences but to maintain compatibility between propositions produced by other programs. Once an incompatibility is discovered, it calls out its own reasoning mechanism, oriented to backtracking on affiliations, and eliminates the incompatibility by modifying the minimum trust set.

3.1.6 *Common Sense Reasoning*

Common sense reasoning is an area of artificial intelligence that aims to help computers understand and interact with people more naturally by gathering all background assumptions and teaching them to the computer. A representative system for common sense reasoning is Cycrop's Cyc system, which operates a logic-based common sense knowledge base.

The Cyc system was developed by Doug Lenat in 1984. The original goal of the project was to encode millions of pieces of knowledge into a machine-usable form to represent human common sense. CycL is the proprietary knowledge representation language of the Cyc project. This knowledge representation language is based on first-order relationships. In 1986, Lennart predicted that if one wanted to complete a huge common sense knowledge system like Cyc, it would involve 250,000 rules and would take 350 man-years to complete. In 1994, the Cyc project became independent from the company, and based on this, Cycorp was established in Austin, Texas, USA.

3.2 Bayesian network

Bayesian network is a graphic model for describing the connection probabilities among variables. It provides a natural representation of casual relationship and is often used to explore potential relationships among data. In the network, nodes represent variables and directed links represent dependant relationships between variables. With firm mathematical foundation, Bayesian theory offers method for brief function calculation, describes the coincidence of brief and evidence, and possesses the incremental learning property that brief varies along with the variation of evidence. In data mining, Bayesian networks can deal with incomplete or noisy datasets. It describes correlations among data with probabilistic measurement and thereby solves the problem of data inconsistency. It describes correlations among data with graphical method, which has clear semantic meaning and understandable representation. It also makes prediction and analysis with casual relationships among data. Bayesian network is becoming one of the most remarkable data mining methods due to its nice properties, including unique knowledge representation of uncertain information, capability for handling probability, and incremental learning with prior knowledge.

3.2.1 *History of Bayesian Theory*

The foundational work of Bayesian School is Reverend Thomas Bayes' (1702–1761) "An Essay towards solving a Problem in the Doctrine of Chances". Maybe he felt the work was not perfect enough, this work was published not in his lifetime but posthumously by his friend. As famous mathematician Laplace P. S. educed Law of Succession based on Bayesian method, Bayesian method and theory began to be recognized. In the 19th century, because the problem of motivating and constructing prior probabilities was not adequately answered, Bayesian theory was not well accepted at that time. Early in the 20th century, de Finetti B. and Jeffreys H. made significant contributions to Bayesian theory. After World War II, Wald A. proposed statistical decision theory. In this theory, Bayesian method played an important role. Besides, the development of information science also contributed to the reincarnation of Bayesian theory. In 1958, Bayes' paper was republished by Biometrika, the most historical statistical magazine in Britain. In the 1950s, Robbins H. suggested combining empirical Bayesian approach and conventional statistical method. The novel approach attracted the attention of statistical research field and soon showed its merits and became an active research direction.

With the development of artificial intelligence, especially after the rise of machine learning and data mining, Bayesian theory gained much more development and applications. Its connotation has also varied greatly from its origination. In the 1980s, Bayesian networks were used for knowledge representation in expert systems. In the 1990s, Bayesian networks were applied to data mining and machine learning. Recently, more and more papers concerning Bayesian theory were published, which covered most fields of artificial intelligence, including casual reasoning, uncertain knowledge representation, pattern recognition, clustering analysis, and so on. There appears an organization and a journal, ISBA, which focus especially on the progress of Bayesian theory.

3.2.2 Basic Concepts of Bayesian Method

In Bayesian theory, all kinds of uncertainties are represented with probabilities. Learning and reasoning are implemented via probabilistic rules. The Bayesian learning results are distributions of random variables, which show the briefs to various possible results. The foundations of Bayesian School are Bayesian theorem and Bayesian assumption. Bayesian theorem connects prior probabilities of events with their posterior probabilities. Assume the joint probability density of random vectors x and θ is $p(x, \theta)$, and $p(x)$ and $p(\theta)$ give the marginal densities of x and θ, respectively. In common cases, x is an observation vector and θ is an unknown parameter vector. The estimation of parameter θ can be obtained with the observation vector via Bayesian theorem. The Bayesian theorem is as follows:

$$p(\theta|x) = \frac{\pi(\theta)p(x|\theta)}{p(x)} = \frac{\pi(\theta)p(x|\theta)}{\int \pi(\theta)p(x|\theta)d\theta} (\pi(\theta) \text{ is the prior of } \theta) \qquad (3.2)$$

From the formula above, we see that in Bayesian method the estimation of a parameter needs the prior information of the parameter and the information from evidence. In contrast, traditional statistical method, e.g., maximum likelihood, only utilizes the information from evidence. The general process to estimate parameter vector via Bayesian method is described as follows:

(1) Regard unknown parameters as random vectors. This is the fundamental difference between Bayesian method and traditional statistical approach.
(2) Define the prior $\pi(\theta)$ based on previous knowledge of parameter θ. This step is a controversial step and is attacked by conventional statistical scientists.
(3) Calculate posterior density and make estimation of parameters according to the posterior distribution.

In the second step, if there is no previous knowledge to determine the prior $\pi(\theta)$ of a parameter, Bayes suggested to assume uniform distribution to be its distribution. This is called Bayesian assumption. Intuitionally, Bayesian assumption is well accepted. Yet, it encounters problems when no information about prior distribution is available, especially when parameter is infinite. Empirical Bayes (EB) estimator combines conventional statistical method and Bayesian method so that it applies conventional method to gain the marginal density $p(x)$ and then ascertains prior $\pi(\theta)$ with the following formula:

$$p(x) = \int_{-\infty}^{+\infty} \pi(\theta)p(x|\theta)d\theta$$

3.2.3 *Foundation of Probability Theory*

Probability is a branch of mathematics, which focuses on the regularity of random phenomena. Random phenomena are phenomena in which different results appear under the same conditions. Random phenomena include individual random phenomena and substantive random phenomena. The regularity from the observation of substantive random phenomena is called statistical regularity.

Statistically, we conventionally call an observation, a registration, or an experiment about a phenomenon a trial. A random trial is an observation of a random phenomenon. Under the same conditions, random trials may lead to different results. But the sphere of all the possible results is estimable. The result of a random trial is both uncertain and predictable. Statistically, the result of a random trial is called random event, shortly by event.

Random event is the result that will appear or not appear in a random trial. In random phenomenon, the frequency of a mark is the total number that the mark appears in all trials.

Example 3.1. To study the product quality of a factory, we make some random samplings. In each sampling, the number of samples is different. The result of sampling is recorded and presented in Table 3.1.

In the table, the number of products examined is the total number of products examined in one sample. The number of qualified products is the total number of qualified products in the examination. The frequency of qualification is the proportion of qualified products in all the products examined in one sample. From the table, we can easily see the relation between the number of a mark and the frequency of a mark. We can also find a statistical regularity. That is, as the number of products examined increases, the frequency of qualification inclines to 0.9 stably. Or the frequency of qualification wavers around a fixed number $p = 0.9$. So p is the statistical stable center of this series of trials. It represents the possibility of qualification of an examined product. The possibility is called probability.

Definition 3.1. Statistical probability: If in a number of repeated trials the frequency of event A inclines to a constant p stably, it represents the possibility of the appearance of event A, and we call this constant p the probability of event A, shortly by $P(A)$:

$$p = P(A)$$

So a probability is the stable center of a frequency. A probability of any event A is a non-negative real number that is not bigger than 1:

$$0 \leq P(A) \leq 1$$

The statistical definition of probability has a close relation with frequency and is easily understood. But it is a tough problem to find the probability of an arbitrary event with experiments. Sometimes it is even impossible. So we often calculate probability with classical probabilistic method or geometrical probabilistic method.

Definition 3.2. Classical probability: Let a trial have and only have finite N possible results, or N basic events. If event A contains K possible results, we call K/N the probability of event A, shortly $P(A)$:

$$P(A) = K/N \tag{3.3}$$

Table 3.1.   The result of sampling of product quality.

Number of products examined	5	10	50	100	300	600	1000	5000	10000
Number of qualified products	5	8	44	91	272	542	899	4510	8999
Frequency of qualification	1	0.8	0.88	0.91	0.907	0.892	0.899	0.902	0.8997

To calculate a classical probability, we need to know the number of all the basic events. So classical probability is restricted to the cases of finite population. In the case of infinite population or the total number of basic events unknown, geometrical probability model is used to calculate probability. Besides, geometrical probability also gives a general definition of probability.

Geometrical random trial: Assume Ω is a bounded domain of M-dimensional space, and $L(\Omega)$ is the volume of Ω. We consider the random trial that we throw a random point into Ω evenly and assume the following: (a) Random point may fall in any domain of Ω but cannot fall outside of Ω. (b) The distribution of random point in Ω is even, viz. the possibility that random point falls into a domain is proportional to the volume of the domain and is independent of the position or the shape of the domain in Ω. Under the restrictions above, we call a trial a geometrical random trial, where Ω is basic event space.

Event in geometrical random trial: Assume that Ω is the basic event space of geometrical random trial, and A is a subset of Ω that can be measured with volume, where $L(A)$ is the M-dimensional volume of A. Then the event of "random point falls in domain A" is represented with A. In Ω, a subset that can be measured with volume is called a measurable set. Each measurable set can be viewed as an event. The set of all measurable subsets is represented by F.

Definition 3.3. Geometrical probability: Assume Ω is a basic event space of a geometrical random trial, and F is the set of all measurable subsets of Ω. Then the probability of any event A in F is the ratio between the volume of A and that of Ω:

$$P(A) = V(A)/V(\Omega) \tag{3.4}$$

Definition 3.4. Conditional probability: The probability of event A under the condition that event B has happened is denoted by $P(A|B)$. We call it the conditional probability of event A under condition B. $P(A)$ is called unconditional probability.

Example 3.2. There are two white balls and one black ball in a bag. Now we take out two balls in turn. Questions: (a) How much is the probability of the event that a white ball is picked in the first time? (b) How much is the probability of the event that a white ball is picked in the second time when a white ball has been picked in the first time?

Solution: Assume A is the event that a white ball is picked in the first time, and B is the event that a white ball is picked in the second time. Then $\{B|A\}$ is the event that a white ball is picked in the second time when a white ball has been picked in the first time. According to Definition 3.4, we have the following:

(1) No matter under repeated sampling or non-repeated sampling, $P(A) = 2/3$.

(2) When sampling is non-repeated, $P(B|A) = 1/2$; when sampling is repeated, $P(B|A) = P(B) = 2/3$. The conditional probability equals to non-conditional probability.

If the appearance of any event A or B will not affect the probability of the other event, viz. $P(A) = P(A|B)$ or $P(B) = P(B|A)$. We call event A and B independent events.

Theorem 3.1 (Addition Theorem). *The probability of the sum of two mutually exclusive events equals the sum of the probabilities of the two events, that is,*

$$P(A + B) = P(A) + P(B)$$

The sum of probabilities of two mutually inverse events is 1. In other words, if $A + A^{-1} = \Omega$, and A and A^{-1} are mutually inverse, then $P(A) + P(A^{-1}) = 1$, or $P(A) = 1 - P(A^{-1})$.

If A and B are two arbitrary events, then

$$P(A + B) = P(A) + P(B) - P(AB)$$

holds. This theorem can be generalized to the case that involves more than three events:

$$P(A + B + C) = P(A) + P(B) + P(C) - P(AB) - P(BC) - P(CA) + P(ABC)$$

Theorem 3.2 (Multiplication Theorem). *Assume A and B are two mutually independent non-zero events, then the probability of the multiple event equals the multiplication of probabilities of event A and B, that is,*

$$P(A \cdot B) = P(A) \cdot P(B) \quad or \quad P(A \cdot B) = P(B) \cdot P(A)$$

Assume A and B are two arbitrary non-zero events, then the probability of the multiple event equals the multiplication of the probability of event A (or B) and the conditional probability of event B (or A) under condition A (or B):

$$P(A \cdot B) = P(A) \cdot P(B|A) \quad or \quad P(A \cdot B) = P(B) \cdot P(A|B)$$

This theorem can be generalized to the case that involves more than three events. When the probability of multiple event $P(A_1 A_2 \ldots A_{n-1}) > 0$, we have

$$P(A_1 A_2 \ldots A_n) = P(A_1) \cdot P(A_2|A_1) \cdot P(A_3|A_1 A_2) \ldots P(A_n|A_1 A_2 \ldots A_{n-1})$$

If all the events are pairwise independent, we have

$$P(A_1 A_2 \ldots A_n) = P(A_1) \cdot P(A_2) \cdot P(A_3) \ldots P(A_n)$$

3.2.4 *Bayesian Probability*

(1) Prior probability: A prior probability is the probability of an event that is gained from historical materials or subjective judgments. It is not verified and is estimated in the absence of evidence. So it is called prior probability. There are two kinds of prior probabilities. One is objective prior probability, which is calculated according to historical materials; the other is subjective prior probability, which is estimated purely based on the subjective experience when historical material is absent or incomplete.

(2) Posterior probability: A posterior probability is the probability that is computed according to the prior probability and additional information from investigation via Bayesian formula.

(3) Joint probability: The joint probability of two events is the probability of the intersection of the two events. It is also called multiplication formula.

(4) Total probability formula: Assume all the influence factors of event A are $B_1, B_2, \ldots$, and they satisfy $B_i \cdot B_j = \emptyset, (i \neq j)$ and $P(\cup B_i) = 1$, $P(B_i) > 0$, $i = 1, 2, \ldots$, then we have

$$P(A) = \sum P(B_i) P(A|B_i) \tag{3.5}$$

(5) Bayesian formula: Bayesian formula, which is also called posterior probability formula or inverse probability formula, has wide application.

Assume $P(B_i)$ is prior probability, and $P(A_j|B_i)$ is new information gained from the investigation, where $i = 1, 2, \ldots, n$ and $j = 1, 2, \ldots, m$. Then the posterior probability calculated with Bayesian formula is

$$P(B_i|A_j) = \frac{P(B_i)P(A_j|B_i)}{\sum_{k=1}^{m} P(B_k)P(A_j|B_k)} \tag{3.6}$$

Example 3.3. One kind of product is made in a factory. Three work teams (A_1, A_2, and A_3) are in charge of two specifications (B_1 and B_2) of the product. Their daily outputs are listed in Table 3.2.

Table 3.2.   Daily outputs of three teams.

	B1	B2	Total
Team A1	2,000	1,000	3,000
Team A2	1,500	500	2,000
Team A3	500	500	1,000
Total	4,000	2,000	6,000

Now we randomly pick out one from the 6000 products. Please answer the following questions:

1. Calculate the following probabilities with classical probability.

(1) Calculate the probabilities that the picked product comes from the outputs of A_1, A_2, or A_3, respectively:

Solution: $P(A_1) = 3000/6000 = 1/2$

$$P(A_2) = 2000/6000 = 1/3$$

$$P(A_3) = 1000/6000 = 1/6$$

Calculate the probabilities that the picked product belongs to B_1 or B_2, respectively:

Solution: $P(B_1) = 4000/6000 = 2/3$

$$P(B_2) = 2000/6000 = 1/3$$

(2) Calculate the probability that the pick product is B_1 and comes from A_1:

Solution: $P(A_1 \cdot B_1) = 2000/6000 = 1/3$

(3) If the product that comes from A_1 is known, how much is the probability that it belongs to B_1?

Solution: $P(B_1|A_1) = 2000/3000 = 2/3$

(4) If the product belongs to B_2, how much is the probability that it comes from A_1, A_2, or A_3, respectively?

Solution: $P(A_1|B_2) = 1000/2000 = 1/2$

$$P(A_2|B_2) = 500/2000 = 1/4$$

$$P(A_3|B_2) = 500/2000 = 1/4$$

2. Calculate the following probabilities with conditional probability:

(1) If a product comes from A_1, how much is the probability that it belongs to B_1?

Solution: $P(B_1|A_1) = (1/3)/(1/2) = 2/3$

(2) If a product belongs to B_2, how much is the probability that it comes from A_1, A_2, or A_3, respectively?

Solution: $P(A_1|B_2) = (1/6)/(1/3) = 1/2$

$$P(A_2|B_2) = (1/12)/(1/3) = 1/4$$

$$P(A_3|B_2) = (1/12)/(1/3) = 1/4$$

3. Calculate the following probabilities with Bayesian Formula

(1) Known: $P(B_1) = 4000/6000 = 2/3$

$$P(B_2) = 2000/6000 = 1/3$$

$$P(A_1|B_1) = 1/2$$

$$P(A_1|B_2) = 1/2$$

Question: If a product comes from A_1, how much is the probability that it belongs to B_2?

Solution: Calculate joint probabilities:

$$P(B_1)P(A_1|B_1) = (2/3)(1/2) = 1/3$$

$$P(B_2)P(A_1|B_2) = (1/3)(1/2) = 1/6$$

Calculate total probability:

$$P(A_1) = (1/3) + (1/6) = 1/2$$

Calculate posterior probability according to Bayesian formula:

$$P(B_2|A_1) = (1/6) \div (1/2) = 1/3$$

(2) Known: $P(A_1) = 3000/6000 = 1/2$

$$P(A_2) = 2000/6000 = 1/3$$

$$P(A_3) = 1000/6000 = 1/6$$

$$P(B_2|A_1) = 1000/3000 = 1/3$$

$$P(B_2|A_2) = 500/2000 = 1/4$$

$$P(B_2|A_3) = 500/1000 = 1/2$$

Question: If a product belongs to B_2, how much is the probability that it comes from A_1, A_2, or A_3?

Solution: Calculate joint probabilities:

$$P(A_1)P(B_2|A_1) = (1/2)(1/3) = 1/6$$

$$P(A_2)P(B_2|A_2) = (1/3)(1/4) = 1/12$$

$$P(A_3)P(B_2|A_3) = (1/6)(1/2) = 1/12$$

Calculate total probability P(B2):

$$P(B_2) = \sum P(A_i)P(B_2|A_i)$$

$$= (1/2)(1/3) + (1/3)(1/4) + (1/6)(1/2) = 1/3$$

Calculate posterior probability according to Bayesian formula:

$$P(A_1|B_2) = (1/6) \div (1/3) = 1/2$$

$$P(A_2|B_2) = (1/12) \div (1/3) = 1/4$$

$$P(A_3|B_2) = (1/12) \div (1/3) = 1/4$$

3.3 Bayesian Problem Solving

Bayesian learning theory utilizes prior information and sample data to estimate unknown data. Probabilities (joint probabilities and conditional probabilities) are the representation of prior information and sample data in Bayesian learning theory. How to get the estimation of these probabilities (also called probabilistic density estimation) is much controversy in Bayesian learning theory. Bayesian density estimation focuses on how to gain the estimation of distribution of unknown variables (vectors) and its parameters based on sample data and prior knowledge from human experts. It includes two steps. One is to determine prior distributions of unknown variables; the other is to get the parameters of these distributions. If we know nothing about previous information, the distribution is called non-informative prior distribution. If we know the distribution and seek its proper parameters, the distribution is called informative prior distribution. Since learning from data is the most elementary characteristic of data mining, non-informative prior distribution is the main subject of Bayesian learning theory research.

The first step of Bayesian problem solving is to select Bayesian prior distribution. This is a key step. There are two common methods to select prior distribution, namely subjective method and objective method. The former makes use of human experience and expert knowledge to assign prior distribution. The latter is analyzing characters of data to get statistical features of data. It requires sufficient data to get the true distribution of data. In practice, these two methods are often combined.

Several common methods for prior distribution selection are listed in the following. Before we discuss these methods, we give some definitions first.

Let θ be the parameter of a model, $X = (x_1, x_2, \ldots, x_n)$ be observed data, and $\pi(\theta)$ be the prior distribution of θ. $\pi(\theta)$ represents the brief of parameter θ when no evidence exists. $l(x_1, x_2, \ldots, x_n|\theta) \propto p(x_1, x_2, \ldots, x_n|\theta)$ is likelihood function. It represents the brief of unknown data when parameter θ is known. $h(\theta|x_1, x_2, \ldots, x_n) \propto p(\theta|x_1, x_2, \ldots, x_n)$ is the brief of parameter θ after new evidence appears. Bayesian theorem describes the relation between them:

$$h(\theta|x_1, x_2, \ldots x_n) = \frac{\pi(\theta)p(x_1, x_2, \ldots x_n|\theta)}{\int \pi(\theta)p(x_1, x_2, \ldots x_n|\theta)d\theta} \propto \pi(\theta)l(x_1, x_2, \ldots x_n|\theta) \quad (3.7)$$

Definition 3.5 (Kernel of Distribution Density). If $f(x)$, the distribution density of random variable z, can be decomposed as $f(x), = cg(x)$, where c is a constant independent of x, we call $g(x)$ the kernel of $f(x)$, shortly $f(x) \propto g(x)$. If we know the kernel of distribution density, we can determine the corresponding constant according to the fact that the integral of distribution density in the whole space is 1. Therefore, the key to solving the distribution density of a random variable is to solve the kernel of its distribution density.

Definition 3.6 (Sufficient Statistic). To parameter θ, the statistic $t(x_1, x_2, \ldots, x_n)$ is sufficient if the posterior distribution of θ, $h(\theta|x_1, x_2, \ldots, x_n)$, is always a function of θ and $t(x_1, x_2, \ldots, x_n)$ in spite of its prior distribution.

This definition clearly states that the information of θ in data can be represented by its sufficient statistics. Sufficient statistics are connections between posterior distribution and data. In the following, we give out a theorem to judge whether a statistic is sufficient.

Theorem 3.3 (The Neyman–Fisher Factorization Theorem). *Let $f\theta(\mathbf{x})$ be the density or mass function for the random vector $\mathbf{x}$, parametrized by the vector θ. The statistic $t = T(\mathbf{x})$ is sufficient for θ if and only if there exist functions $a(\mathbf{x})$ (not depending on θ) and $b\theta(t)$ such that $f\theta(\mathbf{x}) = a(\mathbf{x})b\theta(t)$ for all possible values of $\mathbf{x}$.*

3.3.1 *Common Methods for Prior Distribution Selection*

1. Conjugate family of distributions

Raiffa and Schaifeer suggested using conjugate distributions as prior distributions, where the posterior distribution and the corresponding prior distribution are the same kind of distribution. The general description of conjugate distribution is as follows:

Definition 3.7. Let the conditional distribution of samples $x_1, x_2, \ldots, x_n$ under parameters θ be $p(x_1, x_2, \ldots, x_n | \theta)$. If the prior density function $\pi(\theta)$ and its resulting posterior density function $\pi(\theta | x)$ are in the same family, the prior density function $\pi(\theta)$ is said to be conjugate to the conditional distribution $p(x | \theta)$.

Definition 3.8. Let $P = \{p(x | \theta) : \theta \in \Theta\}$ be the density function family with parameters θ. $H = \pi(\theta)$ is the prior distribution family of θ. If for any given $p \in P$ and $\pi \in H$, the resulting posterior distribution $\pi(\theta | x)$ is always in family H, H is said to be the conjugate family to P.

When the density functions of data distribution and its prior are all exponential functions, the resulting function of their multiplication is the sample kind of exponential function. The only difference is a factor of proportionality. So we have the following:

Theorem 3.4. *If for random variable Z, the kernel of its density function $f(x)$ is an exponential function, the density function belongs to a conjugate family.*

All the distributions with exponential kernel function compose an exponential family, which includes binary distribution, multinomial distribution, normal distribution, Gamma distribution, Poisson distribution, and Dirichlet distribution.

Conjugate distributions can provide a reasonable synthesis of historical trials and a reasonable precondition for future trials. The computation of non-conjugate distribution is rather difficult. In contrast, the computation of conjugate distribution is easy, where only multiplication with prior is required. So, in fact, the conjugate family makes firm foundation for practical application of Bayesian learning.

2. Principle of maximum entropy

Entropy is used to quantify the uncertainty of an event in information theory. If a random variable x takes two different possible values, namely a and b, comparing the following two cases:

(1) $p(x = a) = 0.98$, $p(x = b) = 0.02$
(2) $p(x = a) = 0.45$, $p(x = b) = 0.55$

Obviously, the uncertainty of case 1 is much less than that of case 2. Intuitively, we can see that the uncertainty will reach maximum when the probabilities of two values are equal.

Definition 3.9. Let x be a discrete random variable. It takes at most countable values $a_1, a_2, \ldots, a_k, \ldots$, and $p(x = a_i) = p_i$, $i = 1, 2, \ldots$. The entropy of x is $H(x) = -\sum_i p_i \ln p_i$. For continuous random variable x, if the integral $H(x) = -\int p(x) \ln p(x)$ is meaningful, where $p(x)$ is the density of variable x, the integral is called the entropy of a continuous random variable.

According to the definition, when two random variables have the same distribution, they have equal entropy. So entropy is only related to distribution.

Principle of maximum entropy: For non-information data, the best prior distribution is the distribution, which makes the entropy maximum under parameters θ.

It can be proved that the entropy of a random variable, or vector, reaches maximum if and only if its distribution is uniform. Hence, the Bayesian assumption, which assumes non-information prior to being uniform, fits the principle of maximum entropy. It makes the entropy of a random variable, or vector, maximum. Following is the proof of the case of limited valued random variable.

Theorem 3.5. *Let random variable x take limited values $a_1, a_2, \ldots, a_n$. The corresponding probabilities are $p_1, p_2, \ldots, p_n$. The entropy $H(x)$ is maximum if and only if $p_1 = p_2 = \cdots = p_n = 1/n$.*

Proof. Consider $G(p_1, p_2, \ldots, p_n) = -\sum_{i=1}^{n} p_i \ln p_i + \lambda(\sum_{i=1}^{n} p_i - 1)$. To find its maximum, we let partial derivative of G respective to p_i to be 0 and get the following equations:

$$0 = \frac{\partial G}{\partial p_i} = -\ln p_{i-1} + \lambda \quad (i = 1, 2, \ldots, n)$$

Solve the equations, we get $p_1 = p_2 = \cdots = p_n$. Since $\sum_{i=1}^{n} p_i = 1$, we have $p_1 = p_2 = \cdots = p_n = 1/n$. Here the corresponding entropy is $-\sum_{i=1}^{n} \frac{1}{n} \ln \frac{1}{n} = \ln n$.

For continuous random variables, the result is the same.

From the above, when there is no information to determine prior distribution, the principle of maximum distribution is a reasonable choice for prior selection. There are many cases where no information is available to determine prior, so Bayesian assumption is very important in these cases. $\qquad\square$

3. Jeffrey's principle

Jeffrey had made significant contribution to prior distribution selection. He proposed an invariance principle, which well solved a conflict in Bayesian assumption and gave out an approach to find prior density. Jeffrey's principle is composed of two parts: one is a reasonable requirement for prior distribution; the other is giving out a concrete approach to find a correct prior distribution fitting the requirement.

There is a conflict in Bayesian assumption: If we choose uniform as the distribution of parameter θ, once we take function $g(\theta)$ as the parameter, it should also obey uniform distribution and vice versa. Yet the above precondition cannot lead to expected result. To solve the conflict, Jeffrey proposed an invariance request. That is, a reasonable principle for prior selection should have invariance.

If we choose $\pi(\theta)$ as the prior distribution of parameter θ, according to invariance principle, $\pi_g(g(\theta))$, the distribution of function $g(\theta)$ should satisfy the following:

$$\pi(\theta) = \pi_g(g(\theta))|g'(\theta)| \tag{3.8}$$

The key point is how to find a prior distribution $\pi(\theta)$ to satisfy the above condition. Jeffrey skillfully utilized the invariance of Fisher information matrix to find a required $\pi(\theta)$.

The distribution of parameter θ has the kernel of the square root of information matrix $I(\theta)$, viz. $\pi(\theta) \propto |I(\theta)|^{1/2}$, where $I(\theta) = E(\frac{\partial \ln p(x_1,x_2,...x_n;\theta)}{\partial \theta})(\frac{\partial \ln p(x_1,x_2,...x_n;\theta)}{\partial \theta})'$. The concrete deriving process is not presented here. Interested readers can find them in related references. It is noted that Jeffrey's principle is just a principle of finding reasonable prior, while using square root of information matrix as the kernel of prior is a concrete approach. They are different. In fact, we can seek other concrete approaches to embody the principle.

3.3.2 Computational Learning

Learning is that a system can improve its behavior after running. Is the posterior distribution gained via Bayesian formula better than its corresponding prior? What is its learning mechanism? Here we analyze normal distribution as an example to study the effect of prior information and sample data by changing parameters.

Let $x_1, x_2, \ldots, x_n$ be a sample from normal distribution $N(\theta, \sigma_1^2)$, where σ_1^2 is known and θ is unknown. To seek $\tilde{\theta}$, the estimation of θ, we take another normal distribution as the prior of θ. That is,

$$\pi(\theta) = N(\mu_0, \sigma_0^2)$$

The resulting posterior distribution of θ is also a normal distribution:

$$h(\theta|\bar{x}_1) = N(\alpha_1, d_1^2)$$

where

$$\bar{x}_1 = \sum_{i=1}^{n} \frac{x_i}{n}, \quad \alpha_1 = \left(\frac{1}{\sigma_0^2}\mu_0 + \frac{n}{\sigma_1^2}\bar{x}_1\right) \bigg/ \left(\frac{1}{\sigma_0^2} + \frac{n}{\sigma_1^2}\right), \quad d_1^2 = \left(\frac{1}{\sigma_0^2} + \frac{n}{\sigma_1^2}\right)^{-1}.$$

Take α_1, the expectation of the posterior $h(\theta|\bar{x})$ as the estimation of θ, we have

$$\tilde{\theta} = E(\theta|\bar{x}_1)\left(\frac{1}{\sigma_0^2}\mu_0 + \frac{n}{\sigma_1^2}\bar{x}_1\right) \cdot d_1^2 \tag{3.9}$$

Therefore, $\tilde{\theta}$, the estimation of θ, is the weighted average of μ_0, the expectation of prior, and $\bar{x}_1$, the sample mean. σ_0^2 is the variance of $N(\mu_0, \sigma_0^2)$, so its reciprocal, $1/\sigma_0^2$, is the precision of μ_0. Similarly, σ_1^2/n is the variance of sample mean $\bar{x}$, so its reciprocal is the precision of $\bar{x}_1$. Hence, we see that $\tilde{\theta}$ is the weighted average of μ_0 and $\bar{x}_1$, where the weights are their precisions, respectively. The smaller the variance, the bigger the weight. Besides, the bigger the sample size n, the smaller the variance σ_1^2/n, or the bigger the weight of the sample mean. This means that when n is quite large, the effect of prior mean will be very small. The above analysis illustrates that the posterior from Bayesian formula integrates the prior information and sample data. The result is more reasonable than that based on merely prior information or sample data. The learning mechanism is effective. The analysis based on other conjugate prior distributions leads to similar results.

According to the previous discussion, with the conjugate prior, we can use the posterior information as the prior of the next computation and seek the next posterior by integrating more sample information. If we repeat this process time after time, can we get posterior increasingly close to reality? We study this problem in the following:

Let new sample $x_1, x_2, \ldots, x_n$ be from normal distribution $N(\theta, \sigma_2^2)$, where σ_2^2 is known and θ is unknown. If we use previous posterior $h(\theta|\bar{x}_1) = N(\alpha_1, d_1^2)$ as the prior of next round of computation, then the new posterior is $h_1(\theta|\bar{x}_2) = N(\alpha_2, d_2^2)$, where

$$\bar{x}_2 = \sum_{i=1}^{n} \frac{x_i}{n}, \quad \alpha_2 = \left(\frac{1}{d_1^2}\alpha_1 + \frac{n}{\sigma_2^2}\bar{x}_2\right) \Big/ \left(\frac{1}{d_1^2} + \frac{n}{\sigma_2^2}\right), \quad d_2^2 = \left(\frac{1}{d_1^2} + \frac{n}{\sigma_2^2}\right)^{-1}$$

Now,
$\alpha_2 = (\frac{1}{\sigma_0^2}\mu_0 + \frac{n}{\sigma_1^2}\bar{x})/(\frac{1}{\sigma_0^2} + \frac{n}{\sigma_1^2})$ uses the expectation $h_1(\theta|\bar{x}_2)$ as the estimation of θ.

Since $\alpha_1 = (\frac{1}{\sigma_0^2}\mu_0 + \frac{n}{\sigma_1^2}\bar{x}_1) \cdot d_1^2$, we have

$$\alpha_2 = \left(\frac{1}{d_1^2}\alpha_1 + \frac{n}{\sigma_2^2}\bar{x}_2\right) \cdot d_2^2 = \left(\frac{1}{\sigma_0^2}\mu_0 + \frac{n}{\sigma_1^2}\bar{x}_1 + \frac{n}{\sigma_2^2}\bar{x}_2\right) \cdot d_2^2$$

$$= \left(\frac{1}{\sigma_0^2}\mu_0 + \frac{n}{\sigma_1^2}\bar{x}_1\right) \cdot d_2^2 + \frac{n}{\sigma_2^2}\bar{x}_2 \cdot d_2^2 \tag{3.10}$$

and $\frac{n}{\sigma_2^2} > 0$, so $d_2^2 = (\frac{1}{d_1^2} + \frac{n}{\sigma_2^2})^{-1} = (\frac{1}{\sigma_0^2} + \frac{n}{\sigma_1^2} + \frac{n}{\sigma_2^2})^{-1} < d_1^2 = (\frac{1}{\sigma_0^2} + \frac{n}{\sigma_1^2})^{-1}$ In α_2, $(\frac{1}{\sigma_0^2}\mu_0 + \frac{n}{\sigma_1^2}\bar{x}_1) \cdot d_2^2 < \alpha_1$.

It is clear that because of the addition of a new sample, the proportion of the original prior and old sample declines. According to equation (3.10), with the continuous increase of new sample (here we assume the sample size remains invariant), we have

$$a_m = \left(\frac{1}{\sigma_0^2}\mu_0 + \frac{n}{\sigma_1^2}\bar{x}_1 + \frac{n}{\sigma_2^2}\bar{x}_2 + \cdots + \frac{n}{\sigma_m^2}\bar{x}_m \right) \cdot d_m^2$$

$$= \left(\frac{1}{\sigma_0^2}\mu_0 + \sum_{k=1}^{m} \frac{n}{\sigma_k^2}\bar{x}_k \right) \cdot d_m^2 \quad (k = 1, 2, \ldots, m) \tag{3.11}$$

From equation (3.11), if the variance of new samples is the same, they are equal to a sample with the size of $m \times n$. The above process weighted all the sample means with their precisions. The higher the precision, the bigger the weight. If the prior distribution is estimated precisely, we can use less sample data and only need a little computation. This is especially useful in a situation where the sample is hard to be collected. It is also the point that Bayesian approach outcomes other methods.

Therefore, the determination of prior distribution in Bayesian learning is extremely important. If there is no prior information and we adopt non-information prior, with the increase of the sample, the effect of the sample will become more and more salient. If the noise of the sample is small, the posterior will be increasingly close to its true value. The only matter is that large computation is required.

3.3.3 *Steps of Bayesian Problem-Solving*

The steps of Bayesian problem-solving can be summarized as follows:

(1) Define random variables. Set unknown parameters as random variables or vectors, shortly by θ. The joint density $p(x_1, x_2, \ldots, x_n; \theta)$ of sample $x_1, x_2, \ldots, x_n$ is regarded as the conditional density of $x_1, x_2, \ldots, x_n$ with respect to θ, shortly by $p(x_1, x_2, \ldots, x_n | \theta)$ or $p(D|\theta)$.
(2) Determine prior distribution density $p(\theta)$. Use conjugate distribution. If there is no information about prior distribution, then use Bayesian assumption of non-information prior distribution.
(3) Calculate posterior distribution density via Bayesian theorem.
(4) Make inference of the problem with the resulting posterior distribution.

Take the case of single variable and single parameter, for example. Consider the problem of thumbtack throwing. If we throw a thumbtack up in the air, the thumbtack will fall down and reset at one of the two states: on its head or on its tail. Suppose we flip the thumbtack $N + 1$ times. From the first N observations, how can we get the probability of the case head in the $N + 1$th throwing?

$$\alpha_m = \left(\frac{1}{\sigma_0^2}\mu_0 + \frac{n}{\sigma_1^2}\bar{x}_1 + \frac{n}{\sigma_2^2}\bar{x}_2 + \cdots + \frac{n}{\sigma_m^2}\bar{x}_m \right) \cdot d_m^2$$

$$= \left(\frac{1}{\sigma_0^2}\mu_0 + \sum_{k=1}^{m} \frac{n}{\sigma_k^2}\bar{x}_k \right) \cdot d_m^2 \quad (k = 1, 2, \ldots, m)$$

Step 1. Define a random variable Θ. The value θ corresponds to the possible value of the real probability of head. The density function $p(\theta)$ represents the uncertainty of Θ. The variable of the ith result is X_i $(i = 1, 2, \ldots, N+1)$, and the set of observation is $D = \{X_1 = x_1, \ldots, X_n = x_n\}$. Our objective is to calculate $p(x_{N+1}|D)$.

Step 2. According to Bayesian theory, we have

$$p(\theta|D) = \frac{p(\theta)p(D|\theta)}{p(D)},$$

$$p(D) = \int p(D|\theta)p(\theta)d\theta$$

where $p(D|\theta)$ is the binary likelihood function of the sample. If θ, the value of Θ, is known, the observation value in D is independent, and the probability of head (tail) is θ, the probability of tail is $(1 - \theta)$, then

$$p(\theta|D) = \frac{p(\theta)\theta^h(1 - \theta)^t}{p(D)} \tag{3.12}$$

where h and t are the times of head and tail in the observation D, respectively. They are sufficient statistics of sample binary distribution.

Step 3. Seek the mean of Θ as the probability of case head in the $N + 1$th toss:

$$p(X_{N+1} = heads|D) = \int p(X_{N+1} = heads|\theta)p(\theta|D)d\theta$$

$$= \int \theta \cdot p(\theta|D)d\theta \equiv E_{p(\theta|D)}(\theta) \tag{3.13}$$

where $E_{p(\theta|D)}(\theta)$ is the expectation of θ under the distribution $p(\theta|D)$.

Step 4. Assign prior distribution and supper parameters for Θ.

A common method for prior assignment is to assume prior distribution first and then determine proper parameters. Here we assume the prior distribution is

Beta distribution:

$$p(\theta) = \text{Beta}(\theta|\alpha_h, \alpha_t) \equiv \frac{\Gamma(\alpha)}{\Gamma(\alpha_h)\Gamma(\alpha_t)}\theta^{\alpha_h-1}(1-\theta)^{\alpha_t-1} \tag{3.14}$$

where $\alpha_h > 0$ and $\alpha_t > 0$ are parameters of Beta distribution, and $\alpha = \alpha_h + \alpha_t$, and $\Gamma(\cdot)$ is the Gamma function. To distinguish with parameter θ, σh and α_t are called "Supper Parameters". Since Beta distribution belongs to conjugate family, the resulting posterior is also Beta distribution:

$$
\begin{aligned}
p(\theta|D) &= \frac{\Gamma(\alpha + N)}{\Gamma(\alpha_h + h)\Gamma(\alpha_t + t)}\theta^{\alpha_h+h-1}(1-\theta)^{\alpha_t+t-1} \\
&= \text{Beta}(\theta|\alpha_h + h, \alpha_t + t)
\end{aligned}
\tag{3.15}
$$

To this distribution, its expectation of θ has a simple form:

$$\int \theta \cdot Beta(\theta|\alpha_h, \alpha_t)d\theta = \frac{\alpha_h}{\alpha} \tag{3.16}$$

Therefore, for a given Beta prior, we get the probability of head in the $N + 1$th toss as follows:

$$p(X_{N+1} = heads|D) = \frac{\alpha_h + h}{\alpha + N} \tag{3.17}$$

There are many ways to determine the supper parameters of prior Beta distribution $p(\theta)$, such as imagined future data and equivalent samples. Other methods can be found in the works of Winkler, Chaloner, and Duncan. In the method of imagined future data, two equations can be deduced from equation (3.17), and two supper parameters α_h and α_t can be solved accordingly.

In the case of single variable multiple parameters (a single variable with multiple possible states), commonly X is regarded as a continuous variable with Gaussian distribution. Assume its physical density is $p(x|\theta)$, then we have

$$p(x|\theta) = (2\pi v)^{-1/2}e^{-(x-\mu)^2/2v^2}$$

where $\theta = \{\mu, v\}$.

Similar to the previous approach on binary distribution, we first assign the prior of parameters and then solve the posterior with the data $D = \{X_1 = x_1, X_2 = x_2, \ldots, X_N = x_N\}$ via Bayesian theorem:

$$P(\theta|D) = p(D|\theta)p(\theta)/p(D)$$

Next, we use the mean of Θ as the prediction:

$$p(x_{N+!}|D) = \int p(x_{N+1}|\theta)p(\theta|D)\mathrm{d}\theta \tag{3.18}$$

For exponential family, the computation is effective and close. In the case of multi-samples, if the observed value of X is discrete, Dirichlet distribution can be used as the prior distribution, which can simplify the computation.

The computational learning mechanism of Bayesian theorem is to get the weighted average of the expectation of prior distribution and the mean of the sample, where the higher the precision, the bigger the weight. Under the precondition that the prior is conjugate distribution, posterior information can be used as the prior in the next round computation so that it can be integrated with further obtained sample information. If this process is repeated time after time, the effect of the sample will be increasingly prominent. Since Bayesian method integrates prior information and posterior information, it can both avoid the subjective bias when using only prior information and avoid numerous blind searching and computation when sample information is limited. Besides, it can also avoid the effect of noise when utilizing only posterior information. Therefore, it is suitable for problems of data mining with statistical features and problems of knowledge discovery, especially problems where the sample is hard to collect or the cost of collecting the sample is high. The key to effective learning with Bayesian method is determining prior reasonably and precisely. Currently, there are only some principles for prior determination, and there is no operable whole theory to determine priors. In many cases, the reasonability and precision of prior distribution are hard to evaluate. Further research is required to solve these problems.

3.4 Naïve Bayesian Learning Model

In naïve Bayesian learning models, training sample I is decomposed into feature vector X and decision class variable C. Here, it is assumed that all the weights in a feature vector are independently given the decision variable. In other words, each weight affects the decision variable independently. Although the assumption to some extent limits the sphere of naïve Bayesian model, in practical applications, naïve Bayesian model can both exponentially reduce the complexity of model construction and can express striking robustness and effectiveness even when the assumption is unsatisfied (Nigam *et al.*, 1998). It has been successfully applied in many data mining tasks, such as classification, clustering, model selection, and so on. Currently, many researchers are working to relax the limitation of independence among variables (Heckerman, 1997) so that the model can be applied more widely.

3.4.1 *Naïve Bayesian Learning Model*

Bayesian theorem tells us how to predict the class of incoming sample given training samples. The rule of classification is maximum posterior probability, which is given in the following equation:

$$P(C_i|A) = P(C_i) * P(A|C_i)/P(A) \tag{3.19}$$

Here A is a test sample to be classified and $P(Y|X)$ is the conditional probability of Y under the condition of X. The probabilities at the right side of the equation can be estimated from training data. Suppose that the sample is represented as a vector of features. If all features are independent for given classes, $P(A|C_i)$ can be decomposed as a product of factors: $P(a_1|C_i) \times P(a_2|C_i) \times \cdots \times P(a_m|C_i)$, where a_i is the ith feature of the test sample. Accordingly, the posterior computation equation can be rewritten as

$$P(C_i|A) = \frac{P(C_i)}{P(A)} \prod_{j=1}^{m} P(a_j|C_i) \tag{3.20}$$

The entire process is called naïve Bayesian classification. In common sense, only when the independent assumption holds, or when the correlation of features is very weak, the naïve Bayesian classifier can achieve the optimal or sub-optimal result. Yet the strong limited condition seems inconsistent with the fact that naïve Bayesian classifier gains striking performance in many fields, including some fields where there is obvious dependence among features. In 16 out of a total of 28 datasets of UCI, naïve Bayesian classifier outperforms the C4.5 algorithms and has similar performance to that of CN2 and PEBLS. Some research works report similar results (Clark & Niblett, 1989). At the same time, researchers have also successfully proposed some strategy to relax the limitation of independence among features (Nigam *et al.*, 1998).

The conditional probability in formula (3.19) can be gained using maximum likelihood estimation:

$$P(v_j|C_i) = \frac{count(v_j \wedge c_i)}{count(c_i)} \tag{3.21}$$

To avoid zero probability, if the actual conditional probability is zero, it is assigned to be $0.5/N$, where N is the total number of examples.

Suppose that there are only two classes, namely class0 and class1, and $a_1, \ldots, a_k$ represent features of test set. Let $b_0 = P(C = 0)$, $b_1 = P(C = 1) = 1 - b_0$, $p_{j0} = P(A_j = a_j | C = 0)$, $p_{j1} = P(A_j = a_j | C = 0)$, then

$$p = P(C = 1 | A_1 = a_1 \wedge \cdots \wedge A_k = a_k) = \left(\prod_{j=1}^{k} p_{j1} \right) b_1 / z \qquad (3.22)$$

$$q = P(C = 0 | A_1 = a_1 \wedge \cdots \wedge A_k = a_k) = \left(\prod_{j=1}^{k} p_{j0} \right) b_0 / z \qquad (3.23)$$

where z is a constant. After taking logarithm on both sides of the above two equations, we subtract the second equation from the first one and get

$$\log p - \log q = \left(\sum_{j=1}^{k} \log p_{j1} - \log p_{j0} \right) + \log b_1 - \log b_0 \qquad (3.24)$$

Here, let $w_j = \log p_{j1} - \log p_{j0}$, $b = \log b_1 - \log b_0$, the above equation is written as

$$\log(1 - p)/p = - \sum_{j=1}^{k} w_j - b \qquad (3.25)$$

After taking exponential on both sides of equation (3.24) and rearranging, we have

$$p = \frac{1}{1 + e^{-\sum_{j=1}^{k} w_j - b}} \qquad (3.26)$$

To calculate this value, we assume that feature A_j has $v(j)$ possible values. Let

$$w_{jj'} = \log P(A_j = a_{jj'} | c = 1)$$

$$\log P(A_j = a_{jj'} | c = 0) \quad (1 \leq j' \leq v(j)) \qquad (3.27)$$

We have

$$P(C(x) = 1) = \frac{1}{1 + e^{-(\sum_{j=1}^{k} \sum_{j'}^{v(j)} I(A_j(x) = a_{jj'}) w_{jj'} - b}} \qquad (3.28)$$

where I is a characteristic function. If φ is true, then $I(\varphi) = 1$; else $I(\varphi) = 0$. In practical computation, equation (3.28) can be calculated similarly to equation (3.21).

In fact, equation (3.28) is a perception function with a sigmoid activation function. The input of this function is the possible values of all features. So, to some extent, naïve Bayesian classifier is equal to a perception model. Further research demonstrated that naïve Bayesian classifier can be generalized to logical regression with numerical features.

Consider equation (3.21). If A_j takes discrete values, $count(A_j = a_j \wedge C = c_i)$ can be calculated directly from training samples. If A_j is continuous, it should be discretized. In unsupervised discretization, a feature is discretized into M equally wide sections, where $M = 10$ commonly. We can also utilize more complicated discretization methods, such as the supervised discretization method.

Let each A_j be a numerical feature (discrete or continuous). The logical regression model is

$$\log \frac{P(C = 1|A_1 = a_1, \ldots A_k = a_k)}{P(C = 0|A_1 = a_1, \ldots A_k = a_k)} = \sum_{j=1}^{k} b_j a_j + b_0 \tag{3.29}$$

After transforming as that of equation (3.25), we have

$$p = \frac{1}{1 + e^{-\sum_{j=1}^{k} b_j a_j - b_0}} \tag{3.30}$$

Obviously, this is also a perception function with a sigmoid activation function. Its inputs are all the feature values. Use function $f_j(\varphi)$ to replace $b_j a_j$. If the sphere of A_j is divided into M parts and the ith part is $[c_{j(i-1)}, c_{ji}]$, the function $f_j(\varphi)$ is

$$b_j a_j = f_j(a_j) = \sum_{i=1}^{M} b_{ji} I(c_{j(i-1)} < a_j \leq c_{ji}) \tag{3.31}$$

where b_{ji} is a constant. According to equations (3.30) and (3.31), we have

$$P(C(x) = 1) = \frac{1}{1 + e^{-(\sum_{j=1}^{k} \sum_{i}^{M} b_{ji} I(c_{j(i-1)} < a_j \leq c_{ji})) - b_0}} \tag{3.32}$$

This is the final regression function. So naïve Bayesian classifier is a non-parametric and nonlinear extension of logical regression. By setting $b_{ji} = (c_{j(i-1)} + c_{ij})/(2b_j)$, we can get a standard logical regression formula.

3.4.2 *Boosting of Naïve Bayesian Model*

In boosting, a series of classifiers will be built, and in each classifier in series, examples misclassified by previous classifier will be given more attention. Concretely, after learning classifier k, the weights of training examples that are misclassified by classifier k will increase, and classifier $k + 1$ will be learned based on the newly weighted training examples. This process will be repeated T times. The final classifier is the synthesis of all the classifiers in series.

Initially, each training example is set with a weight. In the learning process, if some example is misclassified by one classifier, in the next learning round, the corresponding weight will be increased so that the next classifier will pay more attention to it.

The boosting algorithm for binary classification problem is given out by Freund and Scbapire as the AdaBoost Algorithm (Freund & Schapire, 1995).

Algorithm 3.1. AdaBoost Algorithm.

Input:

N training examples $< (x_1, y_1 >, \ldots (x_N, y_N) >$

Distribution of the N training examples, D: w, where w is the weight vector of training example.

T: the number of rounds for training.

1. Initialize:
2. Initial weight vector of training examples: $w_i = 1/N \ i = 1, \ldots, N$
3. for $t = 1$ to T
4.   Given weights w_i^t, find a hypothesis $H^{(t)}: X \to [0, 1]$
5.   Estimation the general error of hypothesis $H^{(t)}$:

$$e^{(t)} = \sum_{i=1}^{N} w_i^{(t)} |y_i - h_i^{(t)}(x_i)|$$

6.   Calculate $\beta^{(t)} = e^{(t)}/(1 - e^{(t)})$
7.   Renew the next round weights of examples with

$$w_i^{(t+1)} = w_i^{(t)} (\beta^{(t)})^{1 - |y_i - h_i^{(t)}(x_i)|}$$

8.   Normalize $w_i^{(t+1)}$, so that they are summed up to 1

$$h(x) = \begin{cases} 1 & if \ \sum_{t=1}^{T} \left(\log \tfrac{i}{\beta^{(t)}} \right) h^{(t)}(x) \geq \frac{1}{2} \sum_{t=1}^{T} \left(\log \tfrac{i}{\beta^{(t)}} \right) \\ 0 & otherwize \end{cases}$$

9. End for
10. Output

Here we assume that all the classifiers are effective. In other words, for each classifier, the examples correctly classified are more than the ones misclassified, $e^t < 0.5$. Hence, $\beta^{(t)} < 1$. When $|y_i - h_i^{(t)}(x_i)|$ increases, $w_i^{(t+1)}$ will increase accordingly. The algorithm fulfills the idea of boosting.

Some notes to the algorithm:

(1) $h^{(t)}(x)$ is calculated via output formula, and the result is either 0 or 1.
(2) The calculation of conditional probability $P(A_j = a_{jj'}|C = c)$ in formula (3.20). If we do not consider the weights, the computational basis for $count(condition)$ is 1. For example, if there are k examples of satisfied condition, then $count(condition) = k$. If we consider the weights, the computational basis for each example is its weight. For example, if there are k examples of satisfied condition, then $count(condition) = \sum_i^k w_i$. In this case, the adjustment of weights embodies the idea of boosting.
(3) The output of the algorithm means that for an incoming input x, according to Step 6 in the algorithm, we can use the result of learning to generate of output by voting.

The final combined hypothesis can be defined as

$$H(x) = \frac{1}{1 + \prod_{t=1}^{T} (\beta^{(t)})^{2r(x)-1}}$$

where $r(x) = \frac{\sum_{t=1}^{T}(\log 1/\beta^t)H^{(t)}(x)}{\sum_{t=1}^{T}(\log 1/\beta^t)}$.

In the following, we demonstrate that after boosting the represent capability of combined naïve Bayesian classifier equals to that of multiple layered perception model with one hidden layer. Let $\alpha = \prod_{t=1}^{T} \beta^t$ and $v^{(t)} = \log \beta^{(t)}/\log \alpha$, then

$$H(x) = \frac{1}{1 + \alpha^{2(\sum_{t=1}^{T} v^{(t)} H^{(t)}(x))-1}} = \frac{1}{1 + e^{\sum_{t=1}^{T} 2\log \beta^{(t)} H^{(t)}(x) - \sum_{t=1}^{T} \log \beta^{(t)}}}$$

The output of combined classifier is the output of a sigmoid function, which takes the outputs of single classifiers and their weights as its parameters. Since a naïve Bayesian classifier equals a perception machine, the combined classifier equals a perception network with a hidden layer.

The boosting naïve Bayesian method for multiple classification problems is as follows:

Algorithm 3.2. Multiple Classification AdaBoost Algorithm.

Input:

N training examples $< (x_1, y_1 >, \ldots (x_N, y_N) >$

Distribution of the N training examples, D: w, where w is the weight vector of training example.

T: the number of rounds for training.

1. Initialize:

    Initial weight vector of training examples $w_i = 1/N, i = (1 \ldots N)$

2. *for $t = 1$ to T*

3.   Given weights w_i^t, find a hypothesis $H^{(t)}: X \to Y$

4.   Estimation the general error of hypothesis $H^{(t)}$:

$$e^{(t)} = \sum_{i=1}^{N} w_i^{(t)} I(y_i \neq h_i^{(t)}(x_i))$$

5.   Calculate $\beta^{(t)} = e^{(t)}/(1 - e^{(t)})$

6.   Renew the next round weights of examples with

$$w_i^{(t+1)} = w_i^{(t)} (\beta^{(t)})^{1 - I(y_i = h_i^{(t)}(x_i))}$$

7.   Normalize $w_i^{(t+1)}$, so that they are summed up to 1

$$h(x) = \arg \max_{y \in Y} \sum_{t=1}^{T} \left(\log \frac{i}{\beta^{(t)}} \right) I(h^{(t)}(x) = y)$$

        where $I(\phi) = 1$ if $\phi = T$; $I(\phi) = 0$ otherwise.

8. End for

9. Output:

3.4.3 *The Computational Complexity*

Suppose a sample in the sample space has f features, and each feature takes v values. The naïve Bayesian classifier deduced by formula (3.27) will have $fv + 1$ parameters. These parameters are accumulatively learned $2fv+2$ times. In each learning process, each feature value of each training example will improve the final precision. So the time complexity for n training examples is $O(nf)$, independent of v. Substantially, this time complexity is optimal. For boosting naïve Bayesian classifier, the time complexity of each round is $O(nf)$. T round training corresponds to $O(Tnf)$. Note that T is a constant. So the entire time complexity is still $O(nf)$.

For naïve Bayesian classifier, the primary computation is counting. Training examples can be processed either sequentially or in batches from disk or tape. So this method is perfectly suited for knowledge discovery on large datasets. Training set is not necessarily loaded to memory entirely, and part of it can be kept in the disks or tapes. Yet the boosting naïve Bayesian model also has the following problems:

(1) From the idea of boosting, when noise exists in training set, boosting method will take it as useful information and amplify its effect with a large weight. This will reduce the performance of boosting. If there are many noise data, boosting will lead to a worse result.
(2) Although theoretically boosting can achieve 0 error rate for training set, in its practical application of naïve Bayesian model, 0 classification error in training set is generally hardly guaranteed.

3.5 Inferred Causation

The possibility of studying causal relationships from raw data has been a dream since the time of Hume. That possibility entered the realm in the mid-1980s. An autonomous intelligent system attempting to build a workable model of its environment cannot rely exclusively on preprogrammed causal knowledge rather it must be able to translate direct observations to cause-and-effect relationships (Pearl, 2000).

3.5.1 *Causal Discovery Framework*

We view the task of causal discovery as an induction game. These mechanisms are organized in the form of an acyclic structure, which is identified from the available observations.

Definition 3.10. Causal structure

A causal structure of a set of variables V is a directed acyclic graph in which each node corresponds to a district element of V, and each link represents a direct functional relationship among the corresponding variables.

In a general functional causal model consists of a set of equations,

$$x_i = f_i(pa_i, u_i), \quad i = 1, \ldots, n \tag{3.33}$$

where pa_i stands for the set of variables that directly determine the value of X_i and U_i represent errors due to omitted factors.

Definition 3.11. Causal model

A causal model is a pair $M = \langle D, \Theta_D \rangle$ that consists of a causal structure D and a set of parameters Θ_D compatible with D The parameters Θ_D assignment a function $x_i = f_i(pa_i, u_i)$ to each $X_i \in V$ and a probability measure $P(u_i)$ to each u_i, where PA_i are the parents of X_i in D and where each U_i is a random disturbance distributed according to $P(u_i)$, independent of all other u.

The assumption of independent disturbances renders the model Markovian in the sense that each variable is independent of all its non-descendants, conditional on its parents in D. The ubiquity of the Markov assumption in human discourse may be reflective of the granularity of the models we deem useful for understanding nature. We can start in the deterministic extreme, where all variables are explicated in microscopic detail and where the Markov condition certainly holds. The Markov condition guides us in deciding when a set of parents *PAi* is considered complete in the sense that it includes *all* the relevant immediate causes of variable *Xi*. It permits us to leave some of these causes out of *PAi* but not if they also affect other variables modeled in the system. If a set *PAi* in a model is too narrow, there will be disturbance terms that influence several variables simultaneously, and the Markov property will be lost. Such disturbances will be treated explicitly as "latent" variables. Once we acknowledge the existence of latent variables and represent their existence explicitly as nodes in a graph, the Markov property is restored.

Once a causal model M is formed, it defines a joint probability distribution $P(M)$ over the variables in the system. This distribution reflects some features of the causal structure. Nature then permits the scientist to inspect a select subset of "observed" variables and to ask questions about $P_{[o]}$, the probability distribution over the observables, but it hides the underlying causal model as well as the causal structure.

3.5.2 *Model Preference*

In principle, since V is unknown, there are an unbounded number of models that would fit a given distribution, each invoking a different set of "hidden" variables and each connecting the observed variables through different causal relationships. Therefore, with no restriction on the type of models considered, the scientist is unable to make any meaningful assertions about the structure underlying the phenomena.

Definition 3.12. Inferred causation

A variable X is said to have a causal influence on a variable Y if a directed path from X to Y exists in every minimal structure consistent with the data.

We regard Definition 3.3 as preliminary because it assumes that all variables are observed. The next few definitions generalize the concept of minimality to structures with unobserved variables.

Definition 3.4 Latent structure

A latent structure is a pair $L = \langle D, O \rangle$, where D is a causal structure over V *and where $O \subseteq V$ is a set of observed variables.*

Definition 3.13. Structure preference

One latent structure $L = \langle D, O \rangle$ is preferred to another $L' = \langle D', O \rangle$, written as $L \preceq L$ if and only if D' can mimic D over O, that is, if and only if for every Θ_D *there exists a* $\Theta'_{D'}$ *such that* $P_{[O]}\left(\langle D', \Theta_D \rangle\right) = P_{[O]}\left(\langle D, \Theta_D \rangle\right)$. Two latent structures are equivalent, $L' \equiv L$ if and only if $L \preceq L'$ and $L \succeq L'$.

Note that the set of independencies entailed by a causal structure imposes limits on its expressive power, that is, its power to mimic other structures. Indeed, L_1 cannot be preferred to L_2 if there is even one observable dependence $\hat{P}y$ that is permitted by L_1 and forbidden by L_2. Thus, tests for preference and equivalence can sometimes be reduced to tests of induced dependencies, which in turn can be determined directly from the topology of the directed acyclic graphs without ever concerning ourselves with the set of parameters.

Definition 3.14. Minimality

A latent structure L is minimal with respect to a class $\mathcal{L}$ of latent structures if and only if there is no member of $\mathcal{L}$ that is strictly preferred to L, that is, if and only if for every $L' \in \mathcal{L}$ we have $L' \equiv L$ whenever $L' \preceq L$.

Definition 3.15. Consistency

A latent structure $L = |D, O|$ is consistent with a distribution $\hat{P}$ over O if D can accommodate some model that generates $\hat{P}$. Clearly, a necessary (and sometimes sufficient) condition for L to be consistent with $\hat{P}$ is that L can account for all the dependencies embodied in $\hat{P}$.

Definition 3.16. Inferred causation

Given $\hat{P}$, a variable C has a causal influence on variable E if and only if there exists a directed path from C to E in every minimal latent structure consistent with $\hat{P}$.

We view this definition as normative because it is based on one of the least disputed norms of scientific investigation. However, as with any scientific inquiry, we make no claims that this definition is guaranteed to always identify stable physical mechanisms in nature. It identifies the mechanisms we can plausibly infer from non-experimental data; moreover, it guarantees that any alternative mechanism will be less trustworthy than the one inferred because the alternative would require more contrived, hind-sighted adjustment of parameters to fit the data.

3.5.3 *Stable Distributions*

The structure of the actual data-generating model does not guarantee would be minimal. In order to rule out such pathological parameterizations, Pearl proposed a restriction on the distribution called stability. This restriction conveys the assumption that all the independencies embedded in P are stable; that is, they are entailed by the structure of the model D and hence remain invariant to any change in the parameters Θ_D.

Definition 3.17. Stability

Let $I(P)$ denote the set of all conditional independence relationships embodied in P.

A causal model $M = \langle D, \Theta_D \rangle$ *generates a* stable distribution if and only if $P(\langle D, \Theta_D \rangle)$ contains no extraneous independencies, that is, if and only if $I(P(\langle D, \Theta_D \rangle)) \subseteq I(P(\langle D, \Theta'_D \rangle))$ for any set of parameters Θ'_D.

The stability condition states that, as we vary the parameters from Θ to Θ', no independence in P can be destroyed; hence the name "stability." Succinctly, P is a stable distribution of M if it "maps" the structure D of M.

The analogy with independencies is clear. Some independencies are structural, that is, they would persist for every functional-distributional parameterization of the graph. Others are sensitive to the precise numerical values of the functions and distributions.

3.6 Three Levels of Causality

Causal inference is a kind of logical inference. According to the reasons and conditions of things, we can judge through logical thinking and derive the results. This inference method is called causal inference. Judea Pearl made a breakthrough in understanding causality by discovering and systematically studying the "Ladder of Causation" (Pearl & Mackenzie, 2018), a framework that highlights the distinct roles of seeing, doing, and imagining, as shown in Figure 3.2.

The first is the bottom layer, which refers to the general observation of the phenomenon of things, and according to the observed phenomenon, fined the relevance. According to Pearl, the current machine learning model can only stay in the first stage. This is because statistical machine learning model can only observe

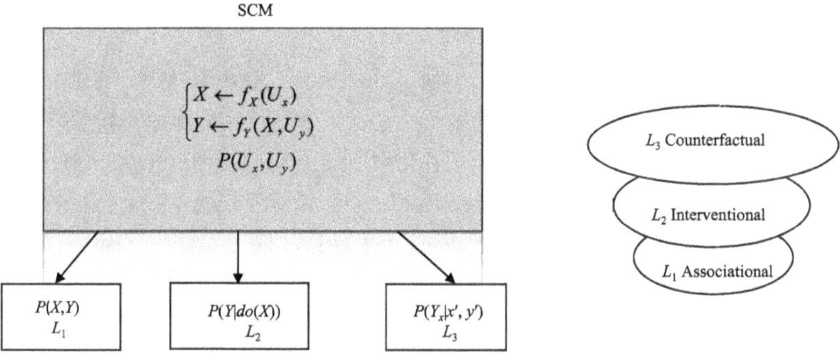

Fig. 3.2.   A framework of causal hierarchy.

phenomena (data) and fit a possible probability distribution. The result of this fitting is that the model can only find the correlation between two variables rather than the real causality.

The middle layer is intervention, that is, through the change of variables, to study the impact of this variable on the results, including whether to change the nature of the results and the intensity of the change. Finally, it is the counterfactual, that is, controlling other variables through simulation, only turning over the studied variables and exploring the possible development. In a sense, counterfactuals are about a fictional world or unrealized possibility because their antecedent is the opposite of the fact: they are about what might happen, but not. Therefore, on the surface, they seem to be beyond the scope of normal scientific research. Counterfactual is so closely related to causality that it is impossible to think with causality instead of counterfactual reasoning. No matter how large the dataset is, the causal conclusion is not only drawn from the observed statistical laws. Instead, we must use all our clues and imagination to create reasonable causal hypotheses and then analyze these hypotheses to see if they are reasonable and how to test them through data. Just piling up more numbers is not a shortcut to gain causal insight.

Causal inference is a logical way to deduce a certain positive result from the cause of things. Through causal inference, human beings can solve the problems in life and science. To express the causality in the language in mathematical form, Pearl proposed the causality diagram. A graphic simulation that represents the causal relationship between variables. A point represents a variable, and an edge with a point represents the causal effect of the variable. This kind of point-edge graph is called causality graph. Pearl thinks that the cause-and-effect diagram can well show the cause-and-effect relationship. If each edge has a weight, it can represent the influence intensity of different factors on this result.

3.7 Causal Diagrams

We introduce graph models into causal inference, and it should be noted that the learning and inference tasks in probabilistic graphical model (PGM) are not the same under the causal framework. Under the causal framework, we estimate the average causal effect and ultimately use a causal model to make decisions.

Shown in Figure 3.3 is a causal diagram. Under the causal framework, we estimate the average causal effect and ultimately use a causal model to make decisions. Figure 3.3 is a directed acyclic graph (DAG), which is often referred to as a DAG graph. Composed of 3 nodes and 3 edges, each node represents a random variable (L, A, Y), and each edge has a direction. We follow convention to represent chronological order from left to right, therefore, L precedes A and Y in time, and A precedes Y.

"Directed" refers to each edge having a direction: because the arrow points from L to A, L is a trigger for A, not the opposite. "No loop" refers to the absence of a closed loop: no variable can be its own trigger, or it can indirectly become its own trigger through other variables.

The arrow from A to Y indicates that we know that in at least one individual, A has a direct causal effect on Y (i.e., there are no other variables in between). Similarly, if there is no arrow between A and Y, then for any individual, A has no direct causal effect on Y. The arrow in the figure points from L to A, indicating that the severity of the condition will affect the probability of receiving a heart transplant. A standard causal diagram does not distinguish whether causal effects are beneficial or harmful. Meanwhile, variable Y has two triggers, and our causal diagram does not indicate whether these two triggers have an interaction.

In a causal-directed acyclic graph, if the direct cause of variable A is controlled, then excluding the variable A as the cause, A is independent of the other variables in the graph. This is one of the basic definitions of causal-directed acyclic graphs. This assumption is called the causal Markov assumption, which implies that in a causal DAG graph, if any two variables have a common cause, then that common cause must also appear in the same graph.

Fig. 3.3. Causal diagrams.

Definition 3.18. Causal diagram

If the DAG graph satisfies the following conditions, it can be considered a causal DAG graph: (1) If there is no arrow pointing from, there is no causal effect; (2) the common cause of any two variables in the graph, even if not measured, must appear in the graph; (3) any variable is the trigger for its downstream variables.

Causal diagrams can be represented as observational studies or as randomized trials.

3.8 Dynamic Uncertain Causality Graph

Over the last two decades or so, graphical models for probabilistic reasoning received a lot of attention. Typical models include Bayesian networks, hidden Markov models, latent tree models, dependency networks, cloud models, and so on. As a newly developed framework of intelligent system, dynamic uncertain causality graph (DUCG) aims to represent uncertain causal knowledge compactly and intuitively provide efficient probabilistic reasoning and make the inference results explanatory (Zhang, 2015).

Dynamic uncertain causality graph has the following advantages:

(1) Represent complex causalities explicitly and easily with graphical symbols including logic gates representing any logic relationship among variables.
(2) Combine separately constructed sub-DUCGs into a whole DUCG so that the maintenance of DUCG knowledge base is easy and each sub-DUCG is simple and well understood.
(3) Reduce the scale of problem for observed evidence significantly, so as to find all possible hypotheses and achieve high inference efficiency.
(4) Deal with not only single-valued cases but also multi-valued cases. The so-called single-valued case means that only the causes of the true state of a child variable are specified, and the false state is just the complement of the true state without specifying its causes separately.

The so-called multi-valued case means that the causes of all states of a child variable can be specified separately. The DUCG involving single-valued cases is called S-DUCG. The DUCG involving multi-valued cases is called M-DUCG. The mixture of them is called DUCG.

The basic idea of S-DUCG is shown in Figure 3.4, in which Figure 3.4(b) reveals the functional mechanism inside Figure 3.4(a). In this example, $X_{1,1}$ and $X_{2,1}$ are in OR relationship in causing $X_{3,1}$. In DUCG, we use the first subscript to index

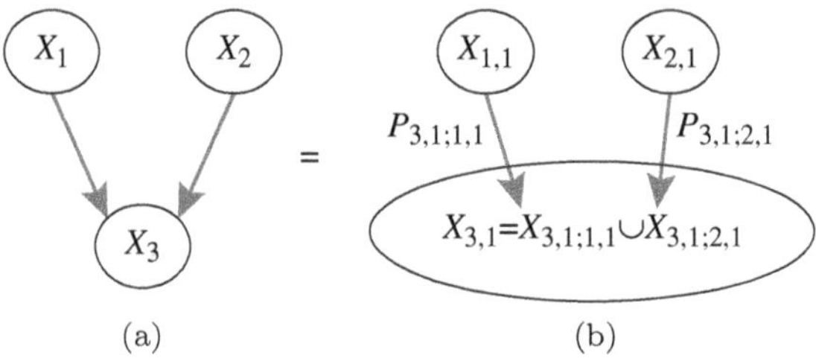

Fig. 3.4.　(a) Example of S-DUCG. (b) Illustration for Figure 3.4(a).

the variable and the second subscript to index the state of the variable. *A comma between them is used to divide them*, e.g., $X_{2,1}$ indicates state 1 of X_2. In the case without confusion, the comma can be ignored, e.g., X_{n1}, X_{ij}, and so on.

In single-valued cases, only the causes of the true state of a child variable are specified. For simplicity, state 1 denotes the true state and state 0 denotes the false state. The false state is just the complement of the true state. This is why only the causes of $X_{3,1}$ are specified in Figure 3.4(b). In this example, $X_{1,1}$ and $X_{2,1}$ are the only causes of $X_{3,1}$.

In S-DUCG, the CPT between a child variable X_n and its parent variables V_i (e.g., is replaced by the linkage events $P_{n1;ij}$ and their occurrence probabilities $P_{n1;ij}$, where $P_{n1;ij}$ is defined as independent of $P_{n1;i'j'}$, $i \neq i'$. In other words, the OR relationship implicitly represented in the CPT as shown in Figure 3.4(b) is explicitly and compactly represented by events $P_{n1;ij}$ and their probabilities $P_{n1;ij} \equiv Pr\{P_{n1;ij}\}$ encoded in the directed arcs, where ";" is used to divide the subscripts of parent event V_{ij} and the subscripts of child event X_{n1}, and $X_{n1;ij}$ is defined as the event that $P_{n1;ij}$ V_{ij} causes X_{n1}. Thus, for the example shown in Figure 3.5(b), we have $X_{3,1} = X_{3,1;1,1} \cup X_{3,1;2,1} = P_{3,1;1,1}X_{1,1} \cup P_{3,1;2,1}X_{2,1}$, and $X_{3,2} = \bar{X}_{3,1} = 1 - X_{3,1}$ ("1" also denotes the complete set).

Figure 3.5 is actually the noisy-OR model in Bayesian network but is more intuitive. Its statistic basis can be illustrated as follows: suppose samples are collected, as shown in Figure 3.5(a), and the corresponding CPT is shown in Figure 3.5(b).

In terms of S-DUCG, these data can be modeled as shown in Figure 3.4(a) by specifying two parameters $P_{3,1;1,1} = 0.3$ and $P_{3,1;2,1} = 0.6$, which means all $P_{n1;ij} = 0$ (null) except P and $P_{3,1;2,1}$. As the relationship between $X_{1,1}$ and $X_{2,1}$ is OR in Figure 3.4 (the default relationship defined in S-DUCG), we have

$$Pr\{X_{3,1}|X_{1,1}X_{2,1}| = Pr\{P_{3,1;1,1} \cup P_{3,1;2,1}\}$$

$$= P_{3,1;1,1} + P_{3,1;2,1} - P_{3,1;1,1}P_{3,1;2,1}$$

$$= 0.3 + 0.6 - 0.3 \times 0.6 = 0.72$$

	$X_{3,1}$	$X_{3,0}$	Total
$X_{1,1}X_{2,0}$	30	70	100
$X_{1,0}X_{2,1}$	120	80	200
$X_{1,1}X_{2,1}$	72	28	100
$X_{1,0}X_{2,0}$	0	200	200
Total	222	378	600

(a)

	$X_{3,1}$	$X_{3,0}$
$X_{1,1}X_{2,0}$	0.3	0.7
$X_{1,0}X_{2,1}$	0.6	0.4
$X_{1,1}X_{2,1}$	0.72	0.28
$X_{1,0}X_{2,0}$	0.0	1.0
Unknown	0.37	0.63

(b)

Fig. 3.5.   (a) Statistic samples and (b) CPT representation in Figure 3.4.

We may use other models if statistic samples cannot be put into the OR model in S-DUCG, logic gates may be employed. The most complex relationship can be modeled by a combined logic gate (Zhang, 2012).

The M-DUCG model is shown in Figure 3.6 and the corresponding CPT is shown in Figure 3.7:

$$a_{3;1} = \begin{pmatrix} a_{3,0;1,0} & a_{3,0;1,1} \\ a_{3,1;1,0} & a_{3,1;1,1} \end{pmatrix} = \begin{pmatrix} 0.75 & 0.6 \\ 0.257 & 0.4 \end{pmatrix}$$

$$a_{3,2} = \begin{pmatrix} a_{3,0;2,0} & a_{3,0;2,1} \\ a_{3,1;2,0} & a_{3,1;21} \end{pmatrix} = \begin{pmatrix} 0.6 & 0.75 \\ 0.4 & 0.25 \end{pmatrix}$$

$$r_{3;1} = 1, \qquad r_{3;2} = 2$$

The learning method can be similar to the existing methods, e.g., EM. As defined in M-DUCG, r3 = r3;1 + r3;2 = 1 + 2 = 3, (r3;1/r3) = (1/3), (r3;2/r3) = (2/3). According to the CPT expression defined in M-DUCG, (rn;i/rn)ank;i j, we can calculate the CPT exactly as shown in Figure 3.7(b). For example,

$$P_r\{X_{3,0}|X_{1,1}X_{2,0}\} = (r_{3;1}/r_3)a_{3,0;1,1} + (r_{3;2}/r_3)a_{3,0;2,0}$$

$$= (1/3)0.6 + (2/3)0.6 = 0.6$$

$$P_r\{X_{3,0}|X_{1,0}X_{2,1}\} = (r_{3;1}/r_3)a_{3,0;1,0} + (r_{3;2}/r_3)a_{3,0;2,1}$$

$$= (1/3)0.75 + (2/3)0.75 = 0.75$$

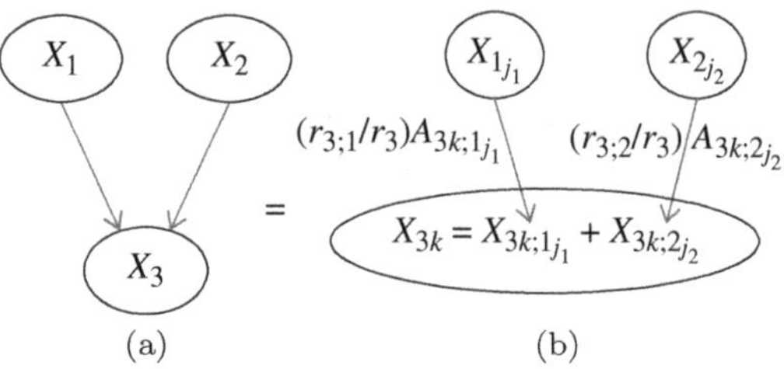

Fig. 3.6.   (a) Example of M-DUCG. (b) Illustration for (a).

	$X_{3,0}$	$X_{3,1}$	Total
$X_{1,1}X_{2,0}$	60	40	100
$X_{1,0}X_{2,1}$	150	50	200
$X_{1,1}X_{2,1}$	70	30	100
$X_{1,0}X_{2,0}$	65	35	100
Total	345	155	500

(a)

	$X_{3,0}$	$X_{3,1}$
$X_{1,1}X_{2,0}$	0.6	0.4
$X_{1,0}X_{2,1}$	0.75	0.25
$X_{1,1}X_{2,1}$	0.7	0.3
$X_{1,0}X_{2,0}$	0.65	0.35
Unknown	0.69	0.31

(b)

Fig. 3.7.   (a) Statistic example. (b) Corresponding CPT.

It is noted that the causes of both $X_{3,0}$ and $X_{3,1}$ are specified separately.

Exercises

3.1 Please explain conditional probability, prior probability, and posterior probability.

3.2 Please describe Bayesian formula and explicate its significance thoroughly.

3.3 Please describe some criterions for prior distribution selection.

3.4 What does 'naïve' mean in naïve Bayesian classification? Please briefly state the main ideas for improving naïve Bayesian classification.

3.5 Please describe the structure of Bayesian network and its construction, and exemplify the usage of Bayesian network.

3.6 What is semi-supervised text mining? Please describe some applications of Bayesian model in Web page clustering.

3.7 In recent years, with the development of Internet technology, Bayesian rules are widely applied. Please exemplify two concrete applications of the Bayesian rules and explain the results.

3.8 Make examples to explain the three levels of causation.

3.9 What is the dynamic uncertain causality graph (DUCG)? How to apply it in medicine?

Chapter 4

Game Theory

Game theory is the study of models of strategic interactions among rational agents. It has applications in many fields of social science, systems science, and computer science. In the 21st century, game theory applies to a wider range of behavioral relations, and it is now an umbrella term for the science of logical decision-making in humans, animals, as well as computers.

4.1 Introduction

In the highly connected world, confrontation and competition between people is a topic that cannot be avoided. Only by facing this topic head-on individuals can have a better chance of success. As people try to succeed, it is necessary to study how to strategically choose their actions in confrontational situations. The science that specializes in studying people's strategic behavior in interactive situations is called game theory. We introduce the main problems and applications of game theory in this chapter.

4.1.1 *History of Game Theory*

Antoine Augustin Cournot was a French philosopher and mathematician who also contributed to the development of economics. In 1838, he published the book *Researches on Mathematical Principles of the Theory of Wealth*, in which Cournot introduced the ideas of functions and probability into economic analysis. He derived the first formula for the rule of supply and demand as a function of price and was the first to draw supply and demand curves on a graph. The Cournot duopoly model developed in his book also introduced the concept of a Nash equilibrium, the reaction function and best-response dynamics.

Game theory emerged as a unique field when John von Neumann published the paper *On the Theory of Games of Strategy* in 1928 (von Neumann, 1928); his original proof used Brouwer's fixed-point theorem on continuous mappings into compact convex sets, which became a standard method in game theory and mathematical economics. Von Neumann's work in game theory culminated in his 1944 book *Theory of Games and Economic Behavior*, co-authored with Oskar Morgenstern. The second edition of this book provided an axiomatic theory of utility, which reincarnated Daniel Bernoulli's old theory of utility (of money) as an independent discipline. This foundational work contains the method for finding mutually consistent solutions for two-person zero-sum games. Subsequent work focused primarily on cooperative game theory, which analyzes optimal strategies for groups of individuals, presuming that they can enforce agreements between them about proper strategies.

In 1950, the first mathematical discussion of the prisoner's dilemma appeared, and an experiment was undertaken by notable mathematicians Merrill M. Flood and Melvin Dresher, as part of the RAND Corporation's investigations into game theory. RAND pursued the studies because of possible applications to global nuclear strategy. Around this same time, John Nash developed a criterion for mutual consistency of players' strategies known as the Nash equilibrium, applicable to a wider variety of games than the criterion proposed by von Neumann and Morgenstern. Nash proved that every finite n-player, non-zero-sum (not just two-player zero-sum) non-cooperative game has what is now known as a Nash equilibrium in mixed strategies.

Game theory experienced a flurry of activity in the 1950s, during which the concepts of the core, the extensive form game, fictitious play, repeated games, and the Shapley value were developed. The 1950s also saw the first applications of game theory to philosophy and political science.

In 1965, Reinhard Selten introduced his solution concept of subgame perfect equilibria, which further refined the Nash equilibrium. Later he would introduce trembling hand perfection as well. In 1994, Nash, Selten, and Harsanyi became Economics Nobel laureates for their contributions to economic game theory.

In the 1970s, game theory was extensively applied in biology, largely as a result of the work of John Maynard Smith and his evolutionarily stable strategy. In addition, the concepts of correlated equilibrium, trembling hand perfection, and common knowledge were introduced and analyzed.

In 1994, John Nash was awarded the Nobel Memorial Prize in the Economic Sciences for his contribution to game theory. Nash's most famous contribution to game theory is the concept of the Nash equilibrium, which is a solution concept for non-cooperative games. A Nash equilibrium is a set of strategies, one for each player, such that no player can improve their payoff by unilaterally changing their strategy.

In 2005, game theorists Thomas Schelling and Robert Aumann followed Nash, Selten, and Harsanyi as Nobel laureates. Schelling worked on dynamic models, early examples of evolutionary game theory. Aumann contributed more to the equilibrium school, introducing equilibrium coarsening and correlated equilibria, and developing an extensive formal analysis of the assumption of common knowledge and of its consequences.

In 2007, Leonid Hurwicz, Eric Maskin, and Roger Myerson were awarded the Nobel Prize in Economics "for having laid the foundations of mechanism design theory". Myerson's contributions include the notion of proper equilibrium and an important graduate text: Game Theory, Analysis of Conflict. Hurwicz introduced and formalized the concept of incentive compatibility.

In 2012, Alvin E. Roth and Lloyd S. Shapley were awarded the Nobel Prize in Economics "for the theory of stable allocations and the practice of market design". In 2014, the Nobel went to game theorist Jean Tirole.

4.1.2 *Basic Concepts of Game Theory*

The basic concepts of game theory consist of seven elements: players, action, information, strategy, payoff, outcome, and equilibrium:

(1) Players: a collection of decision-makers called players.
(2) Action: the collection of feasible moves (decisions, actions, plays,...) that each player can choose to make in each of his possible information states.
(3) Information: the possible information states of each player at each decision time.
(4) Strategy: a procedure or rule for determining how the move choices of all the players collectively determine the possible outcomes of the game.
(5) Payoff: preferences of the individual players over these possible outcomes, typically measured by a utility or payoff function.
(6) Outcome: all results which are interested by game analyst.
(7) Equilibrium: a combination of the best strategies of all participants.

4.1.3 *Applications of Game Theory*

Game theory, also known as multi-person decision theory, is the analysis of situations in which the payoff of a decision-maker depends not only on his own actions but also on those of others. Game theory has applications in several fields, such as economics, politics, law, biology, and computer science.

1. Economics

Game theory is a major method used in mathematical economics and business for modeling competing behaviors of interacting agents. This research usually focuses on particular sets of strategies known as "solution concepts" or "equilibria". A common assumption is that players act rationally. In non-cooperative games, the most famous of these is the Nash equilibrium. A set of strategies is a Nash equilibrium if each represents the best response to the other strategies. If all the players are playing the strategies in a Nash equilibrium, they have no unilateral incentive to deviate, since their strategy is the best they can do given what others are doing.

2. Biology

Unlike those in economics, the payoffs for games in biology are often interpreted as corresponding to fitness. In addition, the focus has been less on equilibria that correspond to a notion of rationality and more on ones that would be maintained by evolutionary forces. The best-known equilibrium in biology is known as the evolutionarily stable strategy (ESS), first introduced in Smith & Price (1973). Although its initial motivation did not involve any of the mental requirements of the Nash equilibrium, every ESS is a Nash equilibrium.

In biology, game theory has been used as a model to understand many different phenomena. It was first used to explain the evolution (and stability) of the approximate 1:1 sex ratios. The 1:1 sex ratios are a result of evolutionary forces acting on individuals who could be seen as trying to maximize their number of grandchildren.

Additionally, biologists have used evolutionary game theory and the ESS to explain the emergence of animal communication. The analysis of signaling games and other communication games has provided insight into the evolution of communication among animals. For example, the mobbing behavior of many species, in which a large number of prey animals attack a larger predator, seems to be an example of spontaneous emergent organization. Ants have also been shown to exhibit feed-forward behavior akin to fashion. Biologists have used the game of chicken to analyze fighting behavior and territoriality.

According to Maynard Smith, in the preface to *Evolution and the Theory of Games*, "paradoxically, it has turned out that game theory is more readily applied to biology than to the field of economic behaviour for which it was originally designed". Evolutionary game theory has been used to explain many seemingly incongruous phenomena in nature. One such phenomenon is known as biological altruism. This is a situation in which an organism appears to act in a way that benefits other organisms and is detrimental to itself. This is distinct from traditional notions of altruism because such actions are not conscious but appear to be evolutionary adaptations to increase overall fitness. Examples can be found in species ranging from vampire bats that regurgitate blood they have obtained from a night's hunting and give it to group members who have failed to feed, to worker bees that care for the queen bee for their entire lives and never mate, to vervet monkeys that warn group members of a predator's approach, even when it endangers that individual's chance of survival. All of these actions increase the overall fitness of a group but occur at a cost to the individual.

Evolutionary game theory explains this altruism with the idea of kin selection. Altruists discriminate between the individuals they help and favor relatives. Hamilton's rule explains the evolutionary rationale behind this selection with the equation $c < b \times r$, where the cost $.33c$ to the altruist must be less than the benefit b to the recipient multiplied by the coefficient of relatedness r. The more closely related the two organisms are, the more the causes of incidences of altruism increase because they share many of the same alleles. This means that the altruistic individual, by ensuring that the alleles of its close relative are passed on through survival of its offspring, can forgo the option of having offspring itself because the same number of alleles is passed on. For example, helping a sibling (in diploid animals) has a coefficient of $1/2$ because (on average) an individual shares half of the alleles in its sibling's offspring. Ensuring that enough of a sibling's offspring survive to adulthood precludes the necessity of the altruistic individual producing offspring. The coefficient values depend heavily on the scope of the playing field; for example, if the choice of whom to favor includes all genetic living things, not just all relatives, we assume the discrepancy between all humans only accounts for approximately 1% of the diversity in the playing field, a coefficient that was $1/2$ in the smaller field becomes 0.995. Similarly, if it is considered that information other than that of a genetic nature (e.g., epigenetics, religion, and science) persisted through time, the playing field becomes larger still, and the discrepancies smaller.

3. Computer Science

Game theory has come to play an increasingly important role in logic and in computer science. Several logical theories have a basis in game semantics. In addition, computer scientists have used games to model interactive computations. Also, game theory provides a theoretical basis to the field of multi-agent systems.

Separately, game theory has played a role in online algorithms, in particular, the k-server problem, which has in the past been referred to as *games with moving cost* and *request-answer games*. Yao's principle is a game-theoretic technique for proving lower bounds on the computational complexity of randomized algorithms, especially online algorithms.

The emergence of the Internet has motivated the development of algorithms for finding equilibria in games, markets, computational auctions, peer-to-peer systems, and security and information markets. Algorithmic game theory and within it algorithmic mechanism design combine computational algorithm design and analysis of complex systems with economic theory.

4.2 Representation of Games

The games studied in game theory are well-defined mathematical objects. To be fully defined, a game must specify the following elements: the players of the game, the information and actions available to each player at each decision point, and the payoffs for each outcome. A game theorist typically uses these elements, along with a solution concept of their choosing, to deduce a set of equilibrium strategies for each player such that, when these strategies are employed, no player can profit by unilaterally deviating from their strategy. These equilibrium strategies determine an equilibrium to the game — a stable state in which either one outcome occurs or a set of outcomes occur with known probability.

In games, players typically have a 'Dominant Strategy', where they are incentivized to choose the best possible strategy that gives them the maximum payoff and stick to it even when the other player/s change their strategies or choose a different option. However, depending on the possible payoffs, one of the players may not possess a 'Dominant Strategy', while the other player might. A player not having a dominant strategy is not a confirmation that another player won't have a dominant strategy of their own, which puts the first player at an immediate disadvantage.

However, there is the chance of both players possessing dominant strategies, when their chosen strategies and their payoffs are dominant, and the combined payoffs form an equilibrium. When this occurs, it creates a dominant strategy equilibrium. This can cause a social dilemma, where a game possesses an equilibrium created by two or multiple players who all have dominant strategies, and the game's solution is different from what the cooperative solution to the game would have been. There is also the chance of a player having more than one dominant strategy. This occurs when reacting to multiple strategies from a second player, and the first player's separate responses having different strategies to each other. This means that there is no chance of a Nash equilibrium occurring within the game. Most cooperative games are presented in the characteristic function form, while the extensive and the normal forms are used to define non-cooperative games.

4.2.1 *Extensive Form*

The extensive form can be used to formalize games with a time sequencing of moves. Extensive form games can be visualized using game trees (as shown in Figure 4.1). Here each vertex (or node) represents a point of choice for a player. The player is specified by a number listed by the vertex. The lines out of the vertex represent a possible action for that player. The payoffs are specified at the bottom of the tree. The extensive form can be viewed as a multi-player generalization of a decision tree. To solve any extensive form game, backward induction must be used. It involves working backward up the game tree to determine what a rational player would do at the last vertex of the tree, what the player with the previous move would do given that the player with the last move is rational, and so on until the first vertex of the tree is reached.

The game pictured consists of two players. The way this particular game is structured (i.e., with sequential decision-making and perfect information), Player 1 "moves" first by choosing either F or U (fair or unfair). Next in the sequence, Player 2, who has now observed Player 1's move, can choose to play either A or R (accept or reject). Once Player 2 has made their choice, the game is considered finished and each player gets their respective payoff, represented in the image as two numbers, where the first number represents Player 1's payoff, and the second number represents Player 2's payoff. Suppose that Player 1 chooses U and then Player 2 chooses A: Player 1 then gets a payoff of "eight" (which in real-world terms can be interpreted in many ways, the simplest of which is in terms of money but could mean things such as eight days of vacation or eight countries conquered or even eight more opportunities to play the same game against other players) and Player 2 gets a payoff of "two".

The extensive form can also capture simultaneous-move games and games with imperfect information. To represent it, either a dotted line connects different vertices to represent them as being part of the same information set (i.e., the players do not know at which point they are) or a closed line is drawn around them.

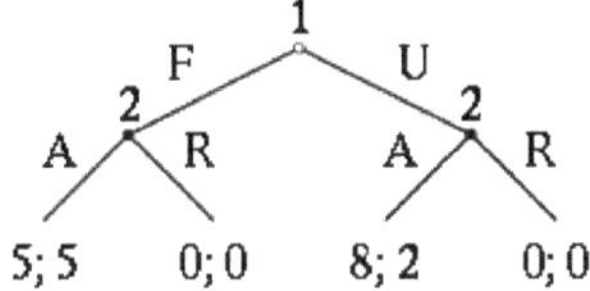

Fig. 4.1.   An extensive form game.

	Player 2 chooses *Left*	Player 2 chooses *Right*
Player 1 chooses *Up*	4, 3	–1, –1
Player 1 chooses *Down*	0, 0	3, 4

Normal form or payoff matrix of a 2–player, 2–strategy game

Fig. 4.2.   Normal game is represented by a matrix.

4.2.2 *Normal Form*

The normal (or strategic form) game is usually represented by a matrix which is as shown in Figure 4.2, the players, strategies, and payoffs (see the example to the right). More generally, it can be represented by any function that associates a payoff for each player with every possible combination of actions. In the accompanying example, there are two players: one chooses the row and the other chooses the column. Each player has two strategies, which are specified by the number of rows and the number of columns. The payoffs are provided in the interior. The first number is the payoff received by the row player (Player 1 in our example); the second is the payoff for the column player (Player 2 in our example). Suppose that Player 1 plays *Up* and that Player 2 plays *Left*. Then Player 1 gets a payoff of 4, and Player 2 gets 3.

When a game is presented in normal form, it is presumed that each player acts simultaneously or, at least, without knowing the actions of the other. If players have some information about the choices of other players, the game is usually presented in extensive form. Every extensive-form game has an equivalent normal-form game, however, the transformation to normal form may result in an exponential blowup in the size of the representation, making it computationally impractical.

4.2.3 *Characteristic Function Form*

In games that possess removable utility, separate rewards are not given rather the characteristic function decides the payoff of each unity. The idea is that the unity that is 'empty', so to speak, does not receive a reward at all. The origin of this form is found in John von Neumann and Oskar Morgenstern's book.

A characteristic function game is a pair (N, v), where $N = \{1, \ldots, n\}$ is the set of players and $v : 2^N \to R$ is the characteristic function. For each subset of players

$C \subseteq N$, $v(C)$ is the amount that the members of C can earn by working together; usually, it is assumed that v is normalized: $v(0) = 0$, non-negative: $v(C) \geq 0$, for any $C \subseteq N$, and monotone: $v(C) \leq v(D)$, for any C, D such that $C \subseteq D$. A coalition is any subset of N. N itself is called the grand coalition.

4.2.4 *Alternative Game Representations*

Alternative game representation forms are used for some subclasses of games or adjusted to the needs of interdisciplinary research. In addition to classical game representations, some of the alternative representations also encode time-related aspects.

4.3 Type of Game Theory

1. Zero-Sum or non-zero-sum

In zero-sum games, the total benefit goes to all players in a game, for every combination of strategies, always adds to zero (more informally, a player benefits only at the equal expense of others). Many games studied by game theorists are non-zero-sum games because the outcome has net results greater or less than zero. Informally, in non-zero-sum games, a gain by one player does not necessarily correspond with a loss by another.

2. Cooperative or non-cooperative game

A game is cooperative if the players are able to form binding commitments externally enforced (e.g., through contract law). A game is non-cooperative if players cannot form alliances or if all agreements need to be self-enforcing (e.g., through credible threats).

3. Perfect information or imperfect information game

A game with perfect information means that all players, at every move in the game, know the previous history of the game and the moves previously made by all other players. In reality, this can be applied to firms and consumers having information about the price and quality of all the available goods in a market. An imperfect information game is played when the players do not know all moves already made by the opponent, such as a simultaneous move game. Most games studied in game theory are imperfect information games.

4. Symmetric or asymmetric game

A symmetric game is a game where each player earns the same payoff when making the same choice. In other words, the identity of the player does not change the resulting game facing the other player. Many of the commonly studied 2×2 games are symmetric. The standard representations of chicken, the prisoner's dilemma, and the stag hunt are all symmetric games. The most commonly studied asymmetric games are games where there are no identical strategy sets for both players. For instance, the ultimatum game and similarly the dictator game have different strategies for each player. It is possible, however, for a game to have identical strategies for both players, yet be asymmetric. For example, the game pictured in this section's graphic is asymmetric despite having identical strategy sets for both players.

5. Sequential or simultaneous game

Sequential games are games where players do not make decisions simultaneously, and players' earlier actions affect the outcome and decisions of other players. This need not be perfect information about every action of earlier players; it might be very little knowledge. Simultaneous games are games where both players move simultaneously, or instead, the later players are unaware of the earlier players' actions.

4.4 Zero-Sum Game

Zero-sum games (more generally, constant-sum games) are games in which choices by players can neither increase nor decrease the available resources. In zero-sum games, the total benefit goes to all players in a game, for every combination of strategies, always adds to zero (more informally, a player benefits only at the equal expense of others), as shown in Figure 4.3. Poker exemplifies a zero-sum game (ignoring the possibility of the house's cut) because one wins exactly the amount one's opponents lose. Other zero-sum games include matching pennies and most classical board games including Go and chess.

Many games studied by game theorists (including the famed prisoner's dilemma) are non-zero-sum games because the outcome has net results greater or less than zero. Informally, in non-zero-sum games, a gain by one player does not necessarily correspond with a loss by another.

	A	B
A	− 1, 1	3, − 3
B	0, 0	− 2, 2

Fig. 4.3. A zero-sum game.

The prisoner's dilemma is a paradox in decision analysis in which two individuals acting in their own self-interests do not produce the optimal outcome. A prime example of game theory, the prisoner's dilemma was developed in 1950 by RAND Corporation mathematicians Merrill Flood and Melvin Dresher. Today, the prisoner's dilemma is a paradigmatic example of how strategic thinking between individuals can lead to suboptimal outcomes for both players. People have developed many methods of overcoming prisoner's dilemmas to choose better collective results despite apparently unfavorable individual incentives.

Constant-sum games correspond to activities like theft and gambling but not to the fundamental economic situation in which there are potential gains from trade. It is possible to transform any constant-sum game into a (possibly asymmetric) zero-sum game by adding a dummy player (often called "the board") whose losses compensate the players' net winnings.

4.5 Nash Equilibrium

Nash equilibrium, also called Nash solution, in game theory, is an outcome in a non-cooperative game for two or more players in which no player's expected outcome can be improved by changing one's own strategy. Nash equilibrium is a strategy combination that ensures that each player's strategy is the optimal response to the strategies of other players. The Nash equilibrium is a key concept in game theory, in which it defines the solution of N-player non-cooperative games. It is named after American mathematician John Nash, who was awarded the 1994 Nobel Prize for Economics for his contributions to game theory.

Definition 4.1. The strategy of a player i is s_i^* with other players' strategy s_{-i}^*, if they meet the following equation:

$$\pi_i(s_i^*, s_{-i}^*) \geq \pi_i(s_i, s_{-i}^*) \quad \text{for } \forall s_i \tag{4.1}$$

then s_i^* is called an optimal response on s_{-i}^*.

Definition 4.2. The strategy vector $s^* = s_1^*, s_2^*, \cdots s_N^*$, if it satisfies the following condition:

$$\pi_i(s_i^*, s_{-i}^*) \geq \pi_i(s_i, s_{-i}^*) \quad \text{for } \forall i \text{ and } s_i \tag{4.2}$$

Then, s^* is called Nash equilibrium.

Equation (4.2) points out take strategy s^* where every player i takes strategy select optimal response.

In order to understand the principle of Nash equilibrium, here we introduce the Cournot model and Bertrand model in monopolistic competition. A French economist, Augstin A. Cournot, has given the duopoly model. According to him, the model has a unique equilibrium when demand curves are linear. The model explains that the two firms choose the output levels in competition with each other. The Cournot model has a continuous strategy. The Cournot game is a non-co-operative game. In order to find Nash equilibrium in the Cournot game, we need the reaction curve. Based on the assumption of Cournot model, economists have given a better solution in terms of reaction curve. The reaction function expresses the output of each duopolistic which is a function of his rivals' output. Bertrand developed his duopoly model in 1883. His model differs from Cournot's in that he assumes that each firm expects that the rival will keep its price constant, irrespective of its own decision about pricing. Thus each firm is faced with the same market demand and aims at the maximization of its own profit on the assumption that the price of the competitor will remain constant. The model may be presented with the analytical tools of the reaction functions.

4.6 Bayesian Nash Equilibrium

4.6.1 Bayesian Game

Bayesian game is a strategic decision-making model which assumes players have incomplete information. Players hold private information relevant to the game, meaning that the payoffs are not common knowledge. Bayesian games model the outcome of player interactions using aspects of Bayesian probability. They are notable because they allowed, for the first time in game theory, the specification of the solutions to games with incomplete information.

Harsanyi defined Bayesian games in the following way: players are assigned by nature at the start of the game a set of characteristics (Harsanyi, 1967/1968). By mapping probability distributions to these characteristics and by calculating the outcome of the game using Bayesian probability, the result is a game whose solution is, for technical reasons, far easier to calculate than a similar game in a non-Bayesian context.

A Bayesian game is defined by (N, A, T, p, u), where it consists of the following elements:

(1) **Set of players, N:** The set of players within the game.
(2) **Action sets, a_i:** The set of actions available to Player i. An action profile $a = (a_1, \ldots, a_N)$ is a list of types, one for each player.

(3) **Type sets, t_i:** The set of types of players i. "Types" capture the private information a player can have. A type profile $t = (t_1, \ldots, t_N)$ is a list of types, one for each player.

(4) **Payoff functions, u:** Assign a payoff to a player given their type and the action profiles. A payoff function $u = (u_1, \ldots, u_N)$ denotes the utilities of player i.

(5) **Prior, p:** A probability distribution over all possible type profiles, where $p(t) = p(t_1, \ldots, t_N)$, is the probability that Player 1 has type t_1 and Player N has type t_N.

There are two important and novel aspects to Bayesian games that were themselves specified by Harsanyi. The first is that Bayesian games should be considered and structured identically to complete information games. Except, by attaching probability to the game, the final game functions as though it were an incomplete information game. Therefore, players can be essentially modeled as having incomplete information and the probability space of the game still follows the law of total probability. Bayesian games are also useful in that they do not require infinite sequential calculations. Infinite sequential calculations would arise where players (essentially) try to "get into each other's heads". For example, one may ask questions and decide "If I expect some action from player B, then player B will anticipate that I expect that action, so then I should anticipate that anticipation" *ad infinitum.* Bayesian games allow for the calculation of these outcomes in one move by simultaneously assigning different probability weights to different outcomes. The effect of this is that Bayesian games allow for the modeling of a number of games that in a non-Bayesian setting would be irrational to compute.

In Bayesian games, there are two strategies as follows:

(1) A pure strategy $s_i \colon \Theta_i \to A_i$ of player i is a mapping from every type player i could have to the action he would play if he had that type. Denote the set of pure strategies of player i as
$S_1 = \{\{C\}, \{D\}\}$
$S_2 = \{\{C \text{ if type I}, C \text{ if type II}\}, \{C \text{ if type I}, D \text{ if type II}\}, \{D \text{ if type I}, C \text{ if type II}\}, \{D \text{ if type I}, D \text{ if type II}\}\}$

(2) A mixed strategy $\sigma_i \colon S_i \to [0; 1]$ of player i is a distribution over his pure strategies.

4.6.2 *Bayesian Nash Equilibrium*

A Bayesian Nash equilibrium of a Bayesian game is a Nash equilibrium of its associated exante normal form game. In a non-Bayesian game, a strategy profile is a

Nash equilibrium if every strategy in that profile is the best response to every other strategy in the profile; i.e., there is no strategy that a player could play that would yield a higher payoff, given all the strategies played by the other players.

An analogous concept can be defined for a Bayesian game, the difference being that every player's strategy maximizes their expected payoff given their beliefs about the state of nature.

We use pure strategies to illustrate the concepts. But they hold the same for mixed strategies:

(1) Player i's *ex ante* expected utility is

$$E_\theta\left[u_i(s(\theta), \theta)\right] = \sum_{\theta_i \in \Theta_i} p(\theta_i) E_{\theta_{-i}}\left[u_i(s(\theta), \theta)|\theta_i\right] \tag{4.3}$$

(2) Player i's best responses to $s_{-i}(\theta_{-i})$ is

$$BR_i = \arg\max_{s_i(\theta_i)\in S_i} E_\theta\left[u_i(s_i(\theta_i), s_{-i}(\theta_{-i}), \theta)\right]$$

$$= \sum_{\theta \in \Theta} p(\theta_i)\left(\arg\max_{s_i(\theta_i)\in S_i} E_{-\theta}\left[u_i(s_i(\theta_i), s_{-i}(\theta_{-i}), |\theta_i)\right]\right) \tag{4.4}$$

(3) A strategy profile $s_i(\theta_i)$ is a Bayesian Nash equilibrium if and only if $\forall_i s_i(\theta_i) \in BR_i$.

Here we give a Bayesian Nash equilibrium example, as shown in Figure 4.4.

As shown in Figure 4.4, Playing D is a dominant strategy for type I player 2; playing C is a dominant strategy for type II player 2. Player 1's expected utility by playing C is $\lambda \times 0 + (1 - \lambda) \times 5 = 5 - 5\lambda$. Player 1's expected utility by playing D is $\lambda X1 + (1 - \lambda)X8 = 8 - 7\lambda > 5 - 5\lambda$. $(D, (D$ if type I, C if type II)) is a Bayesian Nash equilibrium of the game.

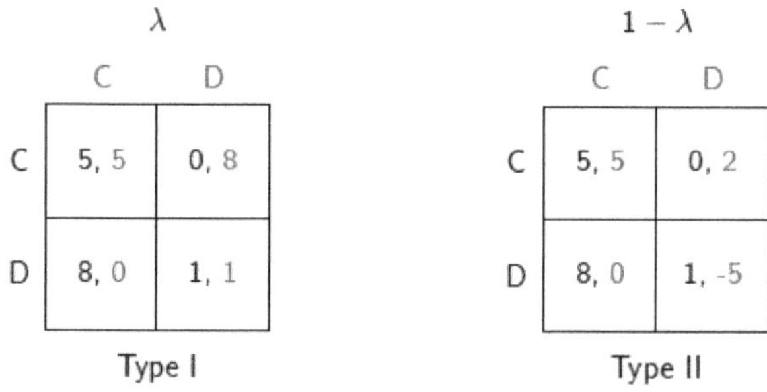

Fig. 4.4.   Bayesian Nash equilibrium example.

4.6.3 *Extensive form Games with Incomplete Information*

Extensive-form games with perfect or imperfect information have the following elements:

(1) set of players,
(2) set of decision nodes,
(3) a player function assigning a player to each decision node,
(4) set of actions for each player at each of her decision nodes,
(5) set of terminal nodes,
(6) a payoff function for each player.

4.7 Mixed Strategy Game

Representation of pure strategy: The set of strategies for player i is $A_i = \{A_{i1}, A_{i2}, \cdots, A_{in_i}\}$, and pure strategy is $a_i \in A_i$.

A mixed strategy assigns a probability to each pure strategy, and a player's set of strategies is a sample space, as shown in Figure 4.5. Use $\Delta(A_i)$ to denote probability distributions in A_i, *that is,*

$$\Delta(A_i) = \{p_i = p_{i1}, p_{i2}, \cdots p_{in_i}\}, \ p_{ij} \geq 0, \ \sum_j p_{ij} = 1\}$$

then a mixed strategy $p_i = p_{i1}, p_{i2}, \cdots p_{in_i} \in \Delta A_i$.

The outcome of a mixed-strategy game $p = (p_1, p_2, \cdots p_N)$, $p_i \in \Delta(A_i)$.
Define $p_{-i} = (p_1, \cdots p_{i-1}, p_{i+1}, \cdots p_N)$ then $p = (p_i, p_{-i})$.
Assuming that each player's decisions are independent, mixed strategy game

$$G = \{N, \Delta(A_1), \ \Delta(A_2) \cdots \Delta(A_N)\}, \ \{U_1, U_2 \cdots U_N\}\}$$

		Player 2	
		L, π_2	$R, 1-\pi_2$
Player 1	U, π_1	1 2	0 4
	$D, 1-\pi_1$	0 5	3 2

Fig. 4.5. Mixed strategy game.

Example 4.1. Suppose probabilities of strategy U and strategy L are 0.4 and 0.5,

$$p = (p_1, p_2) = ((0.4, 0.6), (0.5, 0.5))$$

$$U_1(p) = p_1(U)p_2(L)u_1(U, L) + p_1(U)p_2(R)u_1(U, R)$$

$$+ p_1(D)p_2(L)u_1(D, L) + p_1(D)p_2(D)u_1(D, R) = 1.1$$

$$U_2(p) = \cdots = 3.3$$

4.8 Evolutionary Game Theory

Evolutionary game theory is the application of game theory to evolving populations in biology. It defines a framework of contests, strategies, and analytics into which Darwinian competition can be modeled. It originated in 1973 with John Maynard Smith and George R. Price's formalization of contests, analyzed as strategies, and the mathematical criteria that can be used to predict the results of competing strategies.

Evolutionary game theory differs from classical game theory in focusing more on the dynamics of strategy change (Newton, 2018). This is influenced by the frequency of the competing strategies in the population.

Evolutionary game theory has helped explain the basis of altruistic behaviors in Darwinian evolution. It has in turn become of interest to economists, sociologists, anthropologists, and philosophers.

Evolutionary game theory started with the problem of how to explain ritualized animal behavior in a conflict situation; "why are animals so 'gentlemanly or ladylike' in contests for resources?" The leading ethologists Niko Tinbergen and Konrad Lorenz proposed that such behavior exists for the benefit of the species. John Maynard Smith considered that incompatible with Darwinian thought, where selection occurs at an individual level, so self-interest is rewarded while seeking the common good is not. Maynard Smith, a mathematical biologist, turned to game theory as suggested by George Price, though Richard Lewontin's attempts to use the theory had failed.

Maynard Smith realized that an evolutionary version of game theory does not require players to act rationally — only that they have a strategy. The results of a game show how good that strategy was, just as evolution tests alternative strategies for the ability to survive and reproduce. In biology, strategies are genetically inherited traits that control an individual's action, analogous to computer programs. The success of a strategy is determined by how good the strategy is in the presence of competing strategies (including itself) and of the frequency with which those strategies are used.

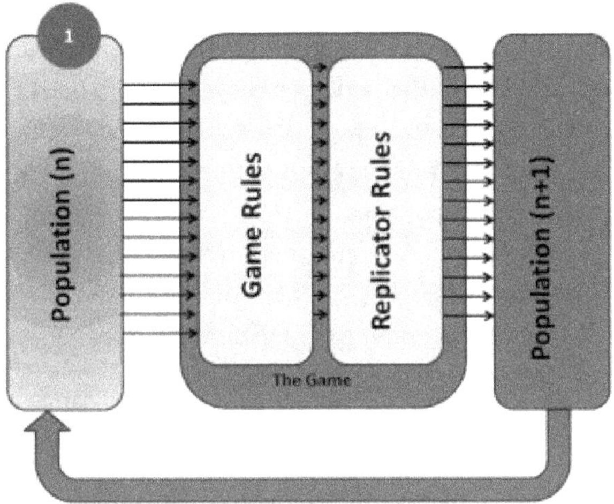

Fig. 4.6.   System model of evolutionary game.

Evolutionary game theory analyses Darwinian mechanisms with a system model with three main components: *population*, *game*, and *replicator dynamics*. The system process has four phases, as shown in Figure 4.6:

(1) The model (as evolution itself) deals with a population (P_n). The population will exhibit variation among competing individuals. In the model, this competition is represented by the game.

(2) The game tests the strategies of the individuals under the rules of the game. These rules produce different payoffs — in units of fitness (the production rate of offspring). The contesting individuals meet in pairwise contests with others, normally in a highly mixed distribution of the population. The mix of strategies in the population affects the payoff results by altering the odds that any individual may meet up in contests with various strategies. The individuals leave the game pairwise contest with a resulting fitness determined by the contest outcome, represented in a *payoff matrix*.

(3) Based on this resulting fitness, each member of the population then undergoes replication or culling determined by the exact mathematics of the replicator dynamics process. This overall process then produces a *new generation $P_{(n+1)}$*. Each surviving individual now has a new fitness level determined by the game result.

(4) The new generation then takes the place of the previous one and the cycle repeats. The population mix may converge to an *evolutionarily stable state* that cannot be invaded by any mutant strategy.

Evolutionary game theory encompasses Darwinian evolution, including competition (the game), natural selection (replicator dynamics), and heredity. Evolutionary game theory has contributed to the understanding of group selection, sexual selection, altruism, parental care, co-evolution, and ecological dynamics. Many counterintuitive situations in these areas have been put on a firm mathematical footing by the use of these models (Hammerstein, 1994).

The common way to study the evolutionary dynamics in games is through replicator equations. These show the growth rate of the proportion of organisms using a certain strategy and that rate is equal to the difference between the average payoff of that strategy and the average payoff of the population as a whole. Continuous replicator equations assume infinite populations, continuous time, complete mixing, and that strategies breed true. Some attractors (all global asymptotically stable fixed points) of the equations are evolutionarily stable states. A strategy which can survive all "mutant" strategies is considered evolutionarily stable. In the context of animal behavior, this usually means such strategies are programmed and heavily influenced by genetics, thus making any player or organism's strategy determined by these biological factors.

4.9 Game Dynamics

Game dynamics is a mathematical model that studies the strategy choices and outcome evolution of participants in the game process. It is not only used for studying game theory but also widely applied in fields, such as economics, political science, and ecology. In the following, we first introduce the basic concepts of game dynamics and then discuss the MDA model, which is a formal approach to bridge the gap between game design and development, game criticism, and technical game research. Basic concepts of game dynamics contain four aspects.

(1) **Game:** Game refers to the decision-making process conducted by two or more individuals under specific conditions. Each participant will choose strategies based on their own interests and goals. The outcome of the game will be influenced by the strategies of the participants.

(2) **Dynamics:** Dynamics refers to the laws and processes of things changing over time. In the game process, the strategies and outcomes of participants will evolve over time, so dynamic methods need to be used for modeling and analysis.

(3) **Strategy:** Strategy refers to the actions and decisions taken by participants in the game process. Each participant will choose the optimal strategy based on their own interests and goals, thereby affecting the outcome of the game.

(4) **Evolution:** Evolution refers to the process in which strategies and outcomes change over time during the game process. The strategies of participants will evolve according to environmental changes, thereby affecting the outcome of the game.

4.9.1 *MDA Model*

From 2001 to 2004, Hunicke, etc. presented MDA as a formal approach to game design and game research at the Game Developers Conference held in San Jose (Hunicke *et al.*, 2004). Games are created by designers/teams of developers and consumed by players, as shown in Figure 4.7. They are purchased, used, and eventually cast away like most other consumable goods. The MDA framework formalizes the consumption of games by breaking them into their distinct components and establishing their design counterparts.

In describing the aesthetics of a game, we can list the following eight taxonomies:

(1) Sensation: Game as sense-pleasure.
(2) Fantasy: Game as make-believe.
(3) Narrative: Game as drama.
(4) Challenge: Game as obstacle course.
(5) Fellowship: Game as social framework.
(6) Discovery Game as uncharted territory.
(7) Expression: Game as self-discovery.
(8) Submission: Game as pastime.

MDA is a formal approach to understanding games — one which attempts to bridge the gap between game design and development, game criticism, and technical game research. MDA is clear for the iterative processes of developers, scholars, and researchers alike, making it easier for all parties to decompose, study, and design a broad class of game designs and game artifacts.

Fig. 4.7.   The procedure of game artifacts.

4.9.2 *Applications*

1. Economics

Game dynamics have a wide range of applications in economics, such as market competition, price formation, resource allocation, and other fields. By establishing mathematical models, market trends and the impact of corporate strategies can be predicted.

2. Political Science

Game dynamics are also widely applied in political science, such as international relations, decision-making, election competition, and other fields. By establishing mathematical models, it is possible to predict the impact of policy implementation and the likelihood of election results.

3. Ecology

The application of game dynamics in ecology mainly focuses on studying the interactions and evolutionary processes between species in ecosystems. By establishing mathematical models, the stability of ecosystems and the possibility of species extinction can be predicted.

Game dynamics have extensive applications in many fields, but it also has some limitations. First, game dynamics can only model and analyze simple games involving a few participants and cannot handle complex multi-player games. Second, the establishment of game dynamics models requires making many assumptions and simplifications, which may lead to inaccurate prediction results.

Exercises

4.1 Please explain what game theory is.

4.2 Give examples to illustrate the representation of the game.

4.3 Give an example of what a zero-sum game is.

4.4 Formally define the representation of Nash equilibrium games.

4.5 What is a mixed strategy in game theory? Give an example to illustrate.

4.6 In game theory economics, the "pig game" is a famous example of Nash equilibrium.

Assume there is a big pig and a little pig in the pigsty. One end of the pigsty has a pig food trough, while the other end is equipped with a button to control the supply of pig food. Pressing the button will result in 10 units of pig food entering the trough, but whoever presses the button will first pay a cost of 2 units. The button and pig food trough are in opposite positions, and the pig who presses the button incurs a cost of 2 units and loses the opportunity to eat at the trough first.

If the piglets first go to the trough to eat, due to lack of competition, the speed of eating is average, and the final ratio of food eaten by the big and small pigs is 6:4; if you eat at the edge of the trough at the same time, the feeding speed of the big pig accelerates, and the final profit ratio of the big and small pigs is 7:3; if the big pig comes to the trough to eat first, the big pig will occupy all the remaining pig food, and the final profit ratio of the big and small pigs is 9:1.

Chapter 5

Statistical Learning

Statistical learning is to build a probabilistic statistical model based on data and use the model to predict and analyze the data. Statistical learning deals with the statistical inference problem of finding a predictive function based on data. With the explosion of "Big Data" problems, statistical learning has become a very hot field in many scientific areas, as well as economy, finance, and other business disciplines. People with statistical learning skills are in high demand.

5.1 Introduction

The statistical method is to perform the external quantity of the object to infer the possible regularity of the thing. The scientific regularity is always hidden deeper. At first, some clues are always seen through statistical analysis from its quantitative performance, and then certain hypotheses or doctrines are proposed for further in-depth theoretical research. When theoretical research makes certain conclusions, it often needs to be verified in practice. That is to say, observing some natural phenomena or specially arranged experimental data, whether it is consistent with the theory, to what extent, and which direction may be in the direction, need to be treated by statistical analysis.

Statistics have been greatly developed in the last hundred years. We can roughly describe the process of statistical development using the following framework:

- 1900–1920: Data description
- 1920–1940: The dawn of statistical models
- 1940–1960: Mathematical statistics era
- 1960–1980: Challenges of stochastic model assumptions
- 1980–1990: Relaxation structure model hypothesis
- 1990–1999: Modeling complex data structures.

Among them, from 1960 to 1980, there was a revolution in the field of statistics. It is sufficient to estimate the dependencies from the observed data, as long as the general properties of the set of functions to which the unknown dependencies belong are known. Leading this revolution are the four discoveries of the 1960s:

(1) the principle of regularization on the resolution of ill-posed problems found by Tikhonov, Ivanov, and Philips,
(2) non-parametric statistics found by Parzen, Rosenblatt, and Chentsov,
(3) the law of large numbers in universal function spaces found by Vapnik and Chervonenkis, and its relationship to the learning process,
(4) the complexity of algorithm and its relationship with inductive reasoning discovered by Kolmogorov, Solomonoff, and Chaitin.

These four findings have also become an important basis for statistical learning research. The traditional statistics study mainly on the asymptotic theory, that is, the statistical properties when the sample tends to infinity. The statistical method mainly considers the hypothesis of the test and the data model fit. It relies on an explicit basic probability model.

Statistical learning methods include the hypothesis space of the model, the criteria for model selection, and the algorithm for model learning, which are called the three elements of statistical learning methods. The general steps to implement statistical learning are as follows:

(1) Prepare a limited training dataset.
(2) Obtain the hypothesis space containing all possible models, that is, the set of learning models.
(3) Determine the criteria for model selection, that is, the learning strategy.
(4) Obtain the algorithm to realize the optimal model, that is, the learning algorithm.
(5) Select the optimal model through learning methods.
(6) Use the learned optimal model to predict or analyze new data.

Common statistical methods include regression analysis (multiple regression, autoregressive, etc.), discriminant analysis (Bayesian discriminant, Fisher discriminant, non-parametric discriminant, etc.), cluster analysis (system clustering, dynamic clustering, etc.), and exploration analysis (principal element analysis, correlation analysis, etc.). The support vector machine (SVM) is based on the principle of structural risk minimization of computational learning theory. The main idea is to find a hyperplane as a two-class segmentation in high-dimensional space for two types of classification problems to ensure the minimum classification error rate

(Vapnik *et al.*, 1997). And an important advantage of SVM is that it can handle linear indivisible situations.

Vapnik and his research group have studied machine learning based on finite samples since the 1960s. A complete theory, statistical learning theory, has been established in the 1990s (Vapnik, 1995). Moreover, a new universal learning algorithm Support Vector Machine (SVM) has been proposed. SVM minimizes the probability of classification error based on the structural risk minimization inductive principle. The main idea of SVM is mapping the nonlinear data to a higher-dimensional linear space where the data can be linearly classified by hyperplane (Vapnik *et al.*, 1997). One advantage of SVM is the capacity of disposing linearity non-separable cases.

5.2 Logistic Regression

Logistic regression is estimating the parameters of a logistic model. Assume vector X is a continuous random variable, $X = (x_1, x_2, \ldots x_n)$, and has the following distribution function and density function (see Figure 5.1):

$$F(x) = P(X \supseteq x) = \frac{1}{1 + e^{-(x-\mu)/\gamma}} \tag{5.1}$$

$$f(x) = F'(x) = \frac{e^{-(x-\mu)/\gamma}}{\gamma (1 + e^{-(x-\mu)/\gamma})^2} \tag{5.2}$$

where μ is the position parameter and $\gamma > 0$ is the shape parameter.

The binomial logistic regression model is a classification model represented by the conditional probability distribution $P(Y|X)$. Here, the random variable X takes the value of a real number, and the random variable Y takes the value of 1 or 0. Model parameters can be estimated through supervised learning. The binomial logistic

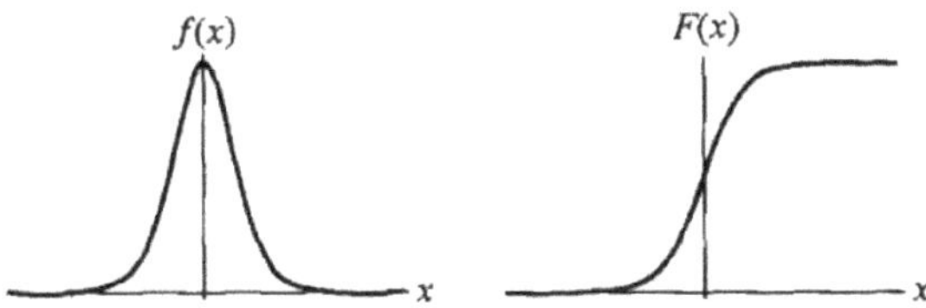

Fig. 5.1. Distribution function and density function in logistic regression.

regression model is the following conditional probability distribution:

$$P(Y = 1|x) = \frac{\exp(w \cdot x + b)}{1 + \exp(w \cdot x + b)} \tag{5.3}$$

$$P(Y = 0|x) = \frac{1}{1 + \exp(w \cdot x + b)} \tag{5.4}$$

where $x \in R^n$ is the input, $Y \in \{0, 1\}$ are the outputs, and $w \in R^n$ and $b \in R$ are parameters. w is called the weight vector, b is called the bias, and $w \cdot x$ is the inner product of w and x.

For a given input instance x, $P(Y = 1|x)$ and $P(Y = 0|x)$ can be obtained according to Equation (5.3) and Equation (5.4). Logistic regression compares the sizes of two conditional probability values and assigns the instance x to the category with the larger probability value.

5.3 Statistical Learning Problem

5.3.1 Empirical Risk

We consider the learning problem as a problem of finding a desired dependence between input and output (or supervisor's response) using a limited number of observations. Learning problems could generally be represented as finding uncertain dependency relationships between variables y and x where the joint probability distribution function $F(x, y)$ is unknown. The selection of the desired function is based on a training set of n independent and identically distributed observations:

$$(x_1, y_1), (x_2, y_2), \ldots, (x_n, y_n) \tag{5.5}$$

Given a set of functions $\{f(x, w)\}$, it aims at choosing the best function to approximate the supervisor's response $f(x, w_0)$, which makes the risk function minimal:

$$R(w) = \int L(y, f(x, w)) dF(x, y) \tag{5.6}$$

where $\{f(x, w)\}$ is the set of expected functions and w is functional general parameters. $L(y, f(x, w))$ measures the loss or discrepancy between the response y of the supervisor to a given input x and the response $f(x, w)$ provided by the learning machine. Different types of learning problems have diverse formal loss functions.

Empirical risk minimization inductive principle is used in the classical methods of learning problems. Empirical risk is defined on the basis of the training set:

$$R_{emp}(w) = \frac{1}{l} \sum_{i=1}^{l} L(y_i, f(x_i, w)) \tag{5.7}$$

Machine learning designs a learning algorithm for minimizing $R_{emp}(w)$.

5.3.2 VC Dimension

Statistical learning theory is inductive learning theory on the basis of a small sample size. One significant concept is VC Dimension (Vapnik–Chervonenkis Dimension) in pattern recognition. The VC dimension of a set of indicator functions $f(x, w)$ is equal to the largest number h of vectors that can be separated into different classes in all the 2^h possible ways using this set of functions (i.e., the VC dimension is the maximum number of vectors that can be shattered by the set of functions). The VC dimension is equal to infinity if there exists a set of vectors that any number of samples can be shattered by the functions $f(x, w)$. The VC dimensions of a set of bounded real functions are defined by transformed indicated functions with thresholds.

The VC dimension of the set of functions (rather than the number of parameters) is responsible for the generalization ability of learning machine. Intuitively, learning machines with high VC dimensions are more complex and powerful. At the moment, there is no universal theory to calculate VC dimension of an arbitrary function set. In n dimension real space *Rn*, the VC dimension of linear classification and linear real function is $n + 1$. However, the VC dimension of $f(x, a) = \sin(ax)$ is infinite. In statistical learning theory, how to calculate VC dimension using theory and experimental method is still a problem.

5.4 Consistency of Learning Processes

5.4.1 Classical Definition of Learning Consistency

In order to construct algorithms for learning from a limited number of observations, we need an asymptotic theory (consistency is an asymptotic concept). We describe the conceptual model for learning processes that are based on the empirical risk minimization inductive principle. The goal of this part is to describe necessary and

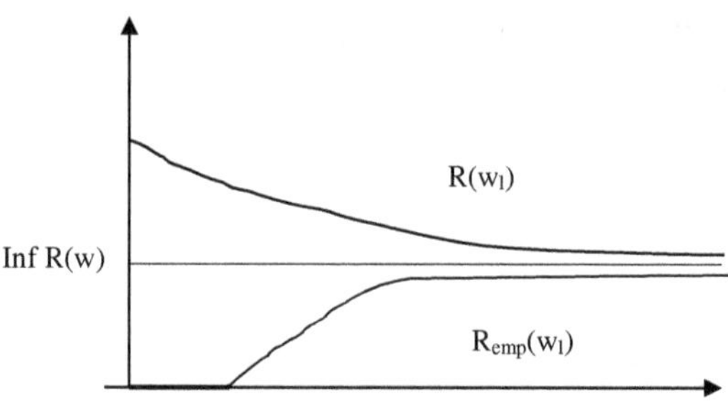

Fig. 5.2.   Consistency of learning process.

sufficient conditions for the consistency of learning processes that minimize the empirical risk.

Definition 5.1. The Empirical Risk Minimization Inductive Principle (Vapnik, 1995).

We say that the principle (method) of the empirical risk minimization (ERM) is consistent for the set of functions $L(y, w)$ and for the probability distribution function $F(y)$ if the following two sequences converge in probability to the same limit (see Figure 5.2):

$$R(w_l) \xrightarrow[l\to\infty]{P} \inf_{w\in\Lambda} R(w) \tag{5.8}$$

$$R_{emp}(w_l) \xrightarrow[l\to\infty]{P} \inf_{w\in\Lambda} R(w) \tag{5.9}$$

In other words, the ERM method is consistent if it provides a sequence of functions $L(y, w_l)$, $l = 1, 2, \ldots$, for which both expected risk and empirical risk converge to the minimal possible value of risk. Equation (5.12) asserts that the values of achieved risks converge to the best possibility, while Equation (5.13) asserts that one can estimate on the basis of the values of empirical risk the minimal possible value of the risk.

5.4.2 *Key Theorem of Learning Theory*

In 1989, Vapnik and Chefvonenkis proposed the key theorem of learning theory as follows (Vapnik & Chervonenkis, 1991).

Theorem 5.1. *Let $L(y, w)$, $w \in \Lambda$ be a set of functions that satisfy the condition*

$$A \le \int L(y, w)dF(y) \le B \quad (A \le R(w) \le B) \tag{5.10}$$

Then for the ERM principle to be consistent in the following sense,

$$\lim_{l\to\infty} P\{\sup_{w\in\Lambda} R(w) - R_{emp}(w)) > \varepsilon\} = 0, \quad \forall \varepsilon > 0 \tag{5.11}$$

It is necessary and sufficient that the empirical risk $R_{emp}(w)$ *converges uniformly to the actual risk* $R(w)$ *over the set* $L(y, w)$, $w \in \Lambda$. *We call this type of uniform convergence uniform one-sided convergence.*

5.4.3 *VC Entropy*

Definition 5.2. Let $A \leq L(y, w) \leq B, \omega \in \Lambda$, be a set of bounded loss functions. Using this set of functions and the training set $z_1, \ldots z_l$, one can construct the following set of l-dimensional vectors:

$$\boldsymbol{q}(w) = (L(z_1, w), \ldots, L(z_l, w)), \quad w \in \Lambda \tag{5.12}$$

This set of vectors belongs to the one-dimensional cube and has a finite minimal ε-net in the metric C (or in the metric L_p).

Let $N = N^\Lambda(\varepsilon; z_1, \ldots, z_l)$ be the number of elements of the minimal ε-net of this set of vectors $\boldsymbol{q}(w)$, $w \in \Lambda$. Note that $N^\Lambda(\varepsilon; z_1, \ldots, z_l)$ is a random variable, since it is constructed using random vectors $z_1, \ldots, z_l$. The logarithm of the random value $N^\Lambda(\varepsilon; z_1, \ldots, z_l)$, $H^\Lambda(\varepsilon; z_1, \ldots, z_l) = \ln N^\Lambda(\varepsilon; z_1, \ldots, z_l)$ is called the random VC entropy of the set of functions.

$A \leq L(y, w) \leq B, w \in \Lambda$ on the sample $z_1, \ldots, z_l$. The expectation of the random VC entropy

$$H^\Lambda(\varepsilon; l) = E H^\Lambda(\varepsilon; z_1, \ldots, z_l)$$

is called the random VC entropy of the set of functions $A \leq L(y, w) \leq B, w \in \Lambda$ on the sample of size l. Here the expectation is taken with respect to the product measure $F(z_1, \ldots, z_l)$.

Theorem 5.2. *For uniform two-sided convergence, it is necessary and sufficient that the equality*

$$\lim_{l\to\infty} \frac{H^\Lambda(\varepsilon, l)}{l} = 0, \quad \forall \varepsilon > 0 \tag{5.13}$$

be valid. In other words, the ratio of the VC entropy to the number of observations should decrease to zero with increasing numbers of observations.

Corollary 5.1. *Under some conditions of measurability on the set of indicator functions* $L(y, w)$, $w \in \Lambda$, *a necessary and sufficient condition for uniform two-sided convergence is* $\lim_{l\to\infty} \frac{H^\Lambda(l)}{l} = 0$, *which is a particular case of equality* (5.9).

Theorem 5.3. *In order for uniform one-sided convergence of empirical means to their expectations to hold for the set of totally bounded functions $L(y, w)$, $w \in \Lambda$, it is necessary and sufficient that for any positive δ, η, and ε, there exists a set of functions $L^*(y, w^*)$, $w^* \in \Lambda^*$ satisfying the following:*

$$L(y, w) - L^*(y, w^*) \geq 0, \quad \forall y$$

$$\int (L(y, w) - L^*(y, w^*))dF(y) \leq \delta \tag{5.14}$$

such that the following holds for the ε-entropy of the set $L^*(y, w^*)$, $w^* \in \Lambda^*$, on samples of size l:

$$\lim_{l \to \infty} \frac{H^{\Lambda^*}(\varepsilon, l)}{l} < \eta$$

According to these key theorems, we study learning theory. Moreover, we describe a sufficient condition for the consistency of the ERM principle by using different methods and functions. On the basis of these functions, three milestones of learning theory are constructed:

(1) We use VC entropy to define the following equation describing a sufficient condition for the consistency of the ERM principle:

$$\lim_{l \to \infty} \frac{H^{\Lambda}(l)}{l} = 0 \tag{5.15}$$

(2) We use the annealed VC entropy to define the following equation describing a sufficient condition for the consistency of the ERM principle:

$$\lim_{l \to \infty} \frac{H^{\Lambda}_{ann}(l)}{l} = 0 \tag{5.16}$$

where the annealed VC entropy $H^{\Lambda}_{ann}(l) = \ln EN^{\Lambda}(z, \ldots, z_l)$.

(3) We use the growth function to define the following equation describing a sufficient condition for the consistency of the ERM principle:

$$\lim_{l \to \infty} \frac{G^{\Lambda}(l)}{l} = 0 \tag{5.17}$$

where the growth function $G^{\Lambda}(l) = \ln \sup_{z,\ldots,z_{l1}} N^{\Lambda}(z, \ldots, z_l)$.

5.5 Structural Risk Minimization Inductive Principle

Statistical learning theory systematically analyzes the relationship between inhomogeneous function set, empirical risk, and actual risk, the bounds on the generalization ability of learning machines (Vapnik, 1995). Here we only consider functions that correspond to the two-class pattern recognition case: For the set of indicator functions (including the function minimizing empirical risk), now choose some η such that $0 \leq \eta \leq 1$. Then for losses taking empirical risk $R_{emp}(w)$ and actual risk $R(w)$ with probability $1 - \eta$, the following bound holds (Burges, 1998):

$$R(w) \leq R_{emp}(w) + \sqrt{\frac{h(\ln(2l/h) + 1) - \ln(\eta/4)}{l}}$$

where h is a non-negative integer called the Vapnik–Chervonenkis (VC) dimension. And l is the total of samples.

It follows that statistical learning's actual risk $R(w)$ has two parts: one is empirical risk $R_{emp}(w)$ defined to be just the measured mean error rate on the training set (for a fixed, finite number of observations); another is VC confidence. The confidence interval reflects the maximal difference between actual risk and empirical risk. Meanwhile, it reflects risk bringing structure complexity. There are some relations between the confidence interval, VC dimension h, and sample number l. The inequality (5.14) could simply be expressed as

$$R(w) \leq R_{emp}(w) + \Phi(h/l) \tag{5.18}$$

According to the second term on the right-hand side of the inequality (5.18), $\Phi(h/l)$ increases while the VC dimension h increases. Therefore, the confidence interval increases while learning machine's complexity and the VC dimension increase. Moreover, the difference between actual risk and empirical risk increases when learning machines use a small sample of training instances.

Note that the bound for the generalization ability of learning machines is a conclusion in the worst case. Furthermore, the bound is not tight in many cases, especially when VC dimension is higher. When $h/l > 0.37$, the bound is guaranteed not tight (Burges, 1998). When VC dimension is infinite, the bound does not exist.

To construct small sample size methods, we use both the bounds for the generalization ability of learning machines with sets of totally bounded nonnegative functions, $0 \leq L(z, w) \leq B$, $w \in \Lambda$ (Λ is abstract parameters set). Each bound is valid with a probability of at least $1 - \eta$:

$$R(w_l) \leq R_{emp}(w) + \frac{B\varepsilon}{2}\left(1 + \sqrt{1 + \frac{4R_{emp}(w_l)}{B\varepsilon}}\right) \tag{5.19}$$

and the bound for the generalization ability of learning machines with sets of unbounded functions:

$$R(w_l) \leq \frac{R_{emp}(w_l)}{(1 - a(p)\tau\sqrt{\varepsilon})_+} \tag{5.20}$$

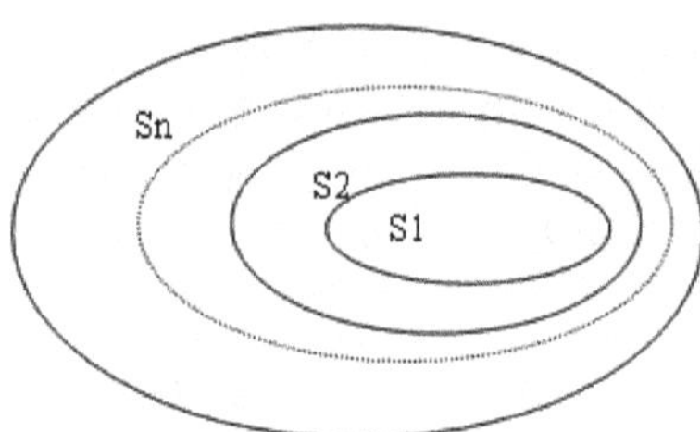

Fig. 5.3.   A structure on the set of functions is determined by the nested subsets of functions.

where

$$a(p) = \sqrt[p]{\frac{1}{2}\left(\frac{p-1}{p-2}\right)^{p-1}} \qquad \varepsilon = 2\frac{\ln N - \ln \eta}{l}$$

There are two methods to minimize the actual risk. One is to minimize the empirical risk. According to upper formula, the upper bound of actual risk decreases while the empirical risk decreases. The other is to minimize the second term on the right-hand side of the inequality (5.15). We have to make the VC dimension a controlling variable. The latter method conformances small sample size.

Let the set S of functions L(z, w) be provided with a structure consisting of nested subsets of functions $S_k = \{L(z, w), w \in \Lambda_k\}$ such that (see Figure 5.3) (Vapnik, 1995):

$$S_1 \subset S_2 \subset \cdots \subset S_n \qquad . \tag{5.21}$$

where the elements of the structure satisfy the following two properties:

(1) The VC dimension hk of each set S_k of functions is finite. Therefore, $h_1 \le h_2 \le$, $\ldots, \le h_n, \ldots$.
(2) Any element S_k of the structure contains either a set of totally bounded functions, $0 \le L(z, w) \le B_k; \alpha \in \Lambda_k$, or a set of functions that satisfy the following inequality for some pair (p, τ_k):

$$\sup_{w \in \Lambda_k} \frac{\left(\int L^p(z, w)dF(z)\right)^{\frac{1}{p}}}{\int L(z, w)dF(z)} \le \tau_k \quad p > 2 \tag{5.22}$$

We call this structure an admissible structure.

For a given set of observations $z_1, \ldots, z_l$, the structural risk minimization (SRM) principle chooses the function L(z, w_{kl}) minimizing the empirical risk in the subset S_k for which the guaranteed risk (determined by the right-hand side of inequality (5.19) or by the right-hand side of inequality (5.20) depending on the circumstances) is minimal.

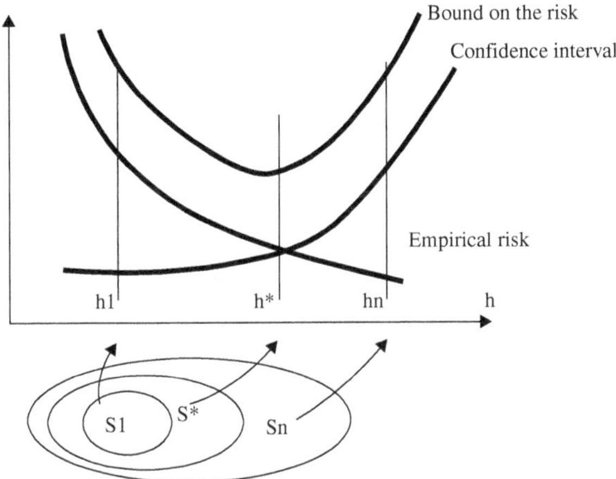

Fig. 5.4.   The bound on the structural risk is the sum of the empirical risk and the confidence interval.

The SRM principle defines a trade-off between the quality of the approximation of the given data and the complexity of the approximating function. As the subset index n increases, the minima of the empirical risks decrease. However, the term responsible for the confidence interval increases. The SRM principle takes both factors into account by choosing the subset Sn for which minimizing the empirical risk yields the best bound on the actual risk, as shown in Figure 5.4 (Vapnik, 1995).

5.6 Support Vector Machine

Support Vector Machine (SVM) is a new type of universal learning machine proposed recently. SVM has extra advantages for pattern classification.

5.6.1 *Linearly Separable Case*

Suppose the training data $(x_1, y_1), \ldots, (x_l, y_l), x \in R_n, y \in \{+1, -1\}$, where l is the sample size and n is the dimension of input data. We can construct a hyperplane to absolutely separate two-class samples for the case where the training data are linearly separable. The hyperplane be described as

$$(w \bullet x) + b = 0 \tag{5.23}$$

where $\bullet$ denote vector dot products. To describe the separating hyperplane, let us use the following form:

$$w \bullet x_i + b \geq 0, \quad \text{if } y_i = +1$$

$$w \bullet x_i + b < 0, \quad \text{if } y_i = -1$$

where w denotes the hyperplane's normal direction, $\frac{w}{\|w\|}$ denotes the unit normal vector, and $\|w\|$ denotes the Euclid modular function.

We say that this set of vectors is separated by the optimal hyperplane (or the maximal margin hyperplane) if the training data is separated without error and the distance between the closest vector to the hyperplane is maximal (see Figure 5.5).

For linearly separable cases, to find the optimal separating hyperplane is to solve the following quadratic programming problem. Given training samples, find a pair consisting of a vector w and a constant (threshold) b such that they minimize the function

$$\min \Phi(w) = \frac{1}{2}\|w\|^2 \tag{5.24}$$

under the constraints of inequality type

$$y_i(w \bullet x_i + b) - 1 \geq 0, \quad i = 1, 2, \ldots, l \tag{5.25}$$

Optimize function $\Phi(w)$ is quadratic form, and the constraint condition is linear. Thus, it is a typical quadratic programming problem that can be solved by Lagrange multiplier method. Introduce Lagrange multipliers $\alpha_i \geq 0, i = 1, 2, \ldots, l$

$$L(w, b, \alpha) = \frac{1}{2}\|w\|^2 - \sum_{i=1}^{l} \alpha_i\{y_i(x_i \bullet w + b) - 1\} \tag{5.26}$$

where the extremum of L is the saddle point of equality (5.26). To find the saddle point, one has to minimize this function over w and b and to maximize it over the non-negative Lagrange multipliers α. At the saddle point, the solutions w^*, b^*, and α^* should satisfy the conditions

$$\frac{\partial L}{\partial b} = \sum_{i=1}^{l} y_i \alpha_i = 0 \tag{5.27}$$

$$\frac{\partial L}{\partial w} = w - \sum_{i=1}^{l} y_i \alpha_i x_i = 0 \tag{5.28}$$

where $\frac{\partial L}{\partial w} = (\frac{\partial L}{\partial w_1}, \frac{\partial L}{\partial w_2}, \cdots \frac{\partial L}{\partial w_l})$.

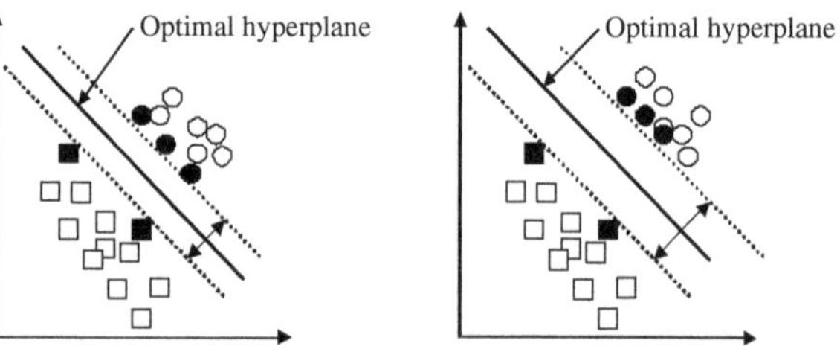

Fig. 5.5.   The optimal separating hyperplane.

Therefore, through solving quadratic programming problem, SVM attains corresponding α^* and w^* satisfying the following equality:

$$w^* = \sum_{i=1}^{l} \alpha_i^* y_i x_i \tag{5.29}$$

and the optimal hyperplane is shown in Figure 5.4.

For transform linearly separable case, the original problem becomes the following problem:

$$\max_{\alpha} W(\alpha) = \sum_{i=1}^{l} \alpha_i - \frac{1}{2} \sum_{i=1}^{l} \sum_{j=1}^{l} \alpha_i \alpha_j y_i y_j x_i \bullet x_j = \Gamma \bullet I - \frac{1}{2} \Gamma \bullet D\Gamma \tag{5.30}$$

satisfying the following constraints:

$$\sum_{i=1}^{l} y_i \alpha_i = 0, \quad \alpha_i \geq 0, \quad i = 1, 2, \ldots, l \tag{5.31}$$

where $\Gamma = (\alpha_1, \alpha_2, \ldots, \alpha_l)$, $I = (1, 1, \ldots, 1)$, and D is an $l \times l$ symmetric matrix, each element is as follows:

$$D_{ij} = y_i y_j x_i \bullet x_j \tag{5.32}$$

This fact follows from the classical Karush–Kuhn–Tucker (KKT) theorem, according to which necessary and sufficient conditions for the optimal hyperplane are that the separating hyperplane satisfies the following conditions:

$$\alpha_i \{ y_i (w \bullet x_i + b) - 1 \} = 0, \quad i = 1, 2, \ldots, l \tag{5.33}$$

According to equality (5.29), only these samples satisfying $\alpha_i > 0$ determine classification results while those samples satisfying $\alpha_i = 0$ do not. We call these samples satisfying $\alpha_i > 0$ support vectors.

We train samples to attain vectors α^* and w^*. Selecting a support vector sample x_i, we attain b^* by the following equality:

$$b^* = y_i - w \bullet x_i \tag{5.34}$$

For a test sample x, calculate the following equality:

$$d(x) = x \bullet w^* + b^* = \sum_{i=1}^{l} y_i \alpha_i^* (x \bullet x_i) + b^* \tag{5.35}$$

According to the sign of $d(x)$ to determine which class x belongs to.

5.6.2 *Linearly Non-Separable Case*

In linearly separable case, decision function is constructed on the basis of Euclid distance, i.e., $K(x_i, x_j) = x_i \bullet x_j = x_i^T x_j$. In linearly non-separable case, SVM maps the input vectors x into a high-dimensional feature space H through some nonlinear mapping (see Figure 5.5). In this space, an optimal separating hyperplane is constructed. The nonlinear mapping $\Phi: R^d \to H$:

$$x \to \Phi(x) = (\phi_1(x), \phi_2(x), \ldots, \phi_{i(x)}, \ldots)^T \tag{5.36}$$

where $\phi_i(x)$ is a real function.

Feature vector $\Phi(x)$ substitutes input vector x, equalities (5.28) and (5.31) are transformed as follows:

$$D_{ij} = y_i y_j \Phi(x_i) \bullet \Phi(x_j) \tag{5.37}$$

$$d(x) = \Phi(x) \bullet w^* + b^* = \sum_{i=1}^{l} \alpha_i y_i \Phi(x_i) \bullet \Phi(x) + b^* \tag{5.38}$$

The optimize function equality (5.26) and decision function equality (5.35) only refer inner product $x_i \bullet x_j$ of training samples. Therefore, we calculate the inner product in high-dimensional space using inner product functions in input space. On the basis of relational theory, a kernel function $K(x_i \bullet x_j)$ is the parallelism of an inner product in a certain space if it satisfies Mercer's condition (Vapnik, 1995).

Using some appropriate inner product function $K(x_i \bullet x_j)$, SVM maps the input space vectors into feature space vectors through some nonlinear mapping. Moreover, the complexity of calculate does not increase. The object function equality (5.26) is transformed to the following equality:

$$\max_{\alpha} W(\alpha) = \sum_{i=1}^{l} \alpha_i - \frac{1}{2} \sum_{i=1}^{l} \sum_{j=1}^{l} \alpha_i \alpha_j y_i y_j K(x_i, x_j) \tag{5.39}$$

Thus, the corresponding classifying function becomes the following equality:

$$d(x) = \sum_{i=1}^{l} y_i \alpha_i^* K(x, x_i) + b^* \tag{5.40}$$

It is called support vector machines (SVM), as shown in Figure 5.6 (Vapnik, 1995).

For a given $K(x, y)$, the corresponding function $\Phi(x)$ exists under certain condition. It is necessary and sufficient that the condition as follows is satisfied.

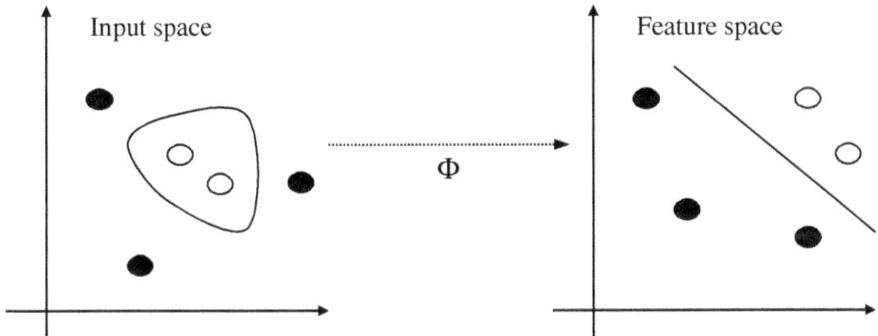

Fig. 5.6.   The SVM maps the input space into a feature space.

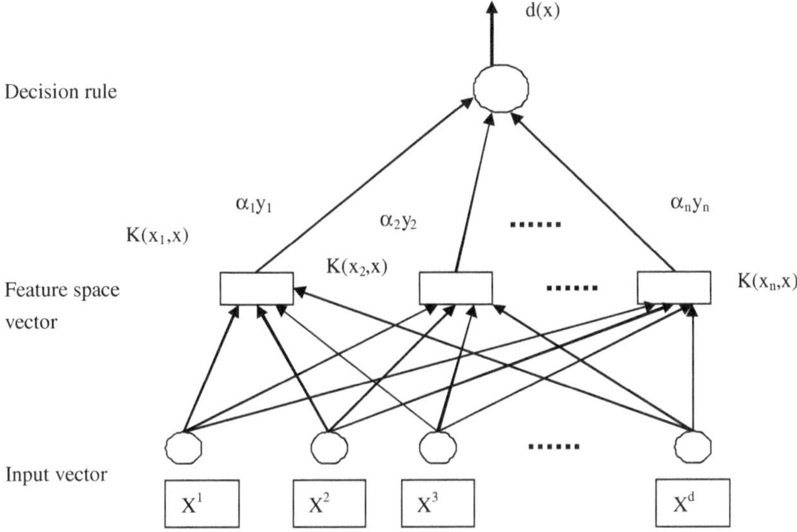

Fig. 5.7.   Support vector machine.

Given an arbitrary function g(x), if $\int_a^b g(x)^2 dx$ is finite,

$$\int_a^b \int_a^b K(x, y)g(x)g(y)dxdy \geq 0 \tag{5.41}$$

Equation (5.41) is valid.

This decision condition is not feasible. It is well known that polynomial functions satisfy Mercer's condition. Therefore, $K(x, y)$ satisfies Mercer's condition if it approaches some polynomial function.

The learning machines that construct decision functions of the type (5.40) are called support vector machines (SVM). In SVM, the complexity of the construction depends on the number of support vectors rather than on the dimensionality of the feature space. The scheme of SVM is shown in Figure 5.7.

In nonlinear case, SVM maps the input space vectors into high-dimensional feature space vectors through some nonlinear mapping defined inner product functions. Thus, SVM can find the generalized optimal separating hyperplane in feature space.

5.7 Kernel Function

Using different functions for the convolution of the inner products' kernel function, one can construct learning machines with different types of nonlinear decision surfaces in input space. At present, prevalent kernel functions are polynomial kernel function, radial basis function, multi-layer perceptron, and dynamic kernel function.

5.7.1 Polynomial Kernel Function

Polynomial kernel function:

$$K(x, x_i) = [(x, x_i) + 1]^d \qquad (5.42)$$

We construct a d-dimensional polynomials decision function of the form

$$f(x, \alpha) = sign\left(\sum_{\text{support vector}} y_i \alpha_i [x_i \bullet x) + 1]^d - b \right)$$

5.7.2 Radial Basis Function

Classical radial basis function (RBF) machines use the following set of decision rules:

$$f(x) = sign\left(\sum_{i=1}^{l} \alpha_i K_\gamma (|x - x_i|) - b \right) \qquad (5.43)$$

where $K\gamma (|x - x_i|)$ depends on the distance $|x - x_i|$ between two vectors. For any fixed γ, the function $K\gamma (|x - x_i|)$ is a non-negative monotonic function. It tends to zero as the training sample's total goes to infinity. The most popular function of this type is

$$K_\gamma (|x - x_i|) = \exp\left\{ -\frac{|x - x_i|^2}{\sigma^2} \right\} \qquad (5.44)$$

To construct the decision rule (5.39), one has to estimate

(1) the value of the parameter γ,
(2) the number N of the centers x_i,

(3) the vectors x_i, describing the centers,
(4) the value of the parameters α_i.

In contrast to classical RBF methods, each center denotes a support vector in this method. Furthermore, all four types of parameters are chosen to minimize the bound on the probability of test error.

5.7.3 *Multi-Layer Perceptron*

Multi-layer perceptron defines the inner product kernel function using a sigmoid function. The number N of hidden units (the number of support vectors) is found automatically. The sigmoid kernel satisfies Mercer conditions as follows:

$$K(x_i, x_j) = \tanh(\gamma \, x_i{}^T x_j - \Theta) \tag{5.45}$$

Using this arithmetic, we avoid local minima problem that puzzles neural network.

5.7.4 *Dynamic Kernel Function*

Amari and Wu proposed a method of modifying a kernel function to improve the performance of a support vector machine classifier (Amari & Wu, 1999). This is based on the structure of the Riemannian geometry induced by the kernel function. U denotes feature mapping $U = \Phi(x)$, then

$$dU = \sum_i \frac{\partial}{\partial x_i} \Phi(x) dx_i$$

$$\|dU\|^2 = \sum_{i,j} g_{ij}(x) dx_i dx_j$$

where

$$g_{ij}(x) = \left(\frac{\partial}{\partial x_i} \Phi(x) \right) \bullet \left(\frac{\partial}{\partial x_j} \Phi(x) \right).$$

The $n \times n$ positive-definite matrix $(g_{ij}(x))$ is the Riemannian metric tensor induced in S. $ds^2 = \sum g_{ij}(x) dx_i dx_j$ is the Riemannian distance. The volume form in a Riemannian space is defined as

$$dv = \sqrt{g(x)} dx_1 \cdots dx_n$$

where $g(x) = \det(g_{ij}(x))$. The factor $g(x)$ represents how a local area is magnified in U under the mapping $\Phi(x)$. Therefore, we call it the magnification factor.

Since $k(x, z) = (\Phi(x) \bullet \Phi(z))$, we can get

$$g_{ij}(x) = \frac{\partial}{\partial x_i \partial z_j} k(x, z)|_{z=x}$$

In particular, for the Gaussian kernel function $k(x, z) = \exp\left\{\frac{|x-z|^2}{2\sigma^2}\right\}$, we have $g_{ij}(x) = \frac{1}{\sigma^2}\delta_{ij}$.

In order to improve the performance of an SVM classifier, Amari and Wu proposed a method of modifying a kernel function. To increase the margin or reparability of classes, we need to enlarge the spatial resolution around the boundary surface in U. Let $c(x)$ be a positive real differentiable function and $k(x, z)$ denote a Gaussian kernel function, then

$$\widetilde{k}(x, z) = c(x)k(x, z)c(z) \tag{5.46}$$

Also it is a kernel function, and

$$\widetilde{g}_{ij}(x) = c_i(x)c_j(x) + c^2(x)g_{ij}$$

where $c_i(x) = \frac{\partial}{\partial x_i}c(x)$. Amari and Wu defined $c(x)$ as

$$c(x) = \sum_{x_i \in SV} h_i e^{\frac{\|x-x_i\|^2}{2\tau^2}} \tag{5.47}$$

where τ is a positive number and hi denotes the coefficient. Around the support vector x_i, we have

$$\sqrt{\widetilde{g}(x)} \approx \frac{h_i}{\sigma^n} e^{\frac{nr^2}{2\gamma^2}} \sqrt{1 + \frac{\sigma^2}{\tau^4}\gamma^2}$$

where $\tau = \|x - x_i\|$ is the Euclid distance between x and x_i. In order to make sure $\sqrt{\widetilde{g}(x)}$ is larger near the support vector x_i and is smaller in other regions, we need

$$\tau \approx \frac{\sigma}{\sqrt{n}} \tag{5.48}$$

In summary, the training process of the new method consists of two steps:

(1) Train SVM with a primary kernel k (Gaussian kernel), then modify training result according to equalities (5.46), (5.47), and (5.48), we attain $\widetilde{k}$.
(2) Train SVM with the modified kernel $\widetilde{k}$.

When SVM uses the new training method, the performance of the classifier is improved remarkably, and the number of support vectors decreases such that it improves the velocity of pattern recognition.

5.8 *K*-Nearest Neighbor Algorithm

Cover and Hart proposed the original proximity algorithm in 1968 to solve classification problems. It is one of the more mature algorithms among machine learning algorithms. The model used by the *K*-nearest neighbor algorithm actually corresponds to the division of the feature space. The KNN algorithm can be used not only for classification but also for regression.

The *K*-nearest neighbor algorithm is to first give a training dataset, which may contain the characteristics and classification of a certain type of item, and then give the characteristics of a certain item, and then classify the items based on the characteristics of each item in the training dataset and the needs. According to the "distance" of the items, find the k items with the closest distance, and then the category to which the most items among the k items belong is the result of the category judgment of the item that needs to be judged.

Advantages of *K*-nearest neighbor algorithm are high accuracy, insensitivity to outliers, and no data input assumptions. Disadvantages of *K*-nearest neighbor algorithm are high computational complexity and high space complexity.

Algorithm 5.1.   *K*-nearest neighbor.

---

1. **Collect data**

   Any method can be used (crawlers, publicly available datasets on the web, etc.).

2. **Prepare data**

   The values required for distance calculation are preferably in a structured data format (usually some matrix or array is used to facilitate the storage of structured data).

3. **Analyze data**

   You can use any method (usually the Matplotlib library is used to draw some graphs to observe the characteristics of the data).

4. **Training algorithm**

   Train a basic parameter model based on the data given in the training set.

5. **Test algorithm**

   Input the feature data from the test set into the generated model, and then record whether the result is the same as the original result to judge the accuracy (good or bad) of the model.

6. **Apply algorithms to applications**

   By making some simple command programs, you can input some characteristics of an item to determine the category of the item.

Exercises

5.1 Write a gradient descent learning algorithm for logistic regression models.

5.2 Compare Empirical Risk Minimization (ERM) inductive principle and Structural Risk Minimization (SRM) inductive principle.

5.3 What is VC dimension's meaning? Why does VC dimension reflect function set's learning capacity?

5.4 What are the three milestones of statistics learning theory? What problem was resolved in each milestone?

5.5 Describe support vector machine's primitive idea and mathematical model.

5.6 Why is statistical learning theory support vector machine's theory foundation, and in which area does it represent?

5.7 Under linearly separable case, given a hyperplane defined as follows:

$$w^T x + b = 0$$

where w denotes the weight vector, b denotes the bias, and x denotes the input vector. If a set of input pattern $\{x_i\}_{i=1}^{N}$ satisfies the following conditions

$$\min_{i=1,2,\ldots,N} |w^T x_i + b| = 1$$

(w, b) is called the canonical pair of a hyperplane. Prove that the conditions of canonical pair conduce distance between bounds of two classifications are $2/\|w\|$.

5.8 Briefly narrate the primitive concept that support vector machines solve non-linearity separable problems.

5.9 A two-layer perceptron's inner product kernel is defined as

$$K(x, x_i) = \tanh(\beta_0 x^T x_i + \beta_1)$$

under which values of β_0 and β_1, the kernel function does not satisfy Mercer's condition.

5.10 What are the advantages and limitations of support vector machine and radial basis function (RBF) network while they respectively solve the following assignments?

(1) pattern recognition

(2) nonlinear regression

5.11 In contrast to other classification methods, what are support vector machine's significant advantages? Please explain theoretically.

Chapter 6

Deep Learning

Deep learning is to imitate the information processing mechanism of human brain multi-layer neural network. It combines low-level features to form more abstract high-level representation attributes to explain the semantics of data, such as images, sounds, and texts.

6.1 Introduction

Deep learning as an exciting new technology has had rich history. Broadly speaking, there have been three waves of development of deep learning: deep learning known as cybernetics in the 1940s–1960s, deep learning known as connectionism in the 1980s–1990s, and the current resurgence under the name deep learning beginning in 2006 (Goodfellow *et al.*, 2016).

The first wave started with cybernetics in the 1940s–1960s, with the development of theories of biological learning. In 1943, McCulloch and Pitts published their papers in a neuron modeling group (McCulloch and Pitts, 1943). In their classic paper, McCulloch and Pitts combined the study of neurophysiology and mathematical logic and described the logic analysis of a neural network. In their model, neurons were assumed to follow the yes-no model law. If the number of such simple neurons is enough and connection weights are appropriately set and synchronously operated, McCulloch and Pitts proved that in principle, any computable function can be calculated with such a network. In Hebb's book *Behavior Histology* (Hebb, 1949), he clearly explained the amendment to the physiological synaptic learning rules. In 1958, Rosenblatt proposed a first model such as the perceptron (Rosenblatt, 1958) allowing the training of a single neuron. Rosenblatt's works are about perceptron, a new method of pattern recognition, and a new method of supervised learning (Rosenblatt, 1962).

The second wave started with the connectionist approach of the 1980–1995 period, with back-propagation to train a neural network with one or two hidden layers. In 1986, Rumelhart, Hinton, and Williams reported the development of back-propagation algorithm. In the same year, the famous book *Parallel Distributed Processing: Exploration of the Microstructure of Cognition*, edited by Rumelhart and McClelland, was published (Rumelhart & McClelland, 1986). This book caused significant impact on the use of back-propagation algorithm. It has become the most common multi-layer perceptron training algorithm. In fact, the back-propagation learning was found independently at the same time in two other places. In the mid-1980s after back-propagation algorithm was found, we found that as early as August 1974 in Harvard University, Werbos in his Ph.D thesis had described it (Werbos, 1974). Werbos' doctoral thesis is the first documentation of effective back-propagation model to describe the gradient calculation, and it can be applied to the general network model, including neural networks as its special case. The basic idea of back-propagation can be further traced back to the book by Bryson and Ho *Applied Optimal Control.*

The current and third wave, deep learning, started around 2006 (Hinton & Salakhutdinov, 2006). Hinton showed that a kind of neural network called a deep belief network could be efficiently trained using a strategy called greedy layer-wise pretraining. This wave of neural network research popularized the use of the term deep learning to emphasize that researchers were now able to train deeper neural networks than had been possible before and to focus attention on the theoretical importance of depth. The third wave began with a focus on new unsupervised learning techniques and the ability of deep models to generalize well from small datasets.

In June 2012, the Google Brain project in *The New York Times* has attracted the public attention. This project was led by Andrew Ng and Jeff Dean at Stanford University. The parallel computation platform with 16000 CPU Cores is served to train the deep neural networks with 10 billion nodes. The mass data is directly loaded into the algorithm and let data speak. The system will automatically learn from the data.

In November 2012, Microsoft publicly demonstrated a fully automatic simultaneous interpretation system in Tianjin, China. The speaker made a speech in English, and the backend computer automatically accomplished the speech recognition, the machine translation from English to Chinese, and the Chinese speech synthesis. The key technology supported backstage is also the deep learning.

On April 24, 2014, Baidu held the fourth technology Open Day in Beijing. During the Open Day, Baidu announced that they officially launched a large data engine, including the core big data capability with the three major components: Open Cloud, Data Factory, and Baidu Brain. Baidu provides big data storage, analysis, and mining techniques to the public through the large data engine. This is the first open big data engine in the world.

Deep learning is a new field in machine learning; its core idea is to simulate the hierarchy abstraction structure of the human brain, analyze large-scale data by unsupervised methods, and reveal the valuable information contained in big data. Deep learning was designed to research big data, which provides a deep thinking brain.

Suppose that there exists a system S with n layers $(S_1, \ldots S_n)$. The system with the input I and the output O can be visually represented as $I => S_1 => S_2 => \ldots . => S_n => O$. If the output O is equal to the input I, namely the information of the input I is not lost after the system transformation. To maintain the same, this means that the input I after each layer S_i does not have any loss of information, that is, in any layer S_i, it is the other kind of representation of original information.

The idea of deep learning is to stack multiple layers, and the output in the current layer is served as the input of the next layer. In this way, the layer-wise representation of the input information can be realized.

In addition, the hypothesis presented above the input is equal to the output is too strict. This limitation can be slightly loosened so that the difference between the input and the output is as far little as possible. This loosened principle can result in other deep learning methods. The above presentation is the basic idea of deep learning.

6.2 Human Brain Visual Mechanism

Human brain visual mechanism is shown in Figure 6.1. For a long time, people have researched the human brain visual system. In 1981, the Nobel Prize in medicine was awarded to David Hubel, Torsten Wiesel, and Roger Sperry. The main contribution of the first two people is in finding the information processing of the vision system: the visual cortex is graded like in Figure 6.2. The low-level V1 extracts the edge features. V2 extracts the shape or object parts, etc. Higher-level regions extract the entire object, the behavior of the object, etc.

In 1958, David Hubel and Torsten Wiesel studied the correspondence between pupillary regions and cerebral cortical neurons at John Hopkins University. They opened a 3 mm hole in the cat's hindbrain skull, inserted electrodes into the hole, and measured the activity of the neurons. Then, in front of the kitten's eyes, they show objects of various shapes and brightness. Also, when each object is displayed, the position and angle at which the object is placed are also changed. They hope that through this method, the kitten will feel the stimulation of different types and different strengths. The reason for doing this experiment is to prove a guess.

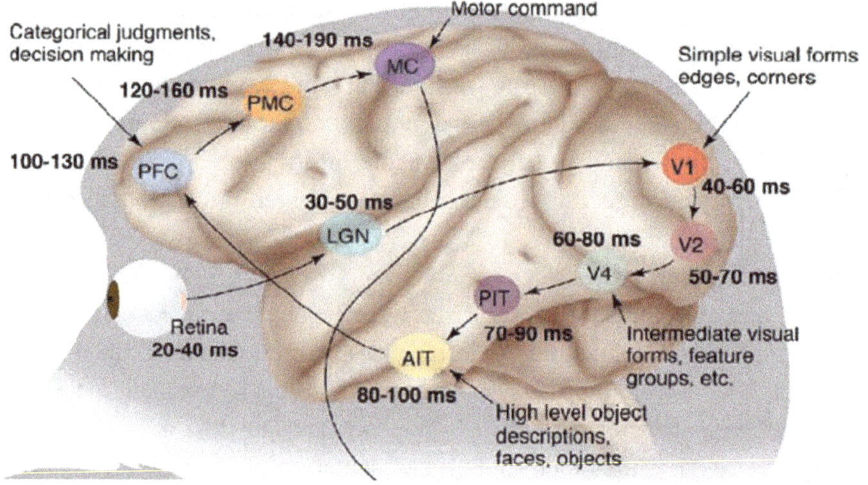

Fig. 6.1.   Human brain visual system.

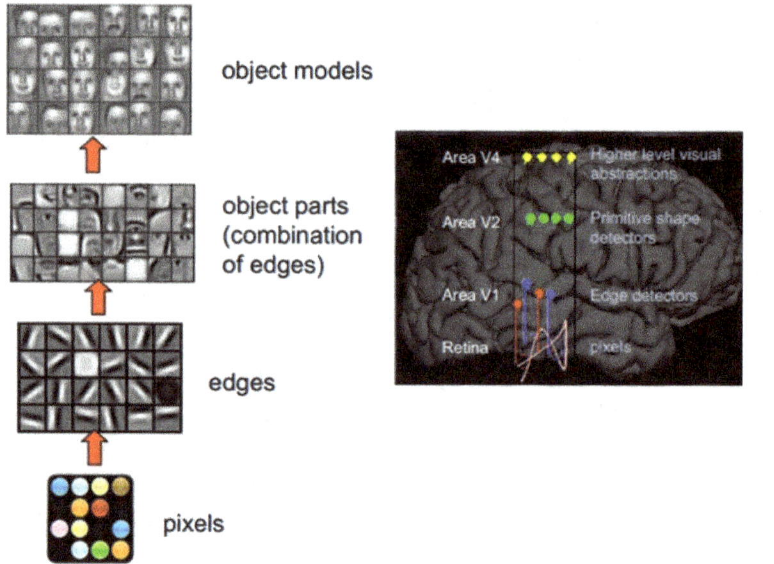

Fig. 6.2.   Visual cortex grading handling.

There is a certain correspondence between different visual neurons located in the posterior cortex and the stimulation of the pupil. Once the pupil is subjected to a certain stimulus, a certain part of the neurons in the posterior cortex will be active.

David Hubel and Torsten Wiesel discovered a neuron called the orientation selective cell. When the pupil finds the edge of the object in front of the eye and the edge points in a certain direction, the neuron cell is active. This discovery has inspired people to think further about the nervous system. The neural-central-brain working process may be an iterative, continuous abstraction process. There are two keywords here: one is abstraction and the other is iteration. From the original signal, do low-level abstraction, and gradually move to high-level abstraction. Human logical thinking often uses highly abstract concepts.

6.3 Autoencoder

Autoencoder is a single hidden layer neural network with the same number of nodes in the input layer and the output layer, as shown in Figure 6.3. The aim of designing an autoencoder is to reconstruct the input signal of a neural network as far as possible. This means that we need to solve the function $h_{W,b}(X) \approx x$, namely the output of the neural network is expectedly equal to the input as expected.

Suppose that there exists a sample set $\{(x^{(1)}, y^{(1)}), \ldots, (x^{(m)}, y^{(m)})\}$ containing m samples. The batch gradient descent method can be used to solve the neural networks. In particular, for a single sample (x, y), the cost function is

$$J(W, b; x, y) = \frac{1}{2}\|h_{W,b}(x) - y\|^2 \tag{6.1}$$

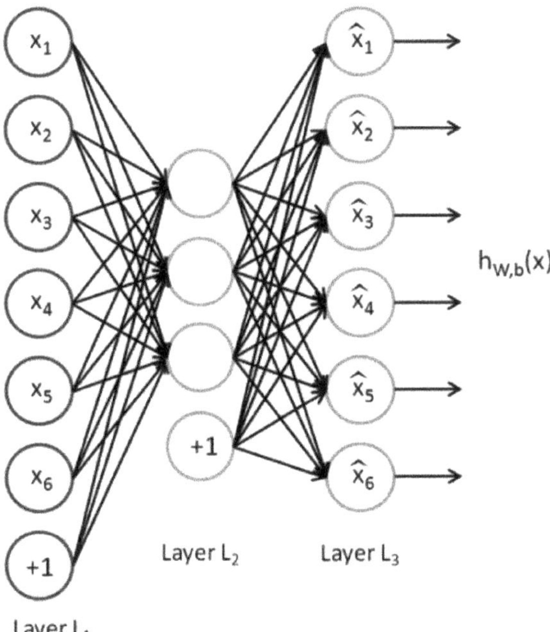

Fig. 6.3.   Autoencoder.

Equation (6.1) is a variance cost function. Given a dataset containing m samples, we can define the entire cost function as follows:

$$J(W, b) = \left[\frac{1}{m} \sum_{i=1}^{m} J(W, b; x^{(i)}, y^{(i)}) \right] + \frac{\lambda}{2} \sum_{l=1}^{n_i-1} \sum_{i=1}^{s_i} \sum_{j=1}^{s_{l+1}} (W_{ji}^{(l)})^2$$

$$= \left[\frac{1}{m} \sum_{i=1}^{m} \left(\frac{1}{2} \| h_{W,b}(x^{(i)}) - y^{(i)} \|^2 \right) \right] + \frac{\lambda}{2} \sum_{l=1}^{n_1-1} \sum_{i=1}^{s_l} \sum_{j=1}^{s_{l+1}} (W_{ji}^{(l)})^2$$

$$(6.2)$$

The first item $J(W, b)$ in equation (6.2) is a mean square error. The second item is a weight decay. Its aim is to reduce the scope of weight to prevent over-fitting.

$a_j^{(2)}(x)$ indicates the activation value of the input vector x on the hidden unit j. The average activation value of the hidden unit j is

$$\hat{\rho}_j = \frac{1}{m} \sum_{i=1}^{m} [a_j^{(2)}(x^{(i)})]$$

$$(6.3)$$

In order to reach a sparsity, the least (most sparse) hidden units will be used to represent the feature of the input layer. When the average activation value of all hidden units is close to 0, the KL distance will be adopted:

$$\sum_{j=1}^{s_2} \rho \log \frac{\rho}{\hat{\rho}_j} + (1 - \rho) \log \frac{1 - \rho}{1 - \hat{\rho}_j}$$

$$(6.4)$$

For ease of writing,

$$\text{KL}(\rho \| \hat{\rho}_j) = \rho \log \frac{\rho}{\hat{\rho}_j} + (1 - \rho) \log \frac{1 - \rho}{1 - \hat{\rho}_j}$$

$$(6.5)$$

So the entire cost function of neural networks can be represented as

$$J_{\text{sparse}}(W, b) = J(W, b) + \beta \sum_{j=1}^{s_2} \text{KL}(\rho \| \hat{\rho}_j)$$

$$(6.6)$$

The error computation formula $\delta_i^{(2)} (\sum_{j=1}^{s_2} W_{ji}^{(2)} \delta_j^{(3)}) f'(z_i^{(2)})$, in back-propagation is modified as

$$\delta_i^{(2)} = \left(\left(\sum_{j=1}^{s_2} W_{ji}^{(2)} \delta_j^{(3)} \right) + \beta \left(-\frac{\rho}{\hat{\rho}_i} + \frac{1 - \rho}{1 - \hat{\rho}_i} \right) \right) f'(z_i^{(2)})$$

$$(6.7)$$

Therefore, the dimension of data can be reduced greatly. Only a few useful hidden units can represent the original data.

6.4 Restricted Boltzmann Machine

In 2002, Hinton at the University of Toronto proposed a machine learning algorithm called Contrastive Divergence (CD) (Hinton, 2002). CD can efficiently train some Markov random model with a simple architecture including restricted Boltzmann machine (RBM) (Smolensky, 1986). This lays the foundation for the birth of deep learning afterward.

The RBM is a single-layer random neural network (generally the input layer is not included in the layer number of neural networks), as shown in Figure 6.4. RBM is essentially a probability graph model. The input layer is fully connected to the output layer, while there aren't connections between the neurons in the same layer. Each neuron either is activated (the value is 1) or is not activated (the value is 0). The activation probability satisfies the sigmoid function. The advantage of RBM is that give a layer the other layer is independent. So it is convenient to randomly sample a layer when the other layer is fixed. Then this process is carried out alternately. Each update of the weights in theory needs all neurons to be sampled infinitely times. Which is called as contrastive divergence (CD). Since CD works too slowly, Hinton proposed an approximation method called as CD-n algorithm where the weights are updated once after sampling n times (Hinton, 2002).

If RBM has n visual units and m hidden units, the vectors v and h are respectively used to represent the states of the visual layer and the hidden layer, where v_i indicates the state of the ith visual unit and h_j indicates the state of the jth hidden unit. So given a set of states v, h, the system energy of RBM can be defined as

$$E(v, h) = -\sum_{i=1}^{n}\sum_{j=1}^{m} v_i W_{ij} h_j - \sum_{i=1}^{n} v_i b_i - \sum_{j=1}^{m} h_j c_j \qquad (6.8)$$

In equation (6.8), W_{ij}, b_i, c_j are the parameters of RBM. They are real numbers, where W_{ij} indicates the connection intension between the visual unit i and the hidden layer j. b_i indicates the bias of the visual unit i. c_j indicates the bias of the hidden

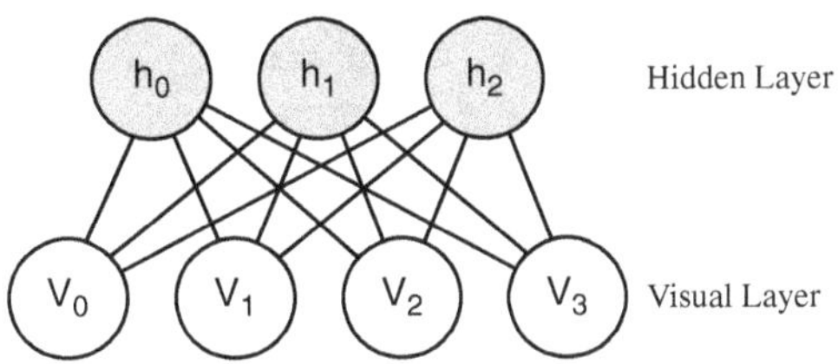

Fig. 6.4.   Architecture of RBM.

unit j. The task of RBM is to solve the value of parameters to fit the given training data.

The states of RBM satisfy the form of the canonical distribution. That is to say, when the parameter is confirmed, the probability of the state v, h is

$$P(v, h) = \frac{1}{Z} \exp(-E(v, h))$$

$$Z = \sum_{v,h} \exp(-E(v, h)) \tag{6.9}$$

Suppose RBM is in the normal temperature $T = 1$, the temperature variable T is omitted. Based on the above definition, when the states for a layer of units are given, the conditional distribution of the states for another layer of units is

$$P(v_i|h) = \sigma \left(\sum_{j=1}^{\cdots} W_{ij} h_j + b_i \right)$$

$$P(h_j|v) = \sigma \left(\sum_{i=1}^{n} v_i W_{ij} + c_j \right) \tag{6.10}$$

In equation (6.10), $\sigma(\cdot)$ is the sigmoid function, and $\sigma(x) = 1/(1+\exp(-x))$. Since the units in the same layer in RBM are not connected to each other, when the states for a layer of units are given, the conditional distribution of the state for another layer of units is independent, namely

$$P(v|h) = \prod_{i=1}^{n} P(v_i|h)$$

$$P(h|v) = \prod_{j=1}^{m} P(h_j|v) \tag{6.11}$$

This means that if we make a Markov chain Monte Carlo (MCMC) sampling on the distribution of RBM (Andrieu *et al.*, 2003), the block Gibbs sampling can be adopted. The block Gibbs sampling starts with an initial state v, h, then $P(.|h)$ and $P(.|v)$ are alternatively used to compute the state transform of all visual units and hidden units. This shows the efficiency advantage of RBM in the sampling.

The unsupervised learning method can be served to RBM through the maximum likelihood principle (Hinton, 2010). Suppose there are training data

$v^{(1)}, v^{(2)}, \ldots, v^{(d)}$, training RBM is equal to maximize the following objective function, where the biases b and c are omitted:

$$L(W) = \sum_{k=1}^{d} \log P(v^{(k)}; W) \equiv \sum_{k=1}^{d} \log \sum_{h} P(v^{(k)}, h; W) \qquad (6.12)$$

In order to use the gradient ascent algorithm in the training process, we need to solve the partial derivatives of the objective function about all parameters. After the algebraic transformation, the final results are

$$\frac{\partial L(W)}{\partial W_{ij}} \propto \frac{1}{d} \sum_{k=1}^{d} v_i^{(k)} P(h_j | v^{(k)}) - \langle v_i h_j \rangle_{P(v, h; W)} \qquad (6.13)$$

Equation (6.13) can be split into two parts: positive phase and negative phase. The positive phase can be computed by equation (6.10) based on the entire training dataset. For the negative phase, $\langle f(x) \rangle_{P(x)}$ is defined as the average value of the function $f(x)$ on the distribution $P(.)$. In RBM, this average value can't be directly represented as a mathematical expression. That is the reason why training RBM is very difficult.

Obviously, the simplest method, MCMC, is used to solve the average value of negative phase. In each MCMC sampling, MCMC needs to make a large number of samples to ensure the samples conform to the objective distribution. So we need to get a large number of samples to accurately approximate the average value. These requirements greatly increase the computational complexity of training RBM. Therefore, MCMC sampling is feasible in theory, but it is not a good choice in efficiency.

Hinton proposed the contrastive divergence (CD) algorithm by considering the state of MCMC starts with the training data (Hinton, 2002). In CD algorithm, the computing process of the average value in the negative phase can be represented as the following. First, each training data are respectively used for the initial states. After a few times of Gibbs sampling on the state transformations, the transformed state is served as the sample to evaluate the average value. Hinton found that just only a time of state transformation can ensure good learning effect in practical application.

The CD algorithm greatly improves the training process of RBM and makes a great contribution to the application of RBM and the rise of deep neural network. The general maximum likelihood learning is similar to minimizing the KL divergence between RBM distribution and training data distribution. In CD algorithm, this principle is inadmissible. Therefore, the essential of CD algorithm is not a maximum likelihood learning (Sutskever, 2010) as the model learned by CD algorithm has a bad model generation capability.

6.5 Deep Belief Networks

In 2006, Hinton proposed a deep belief network (DBN) in his paper (Hinton & Salakhutdinov, 2006), as shown in Figure 6.5. A deep neural network can be viewed as the stack of the multiple RBMs. The training process can be carried out by layer-wise training from the lower layer to the higher layer. The CD algorithm can be served to quickly train RBM. So DBN can avoid the high complexity of directly training neural networks by splitting the entire networks into multiple RBMs. Hinton suggested that the traditional global training algorithm can be used to fine-tune the whole network after the above training way. Then the model will converge to a local optimal point. This training algorithm is similar to the method of initializing the model parameters to a better value by the layer-wise training way and then further training the model through the traditional algorithm. Thus, the problem that the speed of training model is slow can be solved. On the other hand, a lot of experiments

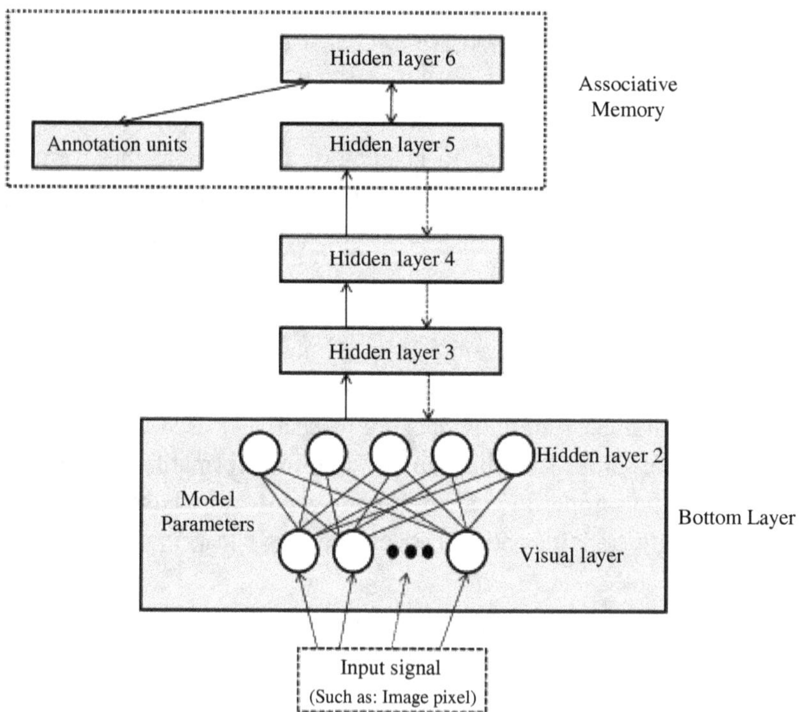

Fig. 6.5. Deep belief networks.

have proven that this method can produce very good parameter initialization values and improve the quality of the final parameters.

In 2008, Tieleman proposed a training algorithm called Persistent Contrastive Divergence (PCD) (Tieleman, 2008), which improves the imperfection of CD algorithm that is unable to maximize the likelihood degree. The efficiency of PCD is the same as CD, and it doesn't destroy the training algorithm of the original objective function (maximum likelihood learning). In Tieleman's paper, a lot of experiments prove that RBM trained by PCD has stronger model generation capability compared with CD. In 2009, Tieleman further improved PCD and proposed a method employing additional parameters to increase the effect of training PCD (Tieleman & Hinton, 2009). He revealed the effect of Markov chain Monte Carlo sampling on training RBM. This lays the foundations for improving the subsequent learning algorithm about RBM. From 2009 to 2010, there existed a series of tempered Markov chain Monte Carlo (Tempered MCMC) RBM training algorithms based on Tieleman's research results.

The left figure in Figure 6.6 is a typical DBN with four layers. It compresses a high-dimensional image single into a representation with 30 dimensions corresponding to the top layer of size 30. We can also use the symmetry architecture of this network to map the 30 dimensions of the compressed signal to the original high-dimensional signal (the middle figure in Figure 6.6). If this network is employed to compress the signal, we can set that the output of the network is equal to the input of the network. Then BP algorithm is used to fine-tune the weights of the network (the right figure in Figure 6.6).

6.6 Convolutional Neural Networks

Convolutional neural network (CNN) is a multiple stage of globally trainable artificial neural network (LeCun *et al.*, 1989). CNN can learn the abstract, essential, and advanced features from the data with a little preprocessing or even the original data. CNN has been widely applied in license plate detection, face detection, handwriting recognition, target tracking, and other fields.

CNN has a better performance in two-dimensional pattern recognition problems than the multi-layer perceptron because the topology of the two-dimensional model is added into the CNN structure, and CNN employs three important structure features: local accepted field, shared weights, sub-sampling ensuring the invariance of the target translation, shrinkage, and distortion for the input signal. CNN mainly consists of the feature extraction and the classifier. The feature extraction contains multiple convolutional layers and sub-sampling layers. The classifier consists of

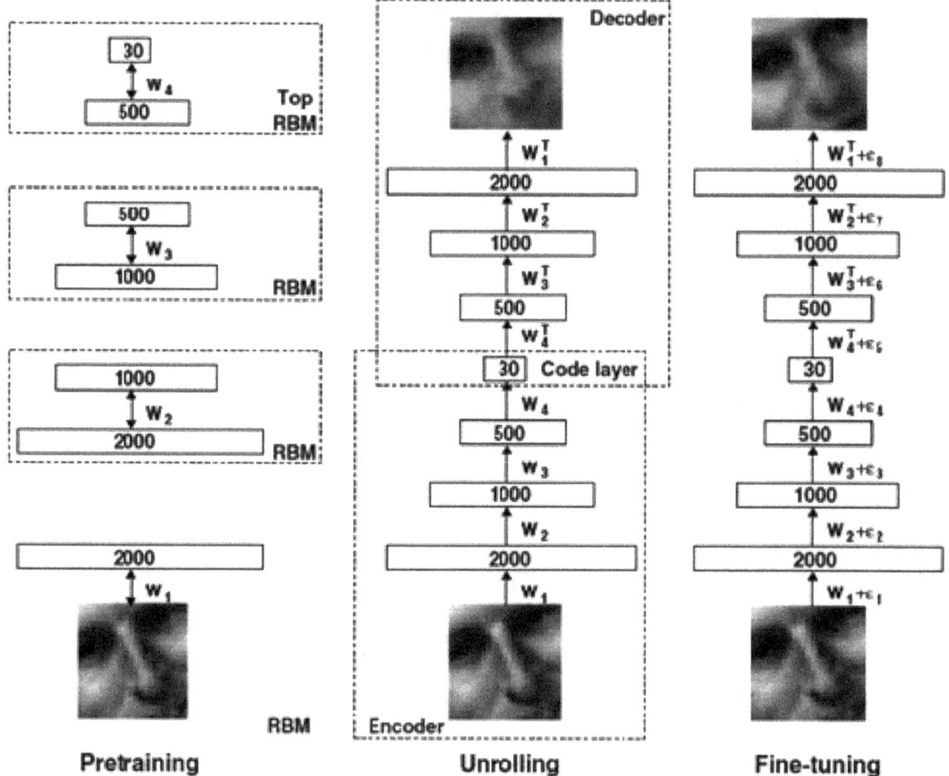

Fig. 6.6.   An example of DBN.

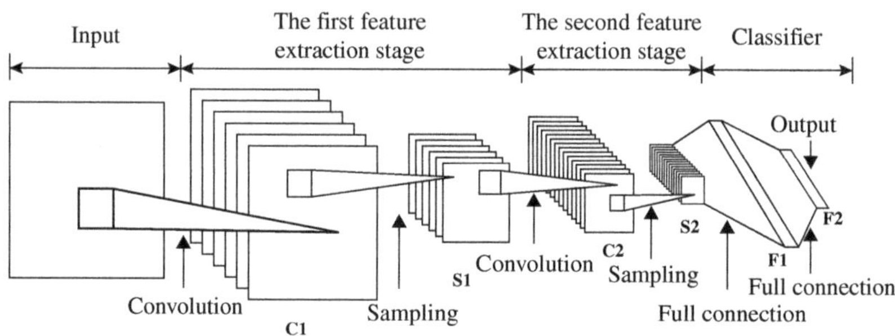

Fig. 6.7.   CNN architecture of recognizing handwriting digits.

one layer or two layers of fully connected neural networks. For the convolutional layer with the local accepted field and the sub-sampling layer with sub-sampling structure, they all have the character of sharing the weights. Figure 6.7 shows CNN served to recognize the handwriting digits.

In Figure 6.7, CNN has seven layers: one input layer, two convolutional layers, two sub-sampling layers, and two fully connected layers. Each input sample in the input layer has $32 \times 32 = 1{,}024$ pixels. C_1 is a convolutional layer with six feature maps, and each feature map contains $28 \times 28 = 784$ neurons. Each neuron in C_1 is connected to the corresponding local accepted field of size 5×5 in the input layer through the convolutional core of size 5×5 and the convolutional step of length 1. So C_1 has the $6 \times 784 \times (5 \times 5 + 1) = 122{,}304$ connections. Each feature map contains the 5×5 weights and biases. Therefore, C_1 has the $6 \times (5 \times 5 + 1) = 156$ trainable parameters.

S_1 is the sub-sampling layer containing six feature maps, and each feature map consists of $14 \times 14 = 196$ neurons. There is a one-to-one correspondence between the feature maps in S_1 and the feature maps in C_1. The window of sub-sampling S_1 is a matrix of size 2×2, and the step size of the sub-sampling defaults to 1. So there are $6 \times 196 \times (2 \times 2 + 1) = 5{,}880$ connections in S_2 layer. Each feature map in S_1 contains a weight and a bias. This means there are 12 trainable parameters.

C_2 is the convolutional layer containing 16 feature maps, and each feature map consists of $10 \times 10 = 100$ neurons. Each neuron in C_2 is connected to the local accepted field of size 5×5 in the k feature maps in S_1 through the k convolutional core of size 5×5, where $k = 6$ if the full connection way is adopted. So there are $41{,}600$ connections in C_2 layer. Each feature map in C_2 contains $6 \times 5 \times 5 = 150$ weights and biases. Which means there are $16 \times (150 + 1) = 2{,}416$ trainable parameters.

S_2 is the sub-sampling layer containing 16 feature maps, and each feature map consists of 5×5 neurons. S_2 contains 400 neurons. There is a one-to-one correspondence between the feature maps in S_2 and the feature maps in C_2. The window of sub-sampling S_2 is a matrix of size 2×2. So there are $16 \times 25 \times (2 \times 2 + 1) = 2{,}000$ connections in S_2 layer. Each feature map in S2 contains a weight and a bias. This means there are 32 trainable parameters in S_2 layer.

F_1 is a full connection layer containing 120 neurons. Each neuron in F_1 is connected to the 400 neurons in S_2. So the number of connections or trainable parameters is $120 \times (400 + 1) = 48{,}120$. F2 is a full connection layer and the output layer, which contains 10 neurons, 1210 connections, and 1210 trainable parameters.

In Figure 6.7, the number of feature maps in the convolutional layer increases layer by layer. This can supplement the loss caused by sampling. On the other hand, the convolutional cores make the convolutional operation on the feature maps of the former layer to produce the current convolutional feature maps. The produced different features expand the feature space and make the feature extraction more comprehensive.

The error back-propagation (BP) algorithm is mostly used to train CNN in a supervised way. In the gradient descent method, the error is back-propagated to constantly adjust the weights and biases. This allows the global sum of squared error on the training samples to reach a minimum value. BP algorithm consists of four processes: the network initialization, the information flow feed-forward propagation, the error back-propagation, the update of the weights, and the biases. In error back-propagation, we need to compute the local variation value of the gradient on the weights and the biases.

In the initial process of training CNN, the neurons in each layer need to be initialized randomly. The initialization of the weights has a great influence on the convergence rate of the network. So the process of how to initialize the weights is very important. There exists a strong relationship between the initialization of the weights and the selection of the activation function. The weights should be assigned by the fastest changing parts of the activation function. The big weights or the small weights in initialization process will lead to a small change in the weights.

In the feed-forward propagation of the information flow, the convolutional layer extracts the primary basic features of the input to make up several feature maps. Then the sub-sampling layer reduces the resolution of the feature maps. After convolution layer and sub-sampling alternatively complete the feature extraction, the network obtains the high-order invariance features of the input. These high-order invariance features will be transmitted to the fully connected neural network and used to classify. After the information transformation and computing in the hidden layer and output layer of the fully connected neural network, which means the network finishes the forward propagation process of the learning. The last results are outputted by the output layer.

The network starts the error back-propagation process when the actual output is not equal to the expected output. The errors are propagated from the output layer to the hidden layer and then propagated from the hidden layer to the sub-sampling layer and convolutional layer in the feature extraction stage. The neurons in each layer start to compute the locally changed value of the weights and biases when they get their output errors. At last, the network will enter the weights update process.

1. Feed-forward propagation of convolutional layer

Each neuron in the convolutional layer extracts the features in the local accepted field of the same location of all feature maps in the former layer. The neurons in the same feature map share the same weight matrix. The convolutional process can be viewed as the convolutional neurons seamlessly scan the former feature maps line by line through the weight matrix. The output, $\mathbf{O}_{(x,y)}^{(l,k)}$, of the neuron located in line x column y in the kth feature map of the lth convolutional layer can be computed by equation (6.14), where $\tanh(\cdot)$ is the activation function:

$$\mathbf{O}_{(x,y)}^{(l,k)} = \tanh\left(\sum_{t=0}^{f-1}\sum_{r=0}^{kh}\sum_{c=0}^{kw}\mathbf{W}_{(r,c)}^{(k,t)}\mathbf{O}_{(x+r,y+c)}^{(l-1,t)} + \mathbf{Bias}^{(l,k)}\right) \qquad (6.14)$$

From equation (6.14), we need to traverse all neurons of the convolutional window in all feature maps of the former layer to compute the output of a neuron in the convolutional layer. The feed-forward propagation of the full connection layer is similar to the convolutional layer. This can be viewed as a convolutional operation on the convolutional weight matrix and the input with the same size.

2. Feed-forward propagation of sub-sampling

There are the same number of feature maps and a one-to-one correspondence relationship between the sub-sampling feature maps and the convolutional feature maps. Each neuron in the sub-sampling layer is connected to the sub-areas with the same size but non-overlapping through the sub-sampling window. The output, $\mathbf{O}_{(x,y)}^{(l,k)}$, of the neuron located in line x column y in the kth feature map of the lth sub-sampling layer can be computed by the following equation:

$$\mathbf{O}_{(x,y)}^{(l,k)} = \tanh\left(\mathbf{W}^{(k)}\sum_{r=0}^{sh}\sum_{c=0}^{sw}\mathbf{O}_{(x\mathrm{sh}+r,y\mathrm{sw}+c)}^{(l-1,k)} + \mathbf{Bias}^{(l,k)}\right) \qquad (6.15)$$

3. Error back-propagation of sub-sampling layer

Error back-propagation starts from the output layer and enters the sub-sampling layer through the hidden layer. The error back-propagation of the output layer needs to first compute the partial derivative of the error about the neurons in the output layer. Suppose the output of the kth neuron is o_k for the training sample d. The expected output of the kth neuron is t_k for the sample d. The error of the output layer for the sample d can be represented as $E = \frac{1}{2}\sum_k(o_k - t_k)^2$. The partial derivative of the error E about the output o_k is $\partial E/\partial o_k = o_k - t_k$. Similarly, we can solve the partial derivatives of the error about all neurons in the output layer. Then we solve the partial derivatives of the error about the input in the output layer. Set $d(o_k)$ is the partial derivative of the error about the input of the kth neuron in the output layer. $d(o_k)$ can be computed by equation (6.16), where $(1 + o_k)(1 - o_k)$ is the partial derivative of the activation function $\tanh(\cdot)$ about the input of the neuron. Then we start to compute the partial derivatives of the error about all neurons in the hidden layer. Suppose the neuron j in the hidden layer is connected to the output neuron through the weight w_{kj}. The partial derivative $d(o_j)$ of the error about the output of the neuron j can be computed by equation (6.16). When we obtain the partial derivative of the error about the output of the hidden layer, the error is back-propagated to the hidden layer. Through a similar process, the error will be back-propagated to the sub-sampling layer. For the convenience of expression, the partial derivatives of the error about the outputs of the neuron are called the output error of the neuron. The partial derivatives of the error about the inputs of the neuron are called the input error of the neuron:

$$d(o_k) = (o_k - t_k)(1 + o_k)(1 - o_k) \tag{6.16}$$

$$d(o_j) = \sum d(o_k)w_{kj} \tag{6.17}$$

There are the same number of feature maps and a one-to-one correspondence relationship between the sub-sampling feature maps and the convolutional feature maps. So it is an intuitive process that the error is propagated from the sub-sampling layer to the convolutional layer. Equation (6.16) is used to compute the input errors of all neurons in the sub-sampling layer. Then the above input errors are propagated to the former layer of the sub-sampling layer. The sub-sampling layer is set to the lth layer, and then the output error of the neuron located in line x column y in the kth feature map of the $(l - 1)$th layer is computed by the following equation:

$$d\big(\mathbf{O}_{(x,y)}^{(l-1,k)}\big) = d\big(\mathbf{O}_{(\lfloor x/sh \rfloor, \lfloor y/sw \rfloor)}^{(l-k)}\big)\mathbf{W}^{(k)} \tag{6.18}$$

All neurons in a feature map of the sub-sampling layer share a weight and a bias. So the local gradient variations of all weights and biases are related to all neurons in the sub-sampling layer. The weight variation $\Delta\mathbf{W}^{(k)}$ and the bias variation $\Delta\mathbf{Bias}^{(l,k)}$ for the kth feature map in the lth sub-sampling layer can be computed by equation (6.19) and equation (6.20). In equation (6.20), f_h and f_w respectively indicate the height and the width of the feature map in the lth sub-sampling layer:

$$\Delta\mathbf{W}^{(k)} = \sum_{x=0}^{fh}\sum_{y=0}^{fw}\sum_{r=0}^{sh}\sum_{c=0}^{sw} \mathbf{O}_{(x,y)}^{(l-1,k)}d\big(\mathbf{O}_{(\lfloor x/sh \rfloor, \lfloor y/sw \rfloor)}^{(l-k)}\big) \tag{6.19}$$

$$\Delta\mathbf{Bias}^{(l,k)} = \sum_{x=0}^{fh}\sum_{y=0}^{fw} d\big(\mathbf{O}_{(x,y)}^{(l,k)}\big) \tag{6.20}$$

4. Error back-propagation of convolutional layer

There exist two error back-propagation ways in the convolutional layer: 'push' and 'pull'. The 'push' way (the left figure in Figure 6.8) can be considered as the neurons in the lth layer actively propagating the errors to the neurons in the $(l–1)$th layer. This is suitable for the serial implementation. There is the 'write conflict' problem in the parallel implementation. The 'pull' way (the right figure in Figure 6.8) can be considered as the neurons in the $(l–1)$th layer actively obtain the errors from

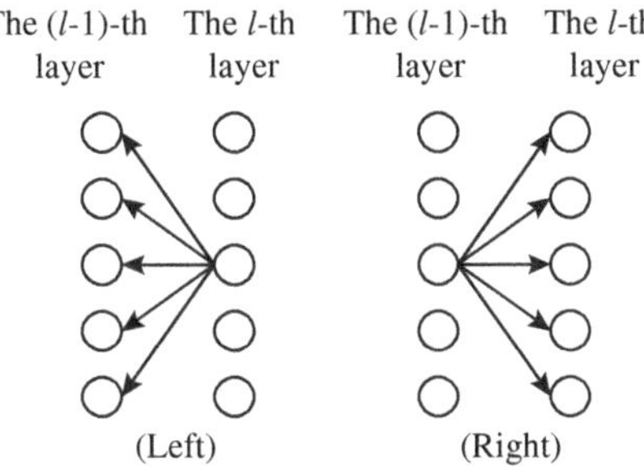

Fig. 6.8.   Two ways of error back-propagation.

the neurons in the *l*th layer. It is difficult to implement the 'pull' way. Due to the border effect of the convolutional operation, we need to confirm which neurons in the feature map of the former layer are connected to the neurons in the current layer.

Here the 'pull' way is used to describe the error back-propagation process of the convolutional layer. Equation (6.16) is employed to compute the input error of each neuron in the convolutional layer. Then the input error is propagated to the neurons in the former layer of the convolutional layer, like the following equation:

$$d\big(O_{(x,y)}^{(l-1,k)}\big) = \sum_{i=0}^{m-1} \sum_{(p,q)\in A} d\big(O_{(p,q)}^{(l,t)}\big)w \tag{6.21}$$

where A is the coordinate set of the neuron located in line x column y in the kth feature map of the *l*th layer and the neurons in the *l*th layer. w indicates the connected weight between the two neurons. The serial computational process that the error in the convolutional layer is back-propagated to the former sub-sampling layer is described as follows:

   for each neuron i in the former layer of the convolutional layer

      Confirm which neurons in the convolutional layer are connected to the neuron i

      'Pull' the error from the relevant neurons in the convolutional layer using equation (6.21)

end for

The local gradient variation of the line r and the column y of the weight matrix describing the connection between the kth feature map in the convolutional layer and the tth feature map in the former layer can be computed by the following equation:

$$\Delta W_{(r,c)}^{(k,t)} = \sum_{x=r}^{fh-kh+r} \sum_{y=c}^{fw-kw+c} d\big(O_{(x,y)}^{(l,k)}\big)O_{(r+x,c+y)}^{(l-1,t)} \tag{6.22}$$

All neurons in a feature map of the convolutional layer share a bias. The computational method is the same as the sub-sampling layer.

With the complication of solving the problem and the higher performance requirement for CNN, the training data need to be more complete and larger. So we need to train the network with stronger capability. Which means the network will have more trainable parameters. The famous ImageNet dataset contains 14,197,122 labeled images (ImageNet, 2014). The CNN containing 650,000 neurons in the library (Krizhevsky *et al.*, 2012) is used to classify the ImageNet dataset. There are 60,000,000 trainable neurons in the above CNN. A large of training data and training parameters will significantly increase the computational time, especially the serial computation will cost a few months of computational time (Ciresan *et al.*, 2010). Therefore, the researchers had started to research the parallel CNN. There exist at least five parallel ways, namely the parallelism of the samples, the parallelism between the former back layers, the parallelism between the feature maps in the same layer, the parallelism between the neurons in the feature map, and the parallelism of the neuron weights (Schmidhuber, 2014).

6.7 Recurrent Neural Networks

A recurrent neural network (RNN) is a class of artificial neural network where connections between nodes form a directed graph along a sequence. This allows it to exhibit temporal dynamic behavior for a time sequence. Unlike feed-forward neural networks, RNNs can use their internal state (memory) to process sequences of inputs. This makes them applicable to tasks, such as unsegmented, connected handwriting recognition, or speech recognition. A typical RNN is shown in Figure 6.9.

RNNs contain input units, and the input set is labeled $\{x_0, x_1, \ldots, x_t, x_{t+1}, \ldots\}$ and the output set of the output units is marked as $\{y_0, y_1, \ldots, y_t, y_{t+1}, \ldots\}$. RNNs also contain hidden units, and we mark their output sets as $\{s_0, s_1, \ldots, s_t, s_{t+1}, \ldots\}$; these hidden units have done the most important work. You will find that in Figure 6.9: there is a one-way flow of information from the input unit to the hidden

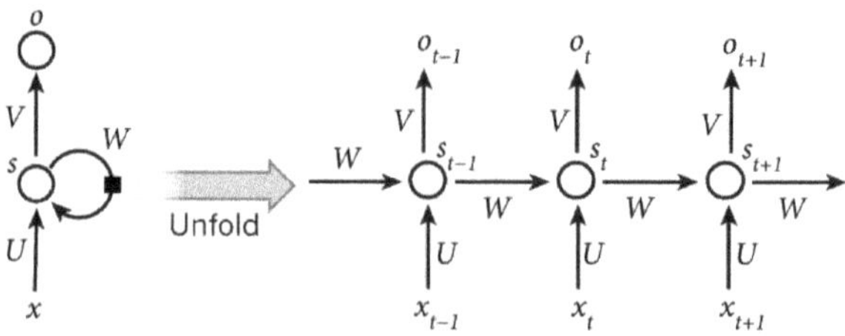

Fig. 6.9. A typical RNN structure.

unit, while another one-way flow of information from the hidden unit to the output unit. In some cases, RNNs break the latter's limitations, and the boot information is returned to the hidden unit from the output unit. These are called "Back Projections", and the input to the hidden layer also includes the state of the last hidden layer, that is, within the hidden layer. Nodes can be self-connected or interconnected.

For example, for a statement containing five words, the expanded network is a five-layer neural network, with each layer representing a word. The calculation process for this network is as follows:

(1) x_t is the input at time step t. For example, x_1 could be a one-hot vector corresponding to the second word of a sentence.
(2) s_t is the hidden state at time step t. It's the "memory" of the network. s_t is calculated based on the previous hidden state and the input at the current step: $s_t = f(Ux_t + Ws_{t-1})$. The function f usually is a nonlinearity, such as tanh (hyperbolic tangent) or ReLU (rectified linear unit). s_{-1}, which is required to calculate the first hidden state, is typically initialized to all zeroes.
(3) o_t is the output at step t. For example, if we wanted to predict the next word in a sentence, it would be a vector of probabilities across our vocabulary: $o_t = \text{softmax}(Vs_t)$.

We can think of the hidden state s_t as the memory of the network. s_t captures information about what happened in all the previous time steps. The output at step o_t is calculated solely based on the memory at time t. As briefly mentioned above, it is a bit more complicated in practice because s_t typically can't capture information from too many time steps back.

Unlike a traditional deep neural network, which uses different parameters at each layer, an RNN shares the same parameters (U, V, W above) across all steps. This reflects the fact that we are performing the same task at each step, just with different inputs. This greatly reduces the total number of parameters we need to learn. Figure 6.9 has outputs at each time step, but depending on the task, this may not be necessary. For example, when predicting the sentiment of a sentence, we may only care about the final output, not the sentiment after each word. Similarly, we may not need inputs at each time step. The main feature of an RNN is its hidden state, which captures some information about a sequence.

6.8 Long Short-Term Memory

RNNs have shown great success in many NLP tasks. At this point, we should mention that the most commonly used type of RNNs are LSTMs, which are much better at capturing long-term dependencies than vanilla RNNs are. LSTMs are essentially the same thing as the RNN; they just have a different way of computing the hidden state.

Long short-term memory (LSTM) is a deep learning system that avoids the vanishing gradient problem. LSTM is normally augmented by recurrent gates called "forget" gates (Gers *et al.*, 2017). LSTM prevents back-propagated errors from vanishing or exploding (Hochreiter & Schmidhuber, 1997). Instead, errors can flow backward through unlimited numbers of virtual layers unfolded in space. That is, LSTM can learn tasks (Schmidhuber, 2015) that require memories of events that happened thousands or even millions of discrete time steps earlier. Problem-specific LSTM-like topologies can be evolved (Bayer *et al.*, 2009). LSTM works even given long delays between significant events and can handle signals that mix low and high-frequency components.

LSTMs are explicitly designed to avoid the long-term dependency problem. Remembering information for long periods of time is practically their default behavior, not something they struggle to learn! All recurrent neural networks have the form of a chain of repeating modules of neural networks. In standard RNNs, this repeating module will have a very simple structure, such as a single tanh layer. Figure 6.10 illustrates the repeating module in a standard RNN containing a single layer.

LSTMs also have this chain-like structure, but the repeating module has a different structure. Instead of having a single neural network layer, there are four, interacting in a very special way, as shown in Figure 6.11. The key to LSTMs is the cell state, the horizontal line running through the top of the diagram. The cell state is kind of like a conveyor belt. It runs straight down the entire chain, with only some minor linear interactions. It's very easy for information to just flow along it unchanged. The LSTM does have the ability to remove or add information to the cell state, carefully regulated by structures called gates. Gates are a way to optionally let information through. They are composed of a sigmoid neural net layer and a point-wise multiplication operation. The sigmoid layer outputs numbers between zero and

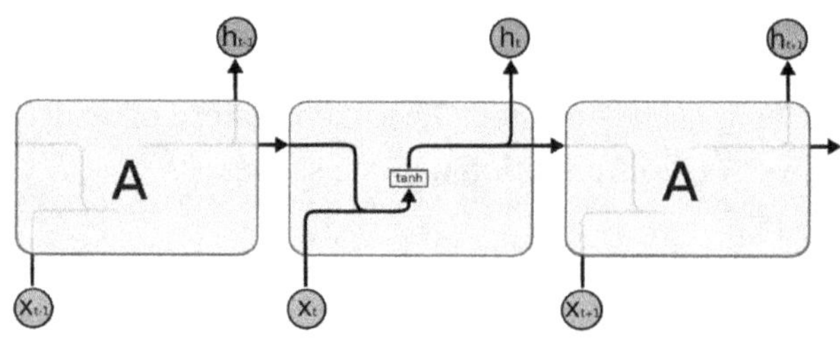

Fig. 6.10.   Standard RNN contains a single layer.

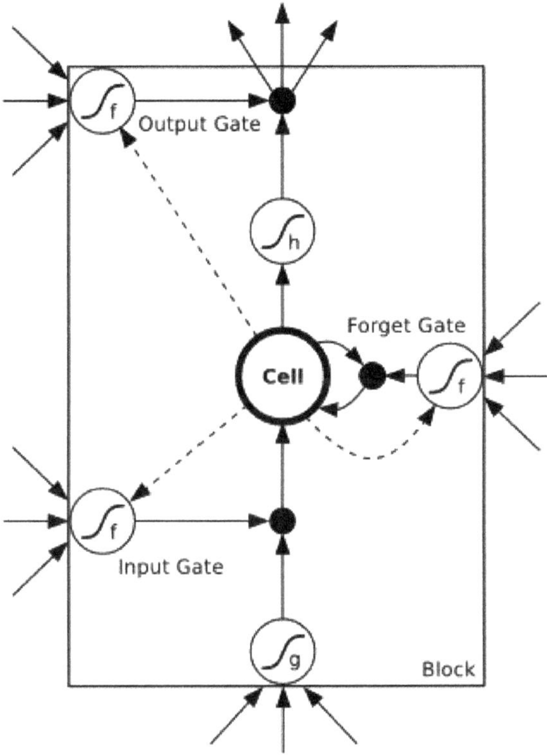

Fig. 6.11.   LSTM block with one cell.

one, describing how much of each component should be let through. A value of zero means "let nothing through," while a value of one means "let everything through!" An LSTM has three of these gates, to protect and control the cell state.

Figure 6.11 shows LSTM block with one cell. The three gates are nonlinear summation units that collect activations from inside and outside the block and control the activation of the cell via multiplications (small black circles). The input and output gates multiply the input and output of the call, while the forget gate multiplies the cell's previous state. No activation function is applied within the cell. The gate activation function 'f' is usually the logistic sigmoid so that the gate activation functions are between 0 (gate closed) and 1 (gate open) The cell input and output activation functions ('g' and 'h') are usually tanh or logistic sigmoid, though in some cases 'h' is the identity function. The weighted 'peephole' connections from the cell to the gates are shown with dashed lines. All other connections within the block are unweighted. The only output from the block to the rest of the network emanates from the output gate multiplication.

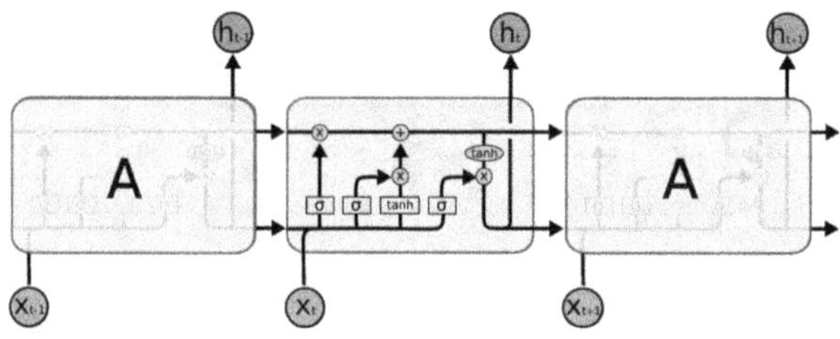

Fig. 6.12. The repeating module in an LSTM.

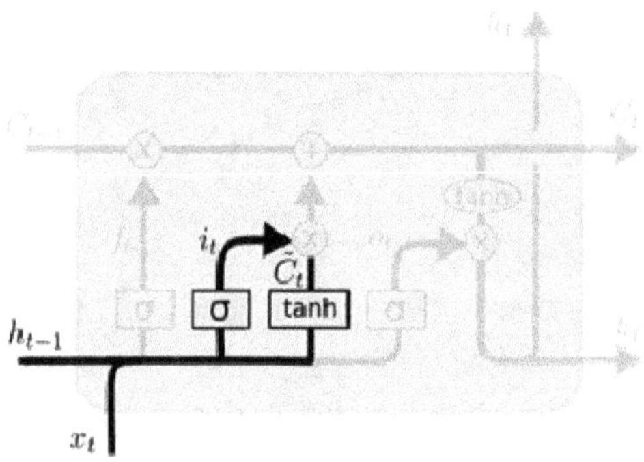

Fig. 6.13. Update a cell state.

Figure 6.12 shows the repeating module in an LSTM. In order to illustrate the principle of LSTM, we have provided a step-by-step LSTM walk-through. The first step in the LSTM is to decide what information we're going to throw away from the cell state. This decision is made by a sigmoid layer called the "forget gate layer." It looks at h_{t-1} and x_t and outputs a number between 0 and 1 for each number in the cell state C_{t-1}. A "1" represents "completely keep this" while a "0" represents "completely get rid of this."

The next step is to decide what new information we're going to store in the cell state. This has two parts. First, a sigmoid layer called the "input gate layer" decides which values we update. Next, a tanh layer creates a vector of new candidate values, $C_{\sim t}$, that could be added to the state. In the next step, we combine these two to create an update to the state, as shown in Figure 6.13:

$$i_t = \sigma\left(W_i \cdot [h_{t-1}, x_t] + b_i\right) \tag{6.23}$$

$$\bar{C}_t = \tanh(W_C \cdot [h_{t-1}, x_t] + b_C) \tag{6.24}$$

In the example of the language model, we'd want to add the gender of the new subject to the cell state, to replace the old one we're forgetting. It's now time to update the old cell state, C_{t-1}, into the new cell state C_t. The previous steps already decided what to do; we just need to actually do it. We multiply the old state by f_t, forgetting the things we decided to forget earlier. Then we add $i_t * C_{\sim t}$. This is the new candidate value, scaled by how much we decided to update each state value. In the case of the language model, this is where we'd actually drop the information about the old subject's gender and add the new information, as we decided in the previous steps:

$$C_t = f_t * C_{t-1} + i_t * \tilde{C}_t \tag{6.25}$$

Finally, we need to decide what we're going to output. This output will be based on our cell state but will be a filtered version. First, we run a sigmoid layer which decides what parts of the cell state we're going to output. Then, we put the cell state through tanh (to push the values to be between -1 and 1) and multiply it by the output of the sigmoid gate so that we only output the parts we decided to (Figure 6.14).

For the language model example, since it just saw a subject, it might want to output information relevant to a verb, in case that's what is coming next. For example, it might output whether the subject is singular or plural so that we know what form a verb should be conjugated into if that's what follows next.

Many applications use stacks of LSTM RNNs (Fernández *et al.*, 2007) and train them by Connectionist Temporal Classification (CTC) (Graves *et al.*, 2006) to find an RNN weight matrix that maximizes the probability of the label sequences in a training set, given the corresponding input sequences. CTC achieves both alignment and recognition. LSTM can learn to recognize context-sensitive languages unlike previous models based on hidden Markov models (HMM) and similar concepts (Gers & Schmidhuber, 2001).

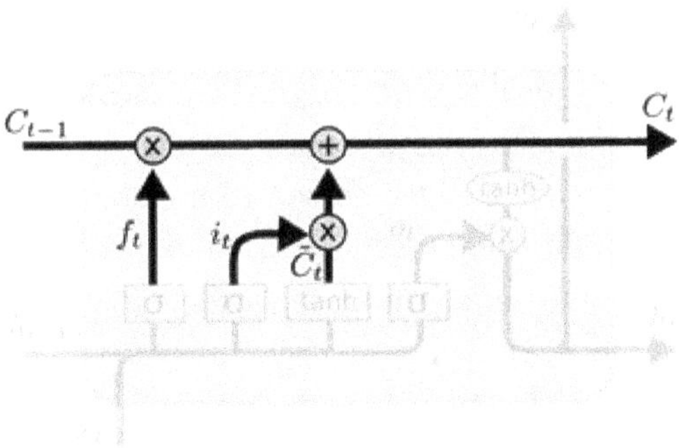

Fig. 6.14.   Generate the output.

6.9 Neural Machine Translation

6.9.1 *Introduction*

The emergence of deep learning has dramatically changed machine translation. Since 2013, neural network-based machine translation has taken translation to a new level in speed and accuracy. At the moment, with fierce competition between technology giants and the study of academic circles, the level of machine translation is still evolving. In recent years, the biggest influence of deep learning on translation comes from neural machine translation (NMT), which greatly improves the accuracy of machine translation.

A decade ago, Google launched machine translation using phrase-based machine translation (PBMT). A few years ago, the Google brain team began using a circular neural network (RNN) to directly learn the mapping between input sequences and output sequences. Phrase-based machine translation (PBMT) translates sentences into words and translates them separately, while Neural Machine Translation (NMT) translates the input as a whole. The advantage of this is that the adjustments that need to be made during translation are much less.

When neural network machine translation technology first appeared, it achieved comparable results with PBMT on medium-sized public datasets. Since then, people engaged in machine translation studies have proposed a number of ways to improve NMT, including using attention to align input and output, split words into smaller units, or mimic external alignment models to deal with uncommon words. Despite this, the performance of NMT is still not enough to become a large-scale deployment of products.

It is not only Google that sees the great value of machine translation. Baidu, Huawei, Ali, and Tencent in China have research, and giants such as Facebook and Microsoft are not falling behind. This competitive situation will maximize the commercial deployment of machine translation and become more "available" to more people. Many novel techniques have been proposed to further improve NMT: using an attention mechanism to deal with rare words, a mechanism to model translation coverage, multi-task, and semi-supervised training to incorporate more data, a character decoder, a character encoder, sub-word units also to deal with rare word outputs, different kinds of attention mechanisms, and sentence-level loss minimization. While the translation accuracy of these systems has been encouraging. Let's take Google GNMT as an example to introduce the working principle of the neural translation machine (Wu *et al.*, 2016).

6.9.2 *Model Architecture*

GNMT follows the common sequence-to-sequence learning framework (Sutskever *et al.*, 2014) with attention (Bahdanau *et al.*, 2015). It has three components: an encoder network, a decoder network, and an attention network. The encoder transforms a source sentence into a list of vectors, one vector per input symbol. Given this list of vectors, the decoder produces one symbol at a time, until the special end-of-sentence symbol (EOS) is produced. The encoder and decoder are connected through an attention module which allows the decoder to focus on different regions of the source sentence during the course of decoding.

Let (X, Y) be a source and target sentence pair. Let $X = x_1, x_2, x_3, \ldots, x_M$ be the sequence of M symbols in the source sentence and let $Y = y_1, y_2, y_3, \ldots, y_N$ be the sequence of N symbols in the target sentence. The encoder is simply a function of the following form:

$$\mathbf{x_1}, \mathbf{x_2}, \ldots, \mathbf{x_M} = Encoder\,RNN(x_1, x_2, x_3, \ldots, x_M) \tag{6.26}$$

In equation (6.26), $\mathbf{x_1}, \mathbf{x_2}, \ldots, \mathbf{x_M}$ is a list of fixed-size vectors. The number of members in the list is the same as the number of symbols in the source sentence (M in this example). Using the chain rule, the conditional probability of the sequence $P(Y|X)$ can be decomposed as

$$P(Y|X) = P(Y|\mathbf{x_1}, \mathbf{x_2}, \mathbf{x_3}, \ldots, \mathbf{x_M})$$

$$= \prod_{i=1}^{N} P(y_i|y_0, y_1, y_2, \ldots, y_{i-1}, ; \mathbf{x_1}, \mathbf{x_2}, \mathbf{x_3}, \ldots, \mathbf{x_M}) \tag{6.27}$$

where y_0 is a special "beginning of sentence" symbol that is prepended to every target sentence.

During inference, we calculate the probability of the next symbol given the source sentence encoding and the decoded target sequence so far:

$$P(y_i|y_0, y_1, y_2, y_3, \ldots, y_{i-1}; \mathbf{x_1}, \mathbf{x_2}, \ldots, \mathbf{x_M}) \tag{6.28}$$

The decoder is implemented as a combination of an RNN network and a softmax layer. The decoder RNN network produces a hidden state $\mathbf{y_i}$ for the next symbol to be predicted, which then goes through the softmax layer to generate a probability distribution over candidate output symbols. A deep stacked long short-term memory network is used for both the encoder RNN and the decoder RNN.

The attention module is similar to Bahdanau *et al.* (2015). More specifically, let y_{i-1} be the decoder-RNN output from the past decoding time step (or use the output from the bottom decoder layer). Attention context $\boldsymbol{a_i}$ for the current time step is computed according to the following formulas:

$$s_t = attentionFunction(y_{i-1}, \mathbf{x}_t) \quad \forall t, \quad 1 \le t \le M$$

$$p_t = \exp(s_t / \sum_{t=1}^{M} \exp(s_t) \quad \forall t, \quad 1 \le t \le M$$

$$a_i = \sum_{t=1}^{M} p_t \cdot \mathbf{x}_t \tag{6.29}$$

where AttentionFunction in GNMT implementation is a feed-forward network with one hidden layer.

In Figure 6.15, on the left is the encoder network, on the right is the decoder network, and in the middle is the attention module. The bottom encoder layer is bi-directional: the pink nodes gather information from left to right while the green nodes gather information from right to left. The other layers of the encoder are uni-directional. Residual connections start from the layer third from the bottom in the encoder and decoder. The model is partitioned into multiple GPUs to speed up training. In fact, we have eight encoder LSTM layers (one bi-directional layer and seven uni-directional layers) and eight decoder layers. With this setting, one model replica is partitioned eight ways and is placed on eight different GPUs typically

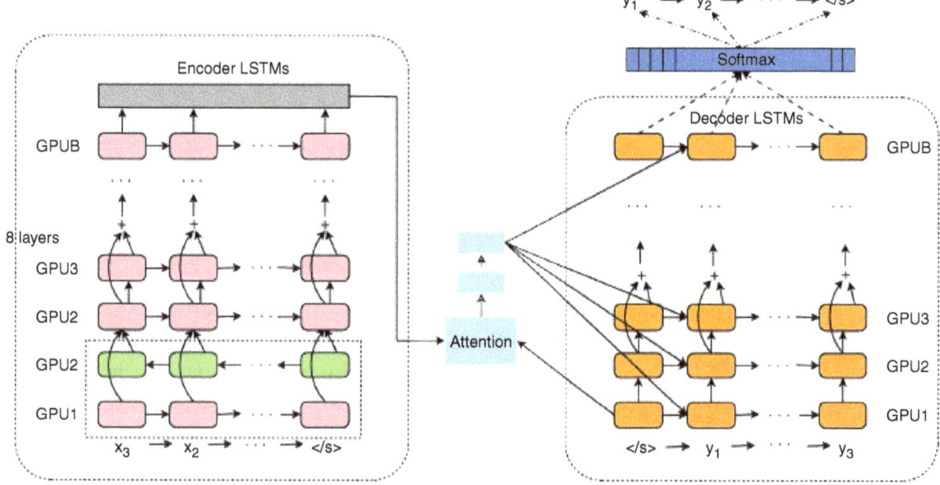

Fig. 6.15.   The model architecture of GNMT.

belonging to one host machine. During training, the bottom bi-directional encoder layers compute in parallel first. Once both are finished, the uni-directional encoder layers can start computing, each on a separate GPU. To retain as much parallelism as possible while running the decoder layers, we use the bottom decoder layer output only for obtaining recurrent attention context, which is sent directly to all the remaining decoder layers. The softmax layer is also partitioned and placed on multiple GPUs. Depending on the output vocabulary size, we either have them run on the same GPUs as the encoder and decoder networks or have them run on a separate set of dedicated GPUs.

6.9.3 *Quantized Inference*

Quantized inference using reduced precision arithmetic is one technique that can significantly reduce the cost of inference for these models, often providing efficiency improvements on the same computational devices. To reduce quantization errors, additional constraints are added to the model during training so that it is quantizable with minimal impact on the output of the model.

GNMT introduced residual connections among the LSTM layers in a stack. More concretely, let $\text{LSTM}i$ and $\text{LSTM}i+1$ be the ith and $(i+1)$th LSTM layers in a stack, whose parameters are W^i and W^{i+1}, respectively. At the tth time step, for the stacked LSTM without residual connections, we have

$$\mathbf{c}_t^i, \mathbf{m}_t^i = \text{LSTM}_i(\mathbf{c}_{t-1}^i, \mathbf{m}_{t-1}^i, \mathbf{x}_t^{i-1}; \mathbf{W}^i)$$

$$\mathbf{x}_t^i = \mathbf{m}_t^i$$

$$\mathbf{c}_t^{i+1}, \mathbf{m}_t^{i+1} = \text{LSTM}_{i+1}(\mathbf{c}_{t-1}^{i+1}, \mathbf{m}_{t-1}^{i+1}, \mathbf{x}_t^i; \mathbf{W}^{i+1}) \tag{6.30}$$

where $\mathbf{x}_t^i$ is the input to LSTM_i at time step t, and m_t^i and c_t^i are the hidden states and memory states of LSTM_i at time step t, respectively. With residual connections between LSTM_i and LSTM_{i+1}, the above equations become

$$\mathbf{c}_t^i, \mathbf{m}_t^i = \text{LSTM}_i(\mathbf{c}_{t-1}^i, \mathbf{m}_{t-1}^i, \mathbf{x}_t^{i-1}; \mathbf{W}^i)$$

$$\mathbf{x}_t^i = \mathbf{m}_t^i + \mathbf{x}_t^{i-1}$$

$$\mathbf{c}_t^{i+1}, \mathbf{m}_t^{i+1} = \text{LSTM}_{i+1}(\mathbf{c}_{t-1}^{i+1}, \mathbf{m}_{t-1}^{i+1}, \mathbf{x}_t^i; \mathbf{W}^{i+1}) \tag{6.31}$$

Residual connections greatly improve the gradient flow in the backward pass, which allows us to train very deep encoder and decoder networks.

In an LSTM stack with residual connections, there are two accumulators: c_t^i along the time axis and $\mathbf{x}_t^i$ along the depth axis. In theory, both of the accumulators are unbounded, but in practice, we noticed their values remain quite small. For quantized inference, we explicitly constrain the values of these accumulators to be within $[-\delta,\delta]$ to guarantee a certain range that can be used for quantization later. The forward computation of an LSTM stack with residual connections is modified as follows:

$$\mathbf{c}_t''^i, \mathbf{m}_t^i = \text{LSTM}_i(\mathbf{c}_{t-1}^i, \mathbf{m}_{t-1}^i, \mathbf{x}_t^{i-1}; \mathbf{W}^i)$$

$$\mathbf{c}_t^i = \max(-\delta, \min(\delta, \mathbf{c}_t''^i))$$

$$\mathbf{x}_t''^i = \mathbf{m}_t^i + \mathbf{x}_t^{i-1}$$

$$\mathbf{x}_t^i = \max(-\delta, \min(\delta, \mathbf{x}_t''^i))$$

$$\mathbf{c}_{c_t}''^{i+1}, \mathbf{m}_t^{i+1} = \text{LSTM}_{i+1}(\mathbf{c}_{t-1}^{i+1}, \mathbf{m}_{t-1}^{i+1}, \mathbf{x}_t^i; \mathbf{W}^{i+1})$$

$$\mathbf{c}_t^{i+1} = \max(-\delta, \min(\delta, \mathbf{c}_t''^{i+1})) \tag{6.32}$$

Let us expand LSTM_i in equation (6.32) to include the internal gating logic. For brevity, we drop all the superscripts i:

$$\mathbf{W} = [\mathbf{W}_1, \mathbf{W}_2, \mathbf{W}_3, \mathbf{W}_4, \mathbf{W}_5, \mathbf{W}_6, \mathbf{W}_7, \mathbf{W}_8]$$

$$\mathbf{i}_t = \text{sigmoid}(\mathbf{W}_1\mathbf{x}_t + \mathbf{W}_2\mathbf{m}_t)$$

$$\mathbf{i}_t' = \tanh(\mathbf{W}_3\mathbf{x}_t + \mathbf{W}_4\mathbf{m}_t)$$

$$\mathbf{f}_t = \text{sigmoid}(\mathbf{W}_5\mathbf{x}_t + \mathbf{W}_6\mathbf{m}_t)$$

$$\mathbf{o}_t = \text{sigmoid}(\mathbf{W}_7\mathbf{x}_t + \mathbf{W}_8\mathbf{m}_t)$$

$$\mathbf{c}_t = \mathbf{c}_{t-1} \odot \mathbf{f}_t + \mathbf{i}_t' \odot \mathbf{i}_t$$

$$\mathbf{m}_t = \mathbf{c}_t \odot \mathbf{o}_t \tag{6.33}$$

When doing quantized inference, we replace all the floating point operations in equations (6.32) and (6.33) with fixed-point integer operations with either 8-bit or 16-bit resolution. The weight matrix W above is represented using an 8-bit integer matrix WQ and a float vector s, as shown in the following:

$$s_i = \max(\text{abs}(\mathbf{W}[i, :]))$$

$$\mathbf{WQ}[i, j] = \text{round}(\mathbf{W}[i, j]/s_i \times 127.0) \tag{6.34}$$

All accumulator values (c_t^i and $\mathbf{x}_t^i$) are represented using 16-bit integers representing the range $[-\delta, \delta]$. All matrix multiplications (e.g., W_1x_t and W_2m_t) in equation (6.33) are done using 8-bit integer multiplication accumulated into larger accumulators. All other operations, including all the activations (sigmoid and tanh) and elementwise operations ($\odot$, $+$) are done using 16-bit integer operations.

We now turn our attention to the log-linear softmax layer. During training, given the decoder RNN network output $\mathbf{y}_t$, we compute the probability vector $\mathbf{p}_t$ over all candidate output symbols as follows:

$$\mathbf{v}_t = \mathbf{W}_s * \mathbf{y}_t$$

$$\mathbf{v}_t' = \max(-\gamma, \min(\gamma, \mathbf{v}_t))$$

$$\mathbf{p}_t = softmax(\mathbf{v}_t') \tag{6.35}$$

In quantized inference, the weight matrix $\mathbf{W}_s$ is quantized into eight bits as in equation (6.34), and the matrix multiplication is done using 8-bit arithmetic. The calculations within the softmax function and the attention model are not quantized during inference.

The challenge of machine translation still exists. GNMT may still make mistakes that humans will never make, such as missing translations and misinterpreting proper nouns or rare words. The translation does not take into account the whole paragraph or even the full text. In short, there are still many areas for improvement in GNMT, but in any case, GNMT represents a major milestone.

Exercises

6.1 What is deep learning? Please list the features of deep learning.

6.2 Compare the three deep learning methods: autoencoder, restricted Boltzmann machine, and deep belief networks.

6.3 Illustrate the architecture of convolutional neural networks.

6.4 Suppose a convolutional neural network is trained on ImageNet dataset (object recognition dataset). This trained model is then given a complete image as an input. What are the output probabilities for this input?

6.5 Error back-propagation starts from the output layer and enters the sub-sampling layer through the hidden layer in CNN. The error back-propagation of the output layer needs to first compute the partial derivative of the error about the neurons in the output layer. How to speed up this processing in CNN?

6.6 The bidirectional RNN is actually just two independent RNNs together. Please construct single-layer static bidirectional RNN network based on TensorFlow, using mnist handwriting as test case.

6.7 Try to build a simple time series prediction module in TensorFlow with LSTM RNN.

6.8 Try to build a neural machine translation (NMT) to translate Chinese to English using an attention model by bi-directional LSTM model.

Chapter 7

Reinforcement Learning

Reinforcement learning is the learning of a mapping from situations to actions so as to maximize a scalar reward of reinforcement signal. The learner does not need to be directly told which actions to take, as in most forms of machine learning, but instead discovers which actions yield the most reward by trying them and finds a balance between exploration and exploitation.

7.1 Introduction

People often learn by interacting with outside environment. Reinforcement learning (RL) is a computational approach to the study of learning from interaction. RL is the learning of a mapping from situations to actions so as to maximize a scalar reward of reinforcement signal. The learner does not need to be directly told which actions to take, as in most forms of machine learning, but instead discovers which actions yield the most reward by trying them. In the most interesting and challenging cases, an action may affect not only the immediate reward but also the next situation, and consequently all subsequent rewards. These two characteristics — trial-and-error and delayed reinforcement — are the two most important distinguishing characteristics of RL.

Reinforcement learning is not defined by characterizing learning methods but by characterizing a learning problem. Any method that is well suited to solving that problem, we consider to be a reinforcement learning method. RL addresses the question of how an autonomous agent that senses and acts in its environment can learn to choose optimal actions to achieve its goals. RL is very different from supervised learning, the kind of learning studied in almost all current research in machine learning, statistical pattern recognition, and artificial neural networks. Supervised learning is learning under the tutelage of a knowledgeable supervisor, or "teacher", which explicitly tells the learning agent how it should respond to training inputs.

RL concerns a family of problems in which an agent evolves while analyzing the consequences of its actions, with a simple scalar signal (the reinforcement) given out by the environment.

The study of RL has developed into an unusually multi-disciplinary field; it includes research specializing in artificial intelligence, psychology, control engineering, statistical research, and so on. RL has particularly rich roots in the psychology of animal learning, from which it takes its name. A number of impressive applications of RL have also been developed. RL has attracted much research in the past decade. Its incremental nature and adaptive capabilities make it suitable for use in various domains, such as automatic control, mobile robotics, and multi-agent system. Particularly with the breakthrough of the mathematical basis of reinforcement learning, the application of RL is increasingly conducted, as one of the hot spots of the current research in the field of machine learning.

The history of reinforcement learning had two main threads, both long and rich, which were pursued independently before intertwining into modern reinforcement learning. One thread concerned learning by trial and error and started in the psychology of animal learning. This thread run through some of the earliest work in artificial intelligence and led to the revival of reinforcement learning in the early 1980s. The other thread concerned the problem of optimal control and its solution using value functions and dynamic programming. For the most part, this thread did not involve learning. Although the two threads were largely independent, the exceptions revolve around a third, less distinct thread concerning temporal-difference methods. All three threads came together in the late 1980s to produce the modern field of reinforcement learning.

In early artificial intelligence, before it was distinct from other branches of engineering, several researchers began to explore trial-and-error learning as an engineering principle. The earliest computational investigations of trial-and-error learning were perhaps by Minsky and Farley and Clark, both in 1954. In his Ph.D. dissertation, Minsky discussed computational models of reinforcement learning and described his construction of an analog machine, composed of components he called SNARCs (Stochastic Neural-Analog Reinforcement Calculators). Farley and Clark described another neural network learning machine designed to learn by trial and error. In the 1960s, one finds the terms "reinforcement" and "reinforcement learning" being widely used in the engineering literature for the first time. Particularly influential was Minsky's paper "Steps Toward Artificial Intelligence" (Minsky, 1961), which discussed several issues relevant to reinforcement learning, including what he called the credit-assignment problem: how do you distribute credit for success among the many decisions that may have been involved in producing it? In 1969, Minsky got Turing Award in computer due to the above contribution.

In 1994 and 1995, the interests of Farley and Clark shifted from trial-and-error learning to generalization and pattern recognition, that is, from reinforcement learning to supervised learning. This began a pattern of confusion about the relationship between these types of learning. Many researchers seemed to believe that they were studying reinforcement learning when they were actually studying supervised learning. For example, neural network pioneers such as Rosenblatt (1958) and Widrow and Hoff (1960) were clearly motivated by reinforcement learning — they used the language of rewards and punishments — but the systems they studied were supervised learning systems suitable for pattern recognition and perceptual learning. Even today, researchers and textbooks often minimize or blur the distinction between these types of learning. Some modern neural network textbooks use the term trial-and-error to describe networks that learn from training examples because they use error information to update connection weights. This is an understandable confusion, but it substantially misses the essential optional character of trial-and-error learning.

The term "optimal control" came into use in the late 1950s to describe the problem of designing a controller to minimize a measure of a dynamical system's behavior over time. One of the approaches to this problem was developed in the mid-1950s by Richard Bellman and colleagues by extending a 19th-century theory of Hamilton and Jacobi. This approach uses the concept of a dynamical system's state and a value function, or "optimal return function" to define a functional equation, now often called the Bellman equation. The class of methods for solving optimal control problems by solving this equation came to be known as dynamic programming (Bellman, 1957). Bellman also introduced the discrete stochastic version of the optimal control problem known as Markov decision processes (MDPs), and Ron Howard devised the policy iteration method for MDPs in 1960. All of these are essential elements underlying the theory and algorithms of modern reinforcement learning.

Finally, the temporal difference and optimal control threads were fully brought together in 1989 with Chris Watkins's development of Q-learning (Watkins *et al.*, 1989). This work extended and integrated prior work in all three threads of reinforcement learning research. By the time of Watkins's work, there had been tremendous growth in reinforcement learning research, primarily in the machine learning subfield of artificial intelligence but also in neural networks and artificial intelligence more broadly. In 1992, the remarkable success of Gerry Tesauro's backgammon playing program, TD-Gammon (Tesauro, 1992), brought additional attention to the field. Other important contributions made in the recent history of reinforcement learning were too numerous to mention in this brief account; we cite these at the end of the individual chapters in which they arise.

7.2 Reinforcement Learning Model

The reinforcement learning problem is meant a straightforward framing of the problem of learning from interaction to achieve a goal. The learner or decision-maker is called the agent. The thing it interacts with, comprising everything outside the agent, is called the environment. These interact continually, with the agent selecting actions and the environment responding to those actions and presenting new situations to the agent. The model of RL is illustrated in Figure 7.1.

More specifically, the agent exists in an environment described by some set of possible state S. It can perform any of a set of possible action A. Each time it performs an action at in some state *st*, the agent receives a real-valued reward r_t that indicates the immediate value of this state–action transition. This produces a sequence of states s_i, actions a_i, and immediate rewards r_i. The task of the agent is to learn a control policy $\pi: S \rightarrow A$ that maximizes the expected sum of these rewards, with future rewards discounted exponentially by their delay. The agent's goal, roughly speaking, is to maximize the total amount of reward it receives over the long run, as shown in Formula (7.1). In learning, the principle of RL is that if the reward is positive, strengthen the action later, otherwise, weaken the action:

$$\sum_{i=0}^{\infty} \gamma^i r_{t+i} \quad 0 < \gamma \leq 1 \tag{7.1}$$

A reinforcement learning task that satisfies the Markov property is called a Markov decision process, or MDP. If the state and action spaces are finite, then it is called a finite Markov decision process (finite MDP). Finite MDPs are particularly important to the theory of reinforcement learning.

Markov decision process: A Markov decision process is defined by a 4-tuples $<S, A, R, P>$, where S is a set of possible states, A is a set of possible actions, R is the reward function ($R: S \times A \rightarrow \mathscr{R}$), and P is the state transition function ($P: S \times A \rightarrow PD(S)$). $R(s, a, s')$ is the immediate reward after transition to state s' from state s by performing action a. $P(s, a, s')$ is the probability that action a in

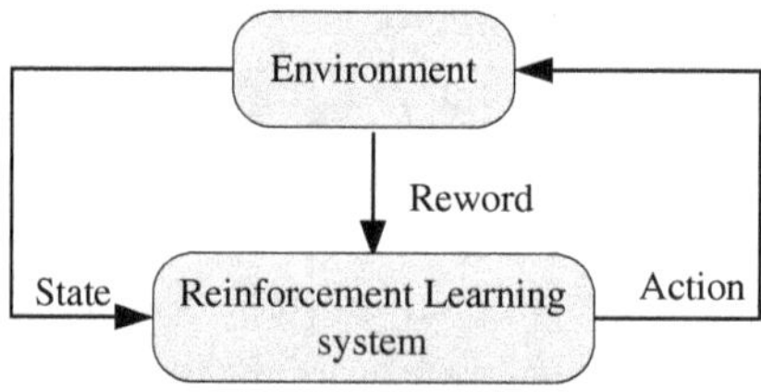

Fig. 7.1. Reinforcement learning model.

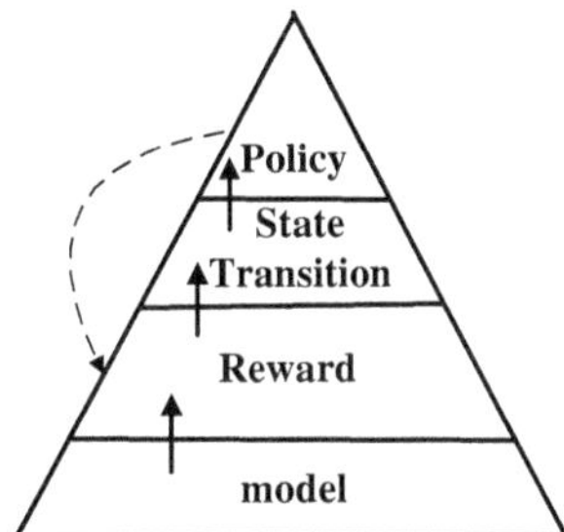

Fig. 7.2.   Four components of RL.

state s will lead to state s'. The task of the agent is to learn a policy, $\pi\colon S \to A$, for selecting its next action at based on the current observed state st, that is, $\pi(st) = at$.

The nature of Markov property is given the state of the MDP at time t is known, transition probabilities to the state at time $t + 1$ are independent of all previous states or actions. If transition function P and reward function R are known, dynamic programming could be used to acquire the optimal policy. However, RL focuses on the solution of optimal policy when P and R are unknown.

To solve this problem, Figure 7.2 shows the relationship of the four basic components of RL: policy π, state transition function (value function) P, reward function R, and environment model. The bottom-up relations of the four elements are the pyramid structure. At any given moment, the policy defines the main method of choice and action of the agent. This policy can be adopted by a group of production rules, or a simple table to find it. As pointed out earlier, under specific circumstances, the policy may need extensive search for a model or plan to process the results. It can also be random. The policy is the main component part of learning agents, as it is sufficient to produce action at any time.

The policy is the decision-making function of the agent, specifying what action it takes in each of the situations that it might encounter. In psychology, this would correspond to the set of stimulus-response rules or associations. This is the core of a reinforcement agent, as suggested in Figure 7.2 because it alone is sufficient to define a full, behaving agent. The other components serve only to change and improve the policy. The policy itself is the ultimate determinant of behavior and performance. In general, it may be stochastic.

The reward function defines the goal of the RL agent. The agent's objective is to maximize the reward that it receives over the long run. The reward function thus defines what are good and bad events for the agent. Rewards are the immediate and defining features of the problem faced by the agent. As such, the reward function must necessarily be fixed. It may, however, be used as a basis for changing the policy.

For example, if an action selected by the policy is followed by a low reward, then the policy may be changed to select some other action in that situation in the future.

Whereas reward indicates what is good in an immediate sense, the transition function specifies what is good in the long run, that is, because it predicts reward. The difference between value and reward is critical to RL. For example, when playing chess, checkmating your opponent is associated with high reward, but winning his queen is associated with high value. The former defines the true goal of the task — winning the game — whereas the latter just predicts this true goal. Learning the value of states, or of state–action pairs, is the critical step in the RL methods.

The fourth and final major component of an RL agent is a model of its environment or external world. This is something that mimics the behavior of the environment in some sense. Not every RL agent uses the model of the environment. Methods that never learn or use a model are called model-free RL methods. Model-free methods are very simple and, perhaps surprisingly, are still generally able to find optimal behavior. Model-based methods just find it faster. The most interesting case is that in which the agent does not have a perfect model of the environment *a priori* but must use learning methods to align it with reality.

The system environment is defined by the environment model. Since the models of P and R functions are unknown, the system can only rely on the immediate reward received by each trial and error to choose policies. The objective is to find a policy that, roughly speaking, maximizes the total reward received. In the simplest formulation, the trade-off between immediate and delayed reward is handled by a discount rate $0 \leq \gamma < 1$. The value of following a policy π from a state s is defined as the expectation of the sum of the subsequent rewards, $r_1, r_2 \ldots$, each discounted geometrically by its delay as follows:

$$R_t = r_{t+1} + \gamma\, r_{t+2} + \gamma^2 r_{t+3} + \cdots = r_{t+1} + \gamma\, R_{t+1} \tag{7.2}$$

$$V^\pi(s) = E_\pi\{R_t|s_t = s\} = E_\pi\{r_{t+1} + \gamma\, V(s_{t+1})|s_t = s\}$$

$$= \sum_a \pi(s,a) \sum_{s'} P^a_{ss'}[R^a_{ss'} + \gamma\, V^\pi(s')] \tag{7.3}$$

Values determine a partial ordering over policies, whereby $\pi_1 \geq \pi_2$ if and only if $V_{\pi_1}(s) \geq V_{\pi_2}(s)$, $\forall s$. Ideally, we seek an optimal policy π^*, one that is greater or equal than all others. All such policies share the same optimal value function. According to Bellman optimality equations, the value function $V^*(s)$ of optimal policy π^* at state s could be defined as follows:

$$V^*(s) = \max_{a \in A(s)} E\left\{r_{t+1} + \gamma\, V^*(s_{t+1})|s_t = s, a_t = a\right\}$$

$$= \max_{a \in A(s)} \sum_{s'} P^a_{ss'}[R^a_{ss'} + \gamma\, V^*(s')] \tag{7.4}$$

Dynamic programming methods involve iteratively updating an approximation to the optimal value function. If the state-transition function P and the expected rewards R are known, a typical example is value iteration, which starts with an arbitrary policy π_0, and then

$$\pi_k(s) = \arg\max_a \sum_{s'} P^a_{ss'}[R^a_{ss'} + \gamma\, V^{\pi_{k-1}}(s')] \tag{7.5}$$

$$V^{\pi_k}(s) \leftarrow \sum_a \pi_{k-1}(s, a) \sum_{s'} P^a_{ss'}[R^a_{ss'} + \gamma\, V^{\pi_{k-1}}(s')] \tag{7.6}$$

In RL, without knowledge of the system's dynamics, we cannot compute the expected value by equations (7.5) and (7.6). It is necessary to estimate the value by iteratively updating an approximation to the optimal value function, and Monte Carlo sampling is one of the basic methods. Keeping policy π, iteratively using equation (7.7) to obtain approximate solutions,

$$V(s_t) \leftarrow V(s_t) + \alpha[R_t - V(s_t)] \tag{7.7}$$

Combining Monte Carlo method and dynamic programming method, equation (7.8) gives the iterative equation of temporal-difference (TD) learning:

$$V(s_t) \leftarrow V(s_t) + \alpha[r_{t+1} + \gamma\, V(s_{t+1}) - V(s_t)] \tag{7.8}$$

7.3 Dynamic Programming

The term dynamic programming (DP) refers to a collection of algorithms that can be used to compute optimal policies given a perfect model of the environment as a Markov decision process (MDP). Classical DP algorithms are of limited utility in RL because of their assumption of a perfect model and their great computational expense, but they are still important theoretically. DP provides an essential foundation for the understanding of the methods presented in the rest of this book. In fact, all of these methods can be viewed as attempts to achieve the same effect as DP, with less computation and without assuming a perfect model of the environment.

First, we consider how to compute the state-value function V^π for an arbitrary policy π. This is called policy evaluation in the DP literature. We also refer to it as the prediction problem. For all s in S,

$$V^\pi(s) = \sum_a \pi(a|s) \times \sum_{s'} \pi(s \rightarrow s'|a) \times (R^a(s \rightarrow s') + \gamma\, (V^\pi(s'))) \tag{7.9}$$

where $\pi(a|s)$ is the probability of taking action a in state s under policy π, and the expectations are subscripted by π to indicate that they are conditional on π being followed. The existence and uniqueness of V^π are guaranteed as long as either

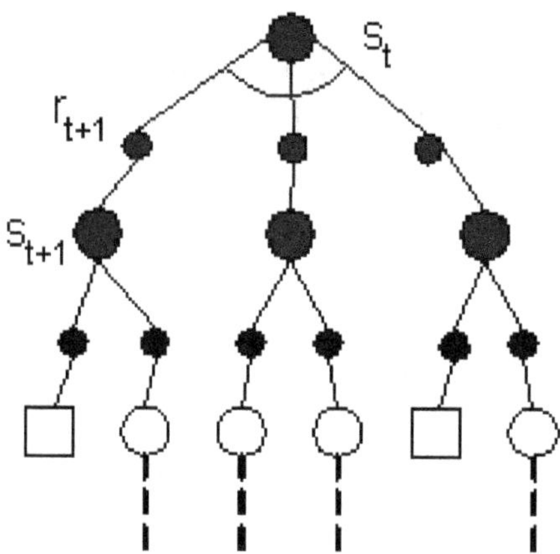

Fig. 7.3. Dynamic programming method.

$\gamma < 1$ or eventual termination is guaranteed from all states under the policy π. Figure 7.3 illustrates the first step of the computing process, and the three subsequence states of s_t are all known. As for policy π, the probability of the action a is $\pi(a|s)$. For each state, the environment may respond to one of the states to s' with a reward r. Bellman equation is adopted to average the probability of these, without weight. It is noted that the value of the starting state must look forward to the discount value of the next state, γ, and the reward obtained with the path. In dynamic programming, if n and m are the number of states and actions, although the total number of the policy is nm, a dynamic planning method can be guaranteed in polynomial time to find the optimal strategy. In this sense, dynamic programming strategy is faster than any of the direct search index class and has polynomial time of action and states. However, if the state is based on the exponential growth of certain variables, of course, there will also appear dimensions of the disaster.

In dynamic programming, iterative equation (7.10) can be derived from equation (7.4):

$$V_{t+1}(s) \leftarrow \underset{a}{\textbf{MAX}} \sum_{s'} P_{ss'}^{a}[\mathfrak{R}_{ss'}^{a} + \gamma V_t(s')] \tag{7.10}$$

$V_t(s) \rightarrow V^*(s)$ when $t \rightarrow \infty$. The award evaluation will be acquired until $|\Delta V|$ is less than a small positive number if repeated and made iteration to each state.

The typical model of dynamic programming model has limited usage, as many problems are difficult to give the integral model of the environment. For example, simulation robot soccer is such a problem, which can be solved by real-time dynamic programming methods. In real-time dynamic programming, the environment model is not required to be given first but to the environment model by testing in a real environment. The use of anti-nerve-state network can be used for the generalization of states. The input unit of the network is the state of the environment s. The output of the network is the evaluation of the state $V(s)$.

DP methods update estimates of the values of states based on estimates of the values of successor states. That is, they update estimates on the basis of other estimates. We call this general idea bootstrapping. Many reinforcement learning methods perform bootstrapping, even those that do not require, as DP requires, a complete and accurate model of the environment. In the following chapter, we explore reinforcement learning methods that do not require a model and do not bootstrap. In the chapter after that, we explore methods that do not require a model but do bootstrap. These key features and properties are separable yet can be mixed in interesting combinations.

7.4 Monte Carlo Methods

Monte Carlo methods are a class of computational algorithms that rely on repeated random sampling to compute their results. Monte Carlo methods are often used when simulating physical and mathematical systems. Due to their reliance on repeated computation and random or pseudo-random numbers, Monte Carlo methods are most suited to calculation by a computer. Monte Carlo methods tend to be used when it is infeasible or impossible to compute an exact result with a deterministic algorithm. Unlike DP, the Monte Carlo methods do not assume complete knowledge of the environment. Monte Carlo methods require only experience — sample sequences of states, actions, and rewards from online or simulated interaction with an environment. Although a model is required, the model need only generate sample transitions, not the complete probability distributions of all possible transitions that are required by dynamic programming (DP) methods. The term Monte Carlo method was coined in the 1940s by physicists working on nuclear weapon projects in the Los Alamos National Laboratory.

Monte Carlo methods are ways of solving the reinforcement learning problem based on averaging sample returns. There is no single Monte Carlo method; instead, the term describes a large and widely used class of approaches. However, these approaches tend to follow a particular pattern:

(1) Define a domain of possible inputs.
(2) Generate inputs randomly from the domain, and perform a deterministic computation on them.
(3) Aggregate the results of the individual computations into the final result.

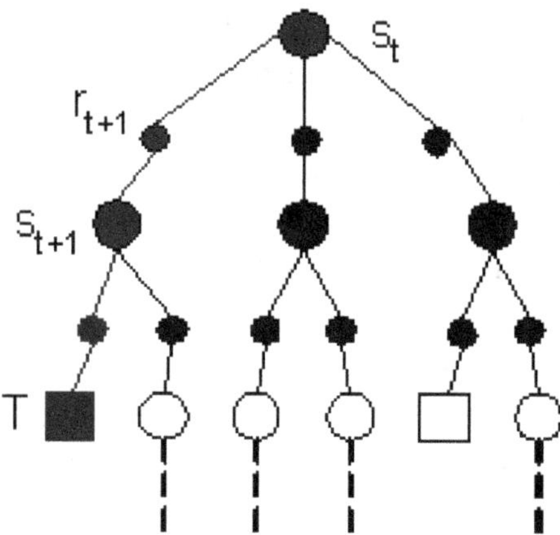

Fig. 7.4. Monte Carlo methods.

Here we use it specifically for methods based on averaging complete returns (as opposed to methods that learn from partial returns, considered in the following chapter). Figure 7.4 gives the return reward by Monte Carlo sampling for one step during learning. Then through iterative learning, the actual obtained rewards are used to approximate the real value function.

Monte Carlo methods do not assume complete knowledge of the environment but learn from online experience. Monte Carlo methods are ways of solving the reinforcement learning problem based on averaging sample returns. Given policy π, compute V^π: subsequence state s_t under policy π, $Rt(st)$ is the long reward return, and add $R_t(s_t)$ to the list R_{si}, $V(s_t) \leftarrow average(Rs_i)$.

The list could use incremental implementation:

$$V(s_t) \leftarrow V(s_t) + \frac{R_t(s_t) - V(s_t)}{N_{s_t} + 1}$$

$$N_{s_t} \leftarrow N_{s_t} + 1$$

(7.11)

Under Monte Carlo control, policy evaluation and improvement use the same random policy as follows:

$$\mathbf{a}^* \leftarrow \arg\max_a Q(s, a)$$

$$\pi(s, a) \leftarrow \begin{cases} 1 - e + \frac{e}{|A(s)|}, & a = a^* \\ \frac{e}{|A(s)|}, & a \neq a^* \end{cases}$$

(7.12)

In learning, if some actions are found to be good, then what action should the agent select in the next decision-making? One consideration is making full use of existing knowledge, by selecting the current best action. But it has a drawback:

maybe some better actions are not found; in contrast, if the agent always tests new actions, it will lead to no progress. The agent faces a trade-off in choosing whether to favor exploration of unknown actions (to gather new information) or exploitation of existing actions that it has already learned will yield high reward (to maximize its cumulative reward). These are two main methods: e-greedy method and genetic simulated annealing. The selection probability of each action is related to its Q value:

$$p(a|s) = \frac{e^{Q(s,a)/T}}{\sum_{a'} e^{Q(s,a')/T}} \tag{7.13}$$

7.5 Temporal-Difference Learning

Temporal-difference (TD) learning is a combination of Monte Carlo ideas and dynamic programming (DP) ideas. Like Monte Carlo methods, TD methods can learn directly from raw experience without a model of the environment's dynamics. TD resembles a Monte Carlo method because it learns by sampling the environment according to some policy. TD is related to dynamic programming techniques because it approximates its current estimate based on previously learned estimates (a process known as bootstrapping). TD learning algorithm is related to the temporal-difference model of animal learning.

As a prediction method, TD learning takes into account the fact that subsequent predictions are often correlated in some sense. In standard supervised predictive learning, one only learns from actually observed values: a prediction is made, and when the observation is available, the prediction is adjusted to better match the observation. TD(0) is the simplest case of temporal-difference learning described in Algorithm 7.1.

TD(0) learning algorithm contains two steps: determine the new action policy according to the current value function and evaluate the action policy by the

---

Algorithm 7.1.   TD(0) Learning Algorithm.

Initialize $V(s)$ arbitrarily, π to the policy to be evaluated

Repeat (for each episode)

    Initialize s

    Repeat (for each step of episode)

    Choose a from s using policy π derived from V (e.g., ε-greedy)

    Take action a, observer r, s'

    $V(s) \leftarrow V(s) + \alpha[r + \gamma V(s') - V(s)]$

    Until s is terminal

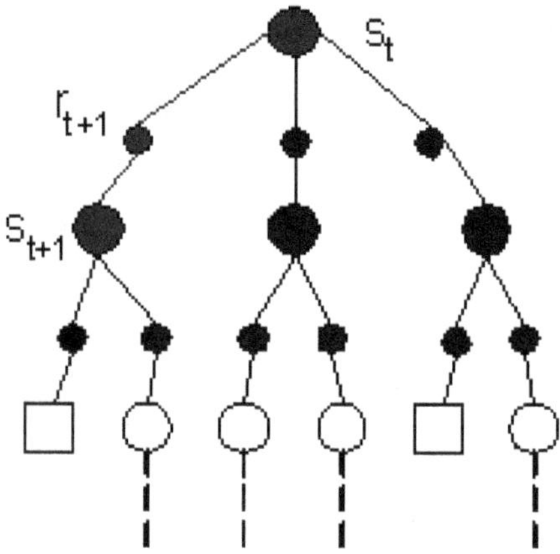

Fig. 7.5. Temporal-difference learning method.

immediate reward under the new action policy. The learning process is as follows:

$$v_0 \to \pi_1 \to v_1 \to \pi_2 \to \cdots \to v^* \to \pi^* \to v^*$$

until the value function and the policy reach a stable value. In TD learning, the computation of value function is shown in Figure 7.5.

To illustrate the general idea of reinforcement learning and contrast it with other approaches, we consider the familiar child's game of tic-tac-toe. Two players take turns playing on a three-by-three board. One player plays ◇s and the other Os until one player wins by placing three marks in a row, horizontally, vertically, or diagonally.

If the board fills up with neither player getting three in a row, the game is a draw. Since a skilled player can play so as never to lose, let us assume that we are playing against an imperfect player, one whose play is sometimes incorrect and allows us to win. For the moment, in fact, let us consider draws and losses to be equally bad for us. How might we construct a player who will find the imperfections in his opponent's play and learn to maximize his chances of winning?

An evolutionary approach to this problem would directly search the space of possible policies for one with a high probability of winning against the opponent. Here, a policy is a rule that tells the player what move to make for every state of the game — every possible configuration of ◇s and Os on the three-by-three board. For each policy considered, an estimate of its winning probability would be obtained by playing some number of games against the opponent. This evaluation would then direct which policy or policies were considered next. A typical evolutionary method would hill-climb in policy space, successively generating and evaluating policies in an attempt to obtain incremental improvements. Or, perhaps, a genetic-style algorithm could be used that would maintain and evaluate a population of policies. Literally, hundreds of different optimization methods could be applied. By directly searching the policy space we mean that entire policies are proposed and compared on the basis of scalar evaluations.

Here is how the tic-tac-toe problem would be approached using reinforcement learning and approximate value functions. First, we set up a table of numbers, one for each possible state of the game. Each number will be the latest estimate of the probability of our winning from that state. We treat this estimate as the state's value, and the whole table is the learned value function. State A has higher value than state B, or is considered "better" than state B, if the current estimate of the probability of our winning from A is higher than it is from B. Assuming we always play ◇s, then for all states with three ◇s in a row, the probability of winning is 1, because we have already won. Similarly, for all states with three Os in a row, or that are "filled up," the correct probability is 0, as we cannot win from them. We set the initial values of all the other states to 0.5, representing a guess that we have a 50% chance of winning.

We play many games against the opponent. To select our moves, we examine the states that would result from each of our possible moves (one for each blank space on the board) and look up their current values in the table. Most of the time, we move greedily, selecting the move that leads to the state with the greatest value, that is, with the highest estimated probability of winning. Occasionally, however, we select randomly from among the other moves instead. These are called exploratory moves because they make us experience states that we might otherwise never see. A sequence of moves made and considered during a game can be diagrammed as in Figure 7.6.

While we are playing, we change the values of the states in which we find ourselves during the game. We attempt to make more accurate estimates of the probabilities of winning. To do this, we "back up" the value of the state after each greedy move to the state before the move, as suggested by the arrows in Figure 7.6. More precisely, the current value of the earlier state is adjusted to be closer to the value of the later state. This can be done by moving the earlier state's value a fraction of the way toward the value of the later state. If we let s_n denote the state before the greedy move, and s_{n+1} the state after the move, then the update to the estimated value of s, denoted $V(s)$, can be written as

$$V(s_n) = S(s_n) + c(V(s_{n+1}) - V(s_n)) \qquad (7.14)$$

where c is a small positive fraction called the step-size parameter, which influences the rate of learning. This update rule is an example of a temporal-difference learning

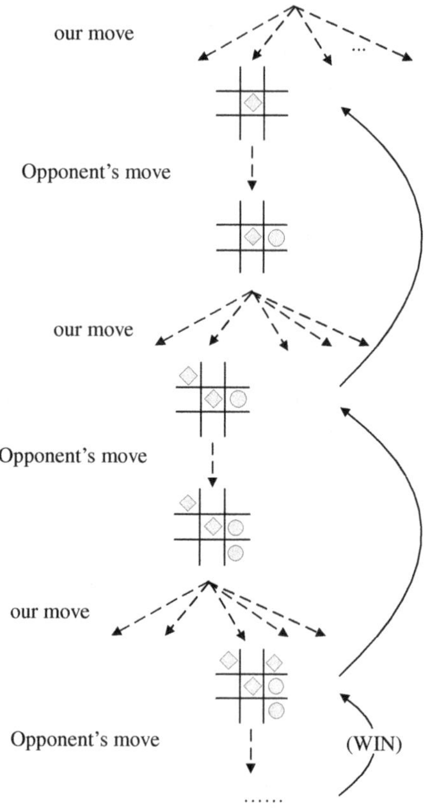

Fig. 7.6. A sequence of tic-tac-toe moves. The solid lines represent the moves taken during a game; the dashed lines represent moves that we (our RL player) considered but did not make.

method, so-called because its changes are based on a difference, $V(s_{n+1}) - V(s_n)$, between estimates at two different times.

The method described above performs quite well on this task. For example, if the step-size parameter is reduced properly over time, this method converges, for any fixed opponent, to the true probabilities of winning from each state given optimal play by our player. Furthermore, the moves then taken (except on exploratory moves) are in fact the optimal moves against the opponent. In other words, the method converges to an optimal policy for playing the game. If the step-size parameter is not reduced all the way to zero over time, then this player also plays well against opponents who slowly change their way of playing.

This example illustrates the differences between evolutionary methods and methods that learn value functions. To evaluate a policy, an evolutionary method must hold it fixed and play many games against the opponent, or simulate many games using a model of the opponent. The frequency of wins gives an unbiased estimate of the probability of winning with that policy and can be used to direct the next policy selection. But each policy change is made only after many games, and only the final outcome of each game is used: what happens during the games is ignored. For example, if the player wins, then all of his behavior in the game is given credit, independently of how specific moves might have been critical to the win. Credit is even given to moves that never occurred! Value function methods, in contrast, allow individual states to be evaluated. In the end, both evolutionary and value function methods search the space of policies, but learning a value function takes advantage of information available during the course of play.

This simple example illustrates some of the key features of reinforcement learning methods. First, there is the emphasis on learning while interacting with an environment, in this case with an opponent player. Second, there is a clear goal, and correct behavior requires planning or foresight that takes into account delayed effects of one's choices. For example, the simple reinforcement learning player would learn to set up multi-move traps for a shortsighted opponent. It is a striking feature of the reinforcement learning solution that it can achieve the effects of planning and looking ahead without using a model of the opponent or conducting an explicit search over possible sequences of future states and actions.

While this example illustrates some of the key features of reinforcement learning, it is so simple that it might give the impression that reinforcement learning is more limited than it really is. Although tic-tac-toe is a two-person game, reinforcement learning also applies in the case in which there is no external adversary, that is, in the case of a "game against nature." Reinforcement learning also is not restricted to problems in which behavior breaks down into separate episodes, like the separate games of tic-tac-toe, with reward only at the end of each episode. It is just as applicable when behavior continues indefinitely and when rewards of various magnitudes can be received at any time.

Finally, the tic-tac-toe player was able to look ahead and know the states that would result from each of its possible moves. To do this, it had to have a model of the game that allowed it to "think about" how its environment would change in response to moves that it might never make. Many problems are like this, but in others even a short-term model of the effects of actions is lacking. Reinforcement learning can be applied in either case. No model is required, but models can easily be used if they are available or can be learned.

7.6 Q-Learning

One of the most important breakthroughs in reinforcement learning was the development of an off-policy TD control algorithm known as Q-learning. Q-learning is

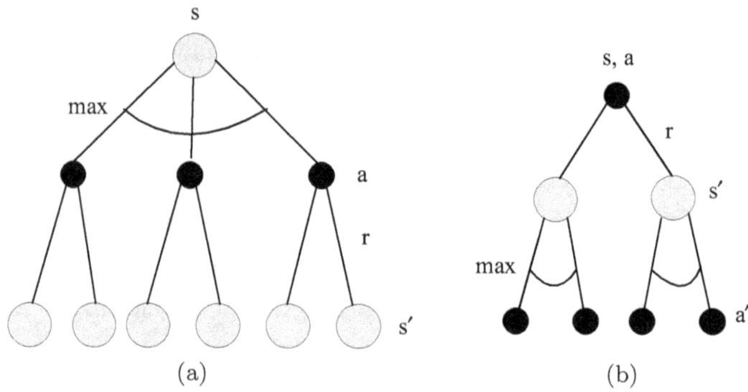

Fig. 7.7.　(a) V^* and (b) Q^* learning trace.

a reinforcement learning technique that works by learning an action-value function that gives the expected utility of taking a given action in a given state and following a fixed policy thereafter. A strength with Q-learning is that it is able to compare the expected utility of the available actions without requiring a model of the environment.

The core of the algorithm is a simple value iteration update. For each state, s, from the state set S, and for each action, a, from the action set A, we can calculate an update to its expected discounted reward with the following expression:

$$Q(s_t, a_t) \leftarrow (1 - c) \times Q(s_t, a_t) + c \times [r_{t+1} + \gamma \, \underset{a}{\mathrm{MAX}} \, Q(s_{t+1}, a) - Q(s_t, a_t)]$$

$$(7.15)$$

where r_t is an observed real reward at time t, c are the learning rates such that $0 \le c \le 1$, and γ is the discount factor such that $0 \le \gamma < 1$. Figure 7.7 illustrates the learning trace of V^* and Q^*.

Q-learning uses tables to store data. This quickly loses viability with increasing levels of complexity of the system it is monitoring/controlling. One answer to this problem is to use an (adapted) artificial neural network as a function approximation, as demonstrated by Tesauro in his backgammon-playing temporal-difference learning research. An adaptation of the standard neural network is required because the required result (from which the error signal is generated) is itself generated at run-time.

Monte Carlo methods perform a backup for each state based on the entire sequence of observed rewards from that state until the end of the episode. The backup of Q-learning, on the other hand, is based on just the next reward, using the value of the state one step later as a proxy for the remaining rewards (bootstrapping method). Thus, RL needs repeated learning to reach optimal policies. We construct

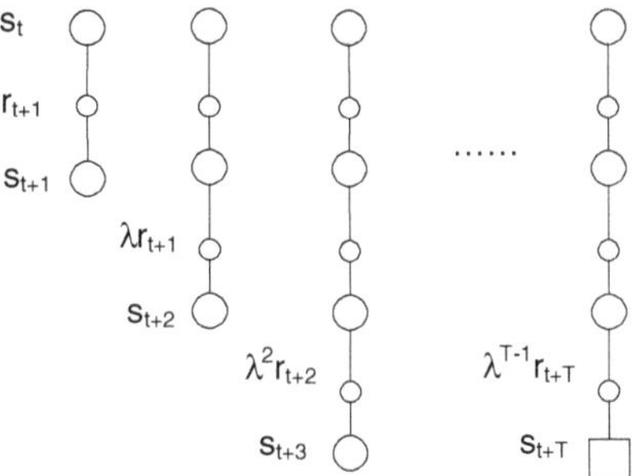

Fig. 7.8. λ-reward function.

Algorithm 7.2. TD(λ) Algorithm.

Initialize $V(s)$ arbitrarily and $e(s) = 0$ for all $s \in S$
Repeat (for each episode)
  Initialize s
  Repeat (for each step of episode)
    $a \leftarrow$ action given by π for s(e.g., ε-greedy)
    Take action a, observer r, s'
    $\delta \leftarrow r + \gamma V(s') - V(s)$
    $e(s) \leftarrow e(s) + 1$
    for all s
      $V(s) \leftarrow V(s) + \alpha \delta e(s)$
      $e(s) \leftarrow \gamma \lambda e(s)$
    $s \leftarrow s'$
  Until s is terminal

---

a λ-reward function Rt' by rewriting (7.8) as shown in equation (7.16). If the system reaches the end state at T step, the value function conforms to equation (7.17). The theoretical meaning of λ-reward function is illustrated in Figure 7.8:

$$R'_t = r_{t+1} + \lambda r_{t+2} + \lambda^2 r_{t+3} + \cdots + \lambda^{T-1} r_{t+T} \tag{7.16}$$

$$V(s_t) \leftarrow V(s_t) + \alpha [R'_t - V(s_t)] \tag{7.17}$$

The TD(λ) algorithm can be understood as one particular way of averaging n-step backups. According to equation (7.17), TD(λ) could be designed. In TD(λ), the value function will be updated by equation (7.17) through $e(s)$. A complete algorithm for online TD(λ) is given in Algorithm 7.2.

Algorithm 7.3. Q-Learning Algorithm.

Initialize $Q(s, a)$ arbitrarily
Repeat (for each episode)
  Initialize s
  Repeat (for each step of episode)
  Choose a from s using policy derived from Q (e.g., ε-greedy)
  Take action a, observer r, s'
    $Q(s, a) \leftarrow Q(s, a) + \alpha [r + \gamma \max_{a'} Q(s', a') - Q(s, a)]$
    $s \leftarrow s'$
Until s is terminal

We could combine the two steps of estimate and evaluation of value function to construct value function of state–action pair, Q function. In Q-learning, the learned action-value function Q directly approximates Q^*, the optimal action-value function, independent of the policy being followed. The policy still has an effect in that it determines which state–action pairs are visited and updated. However, all that is required for correct convergence is that all pairs continue to be updated. This is the minimal requirement in the sense that any method guaranteed to find optimal behavior in the general case must require it. Under this assumption and a variant of the usual stochastic approximation conditions on the sequence of step-size parameters, Q_t has been shown to converge with probability 1 to Q^*. The Q-learning algorithm is shown in procedural form in Algorithm 7.3.

7.7 Function Approximation

RL is a broad class of optimal control methods based on estimating value functions from experience, simulation, or search. Most of the theoretical convergence results for RL algorithms assume a tabular representation of the value function, in which the value of each state is stored in a separate memory location. However, most practical applications have continuous state spaces, or very large discrete state spaces, for which such a representation is not feasible. Thus, generalization is crucial to scaling RL algorithms to real-world problems. The kind of generalization we require is often called function approximation because it takes examples from a desired function (e.g., a value function) and attempts to generalize from them to construct an approximation of the entire function. The mapping relations in RL include S $\rightarrow$ A, S $\rightarrow$ R, S $\times$ A $\rightarrow$ R, S $\times$ A $\rightarrow$ S, and so on. The nature of function approximation in RL is to estimate these mapping relations by parameterized functions.

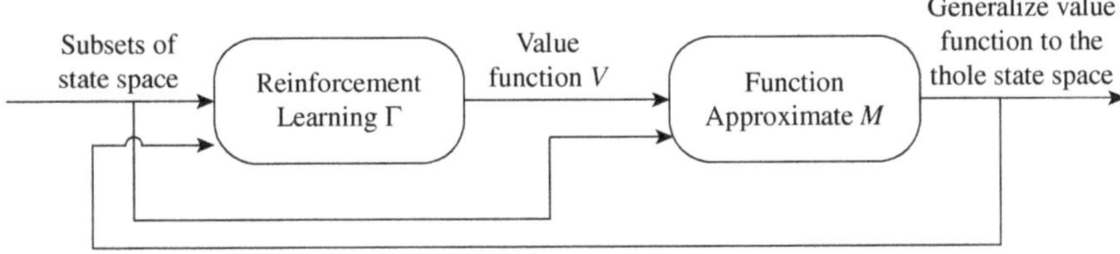

Fig. 7.9.   RL model with function approximation.

Assuming the starting value of value function is V_0, then the sequence of value functions during learning is

$$V_0, \ \Gamma(V_0), \ \Gamma(\Gamma(V_0)), \Gamma(\Gamma(\Gamma(V_0))), \ldots$$

where Γ represents equation (7.8).

Most of the traditional RL algorithms adopt a lookup table to save the value functions. And function approximation adopts parameterized functions to replace the lookup table. The model of RL with function approximation is shown in Figure 7.9. In the model, value function V is the objective function, function V' is the estimated function, and $M: V \to V'$ is the estimated operator. Assuming the starting value of value function is V_0, then the sequence of value functions during learning is

$$V_0, \ M(V_0), \ \Gamma(M(V_0)), \ M(\Gamma(M(V_0))), \ \Gamma(M(\Gamma(M(V_0)))), \ldots$$

Like Q-learning, the equations of RL with function approximation are as follows:

$$Q(s, a) \leftarrow (1 - \alpha)V'(s, a) + \alpha(r(s, a, s') + \max_{a'} V'(s', a')) \qquad (7.18)$$

$$V'(s, a) = M(Q(s, a)) \qquad (7.19)$$

In RL learning with function approximation, two iterative processes work simultaneously. One is the iterative process of value function Γ. The other is the approximation process of value function M. The correctness and convergence of the approximation process M play the key role in RL. Function approximation is an instance of supervised learning, the primary topic studied in machine learning, artificial neural networks, pattern recognition, and statistical curve fitting, such as state aggregation, function interpolation, and artificial neural networks.

Aggregation is an intuitive and applicable technique to solve large-scale problems. In state aggregation, the state space of the Markov chain is partitioned, and the states belonging to the same partition subset are aggregated into one meta-state. The Markov chain is said to be lumpable if the transition process among meta-states is Markovian for every probability distribution of the initial state of the original

Markov chain and weak lumpable if the transition process among meta-states is Markovian only for some initial probability distributions. It is proved that the function approximation with state aggregation is convergent. However, it is possible that the convergent value is not the optimal value. To reach the optimal value, the step could be too long. Thus, it also suffers from the dimension tragedy for large MDP problems.

Function approximation with artificial neural networks has attracted much research currently. Though these new methods could accelerate the speed largely, the convergence could not be ensured. Therefore, the new methods of function approximation which have both convergence and high speed are still one of the most important research in reinforcement learning.

7.8 Reinforcement Learning Applications

Reinforcement learning addresses the question of how an autonomous agent that senses and acts in its environment can learn to choose optimal actions to achieve its goals. In a Markov decision process (MDP), the agent can perceive a set S of distinct states of its environment and has a set A of actions that it can perform. At each discrete time step t, the agent senses the current state s_t, chooses a current action at, and performs it. The environment responds by giving the agent a reward $r_t = Q(s_t, a_t)$ and by producing the succeeding state $s_{t+1} = P(s_t, a_t)$. Here the functions P and Q are part of the environment and are not necessarily known to the agent. In an MDP, the functions P and Q depend only on the current state and action, and not on earlier states or actions. Reinforcement learning is a useful way to solve MDP problems. Reinforcement learning reaches its goal by learning reward function $r_t = Q(s_t, a_t)$ and state transition function $P(s_t, a_t)$. Q-learning acquires the optimal policy by learning $r_t = Q(s_t, a_t)$.

7.8.1 *RoboCup*

RoboCup is an international robotics competition founded in 1993. The aim is to develop autonomous robots with the intention of promoting research and education in the field of artificial intelligence. The name RoboCup is a contraction of the competition's full name, "Robot Soccer World Cup". The following is the application of Q-learning algorithm to simulate robot soccer with three members (2 to 1). The training is aimed at trying to get to the main strategy of awareness in the attack when running. In Figure 7.10, striker A controls the ball in the shoot region. But A has no angle to shoot; teammate B also is in the shoot region, and B has a good shot angle. Thus, A passes the ball to B, and B completes the shot. Then, the cooperation

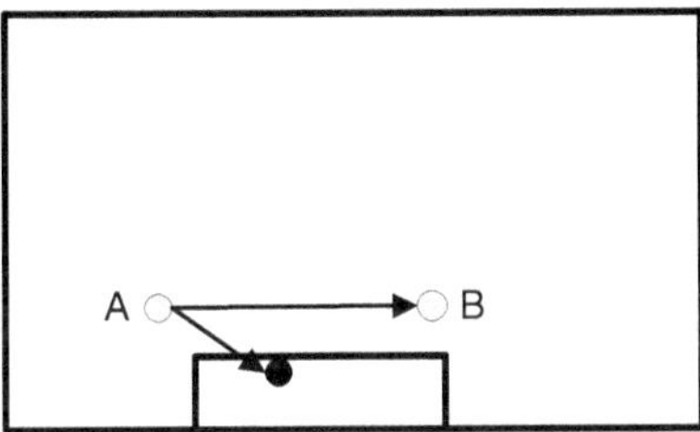

Fig. 7.10.   Robot soccer world cup training, 2 to 1.

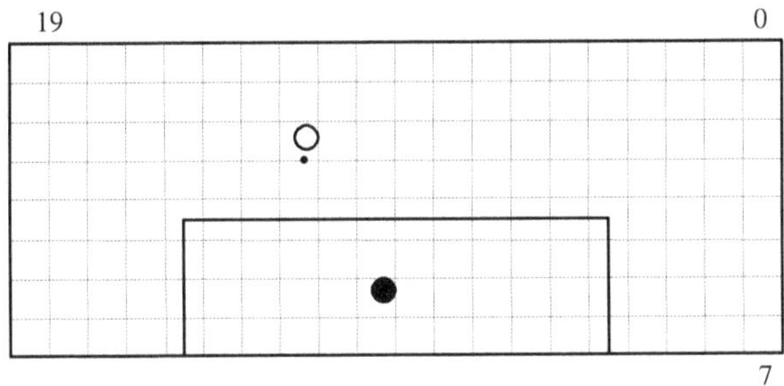

Fig. 7.11.   Position partition.

is very successful. Through the Q-learning approach in the training, the action of
A passing the ball to B is the best action in this state after training a large number
of examples.

Figure 7.11 illustrates the description of states. The attack region is divided
into 20 ∗ 8 small regions. Each small region is a square with a length of 2 m. A
two-dimensional array $A_{i,j}$ $(0 \leq i \leq 19, 0 \leq j \leq 7)$ can be used to describe the
region. The attack state can be described by the location of three agents. Figure 7.11
shows the generalization of the state. The state in the same region can be considered
as similar states. Though the description of the state is not precise, it is a description
of the strategic level that agents can be running in the same strategic region actively.
So, (S_A, S_B, S_G) describes a particular state, in which S_A is the regional code of
offensive team member A, S_B is the regional code of offensive team member B, and
S_G is the regional code of offensive team member G. The regional code is calculated
as follows: $S = i * 8 + j$. And the states are preserved by triples of the three regional
codes.

The optional actions have {Shoot, Pass, Dribble}, described as follows.

Shoot: The strategy is obtained by learning through a strategy based on the probability of a shot.

Dribble: The strategy is always to reduce threat and pass the ball to regions with a high probability of shooting the goal. In order to achieve this strategic objective, the offensive region can be divided into a number of strategic areas. In each strategic area, shot evaluation is recorded with the shooting success rate.

Pass: The strategy is very simple, just passing the ball between any two agents, and do not need to choose the target agent. If the pass fails, then the state of adoption of this strategy is unsuccessful; through this training, an impossible path of passing the ball cannot be adopted.

All states in training include four absorption states. Assume that the offense is in the left half, according to the specifications of the standard soccer server, the four states are played on, goal left, goal kick right, and free kick right. If taking the action and the state reaching the four absorption states, the agent will be given the ultimate reward r. For other actions, the agent will be given the procedure rewards as immediate rewards. For example, the maximum reward value of the goal left is 1, which means shooting the ball to the goal region.

The agent will obtain the ultimate reward by taking several actions through corresponding states. At this time, the state–action pair will get the reward value. The core of Q-learning algorithm is that every state–action pair has its Q-value. And the Q-values will be updated when getting the ultimate rewards. As RoboCup simulation platform adds a smaller random noise in the design of the state transition, the model is non-deterministic MDP. The Q value is updated by the following equation:

$$Q(s, a) = (1 - a)Q(s, a) + a(r + \gamma \max Q(s_{t+1}, a_{t+1})) \tag{7.20}$$

where $a = 0.1, \gamma = 0.95$.

In actual training, the initial Q is 1. After about 20,000 of the training (to reach a state of absorption), the majority of items in Q value have changed and have separated. Table 7.1 is the updated scene of Q-values with different training numbers.

Table 7.1.   Q value.

	Initial value	5,000	10,000	20,000
Shoot	1	0.7342	0.6248	0.5311
Pass	1	0.9743	0.9851	0.993
Dribble	1	0.9012	0.8104	0.7242

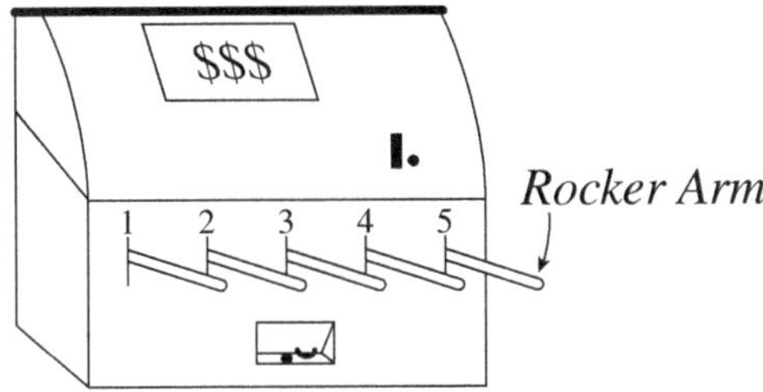

Fig. 7.12. Multi-armed gambling machine.

7.8.2 *Multi-Armed Gambling Machine*

Problem. There are K gambling machines (shown in Figure 7.12), and each machine has a certain probability of spitting out coins, but we don't know what this probability is. The value of the coins spitted out by each gambling machine is also different. Now that there are T chances to choose a gambling machine, how can we choose to maximize the total value of the coins obtained?

Set value:

$$K = 5$$
$$P = [0.1, 0.9, 0.3, 0.2, 0.7]$$
$$V = [5, 3, 1, 7, 4]$$
$$T = 1,000,000$$

The following can be calculated in this case:

- If you choose the 4th gambling machine with the highest expected value every time, the maximum total value you can obtain is 2,800,000.
- If you choose the second gambling machine with the lowest expected value every time, the minimum total value you can obtain is 300,000.
- If a gambling machine is randomly selected, the expected total value that can be obtained is 1,540,000.

The exploration only algorithm is to evenly allocate opportunities to each gambling machine and randomly select gambling machines. The exploitation only algorithm selects the gambling machine with the highest average value currently. However, due to the limited number of attempts, there is a trade-off between exploration and utilization, which is also a major problem faced by reinforcement learning: exploration–exploitation dilemma. There are two algorithms that can be used for K gambling machines: ε-greedy algorithm and softmax.

$$\textbf{Algorithm 7.4.} \quad \varepsilon\text{-Greedy Algorithm.}$$

Input: Number of rocker arms K; Reward function R; Attempt count T; Exploring probability ε.

Procedure

1. $r = 0$;
2. $\forall i = 1, 2, \ldots, K$: $Q(i) = 0$, count$(i) = 0$;
3. For $t = 1, 2, \ldots, T$ do
4.   if rand() $\ll \varepsilon$ then
5. $k = 1, 2, \ldots, K$ random selected
6.   else
7. $K = \arg\max_i Q(i)$
8.   end if
9. $v = R(k)$;
10. $r = r + v$
11. $Q(k) = \frac{Q(k) \times count(k) + v}{count(k) + 1}$
12.   **Count $(k) = $ count$(k) + 1$;**
13.   **End for**

Output: Accumulated rewards r

---

1. ε-Greedy Algorithm (Algorithm 7.4)

The principle of this method is to use the ε Explore with a uniform probability, that is, randomly select a rocker arm. With $1 - \varepsilon$ probability utilization, that is, selecting the rocker arm with the highest current average value. ε the general value is a smaller value of 0.1 or 0.01, but it can also decrease as the number of attempts increase. If the number of attempts is n, then set to $\varepsilon = 1/\sqrt{n}$ is sufficient.

2. Softmax (Algorithm 7.5)

The ε-greedy algorithm method is based on ε. The probability is used to explore and utilize decisions, while the softmax method is based on the Boltzmann distribution:

$$p(k) = \frac{e^{\frac{Q(k)}{\tau}}}{\sum_{i=1}^{K} e^{\frac{Q(k)}{\tau}}} \tag{7.21}$$

Reinforcement learning has received much attention in the past decade. Its incremental nature and adaptive capabilities make it suitable for use in various domains, such as automatic control, mobile robotics, and multi-agent system. A critical problem in conventional reinforcement learning is the slow convergence of the learning process. However, in most learning systems, there usually exists prior

---

Algorithm 7.5. **Softmax** Algorithm.

Input: Number of rocker arms K; Reward function R; Attempt count T; Temperature parameters τ.

Procedure

1. $r = 0$;
2. $\forall i = 1, 2, \ldots, K: Q(i) = 0$, count$(i) = 0$;
3. For $t = 1, 2, \ldots, T$ do
4. $k = 1, 2, \ldots, K$, random select according to equation (7.21)
5. $v = R(k)$;
6. $r = r + v$;
7. $Q(k) = \frac{Q(k) \times count(k) + v}{count(k) + 1}$
8.   count $(k) = $ count$(k) + 1$;
9. end for

Output: Accumulated rewards r

---

knowledge in the form of human expertise or previously learned experience. Therefore, how to integrate other machine learning techniques, such as neural networks and symbol learning technology, to help accelerate the learning speed is an important direction of RL. At present, the main technical difficulty is how to prove and guarantee the convergence of learning algorithm from theoretical aspects. The development of effective models for complex MDP will also be an important direction in the future.

Exercises

7.1 Give a brief description of the main branches of reinforcement learning and its research history.

7.2 Explain the similarities and differences between reinforcement learning model and other machine learning methods.

7.3 Explain the decision process of MDP and its essence.

7.4 Give the basic ideas of Monte Carlo methods and their applications in reinforcement learning.

7.5 Give the basic ideas of temporal-difference (TD) learning and illustrate its process considering playing the game of tic-tac-toe.

7.6 Consider the deterministic grid world shown in Figure 7.13 with the absorbing goal-state G. Here the immediate rewards are shown in Figure 7.13 for the labeled transitions and 0 for all unlabeled transitions. Give the V^* value for

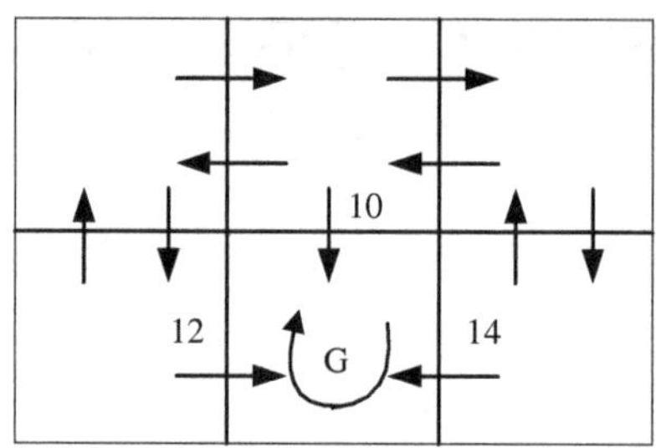

Fig. 7.13. Deterministic grid world.

every state in this grid world. Give the $Q(s, a)$ value for every transition. Finally, show an optimal policy using $\gamma = 0.8$.

7.7 Try to compare ε-greedy algorithm and **softmax** algorithm for multi-armed gambling machine.

Chapter 8

Transfer Learning

Transfer learning is a kind of machine learning in which knowledge learned from a task is reused in order to boost performance on a related task. Transferring information from previously learned tasks to new tasks has the potential to significantly improve learning efficiency.

8.1 Introduction

Transfer learning is related to the psychological literature on transfer of learning, although practical ties between the two fields are limited. Transfer learning stems from many research areas and is driven by many application needs.

8.1.1 *History*

In 1976, Bozinovski and Fulgosi published a paper addressing transfer learning in neural network training (Stevo & Ante, 1976). The paper gives a mathematical and geometrical model of the topic. In 1981, a report considered the application of transfer learning to a dataset of images representing letters of computer terminals, experimentally demonstrating positive and negative transfer learning.

Transfer learning has been applied in cognitive science. Pratt guest-edited an issue of *Connection Science* on the reuse of neural networks through transfer in 1996.

In 1997, Pratt and Thrun guest-edited a special issue of Machine Learning devoted to transfer learning.

In 2005, the Broad Agency Announcement (BAA) 05–29 of Defense Advanced Research Projects Agency (DARPA)'s Information Processing Technology Office (IPTO) gave a new mission of transfer learning: the ability of a system to recognize and apply knowledge and skills learned in previous tasks to novel tasks.

In 2020, Qiang Yang *et at.* published *Transfer Learning* (Yang *et al.*, 2020), which provides a comprehensive introduction to transfer learning.

The application of transfer learning is often not limited to a specific domain, as long as the problem meets the transfer learning scenario, you can try to use transfer learning to solve it. Transfer learning technology can be used in computer vision, text classification, behavior recognition, natural language processing, indoor positioning, video surveillance, public opinion analysis, human-computer interaction, and other fields.

8.1.2 *Important Concepts*

1. Domain

This is the subject of learning. The domain consists of two main parts: the feature space X and the probability distribution $P(X)$ that generates these data. Usually, we use D for a domain and P for a probability distribution. In particular, because migration is involved, it corresponds to two basic domains: the source domain and the target domain. Usually, we use Ds for the source domain and Dt for the target domain.

2. Task

Task is the goal of learning. The task consists of two main parts: the label space Y and the function f corresponding to the label. Usually, we use Y to represent a label space and f($\bullet$) to represent a learning function. Correspondingly, the category spaces of the source and target domains can be represented as Ys and Yt, respectively.

With the above two concepts in mind, we can give a definition of transfer learning.

Definition 8.1 (Transfer Learning). Given the source domain D_s and the source task T_s, the target domain D_t and the target task T_t, the goal of transfer learning is to use the knowledge of D_s and T_s to improve the task learning function $ft(\bullet)$ in the case of $D_s \neq D_t$ or $T_s \neq T_t$ prediction effect.

Transfer learning is to use the knowledge learned in previous tasks, such as data features and model parameters, to assist the learning process in new fields and obtain its own model.

8.1.3 *Similarity Measure*

The similarity coefficient reflects the degree of similarity between objects, reflecting the degree of similarity between samples relative to certain attributes. There are many ways to determine the similarity coefficient. Here are some common methods that can be selected based on actual problems.

Set $O = \{x_1, x_2, \cdots, x_n\}$ to the entirety of the object to be classified, $(x_{i1}, x_{i2}, \cdots, x_{im})$ to represent the feature data of each object. Let $x_i, x_j \in O$, r_{ij} be the similarity coefficient between x_i and x_j, satisfying the following conditions:

(1) $r_{ij} = 1 \Leftrightarrow x_i = x_j$
(2) $\forall x_i, x_j, r_{ij} \in [0, 1]$
(3) $\forall x_i, x_j, r_{ij} = r_{ji}$

The following methods are commonly used to determine the measure of similarity coefficients:

1. Quantitative product method

$$r_{ij} = \begin{cases} 1 & i = j; \\ \frac{1}{M} \sum_{k=1}^{m} x_{ik} x_{jk} & i \neq j \end{cases} \tag{8.1}$$

where M is a positive number and satisfies $M \geq \max\limits_{i \neq j} \left(\sum_{k=1}^{m} x_{ik} x_{jk} \right)$.

2. Angle cosine method

$$r_{ij} = \frac{\left| \sum_{k=1}^{m} x_{ik} x_{jk} \right|}{\sqrt{\left(\sum_{k=1}^{m} x_{ik}^2 \right) \left(\sum_{k=1}^{m} x_{jk}^2 \right)}} \tag{8.2}$$

The cosine between the two vectors is used as the similarity coefficient, and the range is $[-1, 1]$. When the two vectors are orthogonal, the value is 0, indicating that they are completely dissimilar.

3. Correlation coefficient method

$$r_{ij} = \frac{\sum_{k=1}^{m} (x_{ik} - \bar{x}_i)(x_{jk} - \bar{x}_j)}{\sqrt{\sum_{k=1}^{m} (x_{ik} - \bar{x}_i)^2} \sqrt{\sum_{k=1}^{m} (x_{jk} - \bar{x}_j)^2}} \tag{8.3}$$

where

$$\bar{x}_j = \frac{1}{m} \sum_{k=1}^{m} x_{ik}, \quad \bar{x}_j = \frac{1}{m} \sum_{k=1}^{m} x_{jk}.$$

Calculate the correlation between two vectors, in the range of $[-1, 1]$, where 0 means irrelevant, 1 means positive correlation, and -1 means negative correlation.

4. Maximum and minimum method

$$r_{ij} = \frac{\sum_{k=1}^{m} (x_{ik} \wedge x_{jk})}{\sum_{k=1}^{m} (x_{ik} \vee x_{jk})} \tag{8.4}$$

5. Arithmetic mean minimum method

$$r_{ij} = \frac{2\sum_{k=1}^{m} (x_{ik} \wedge x_{jk})}{\sum_{k=1}^{m} (x_{ik} + x_{jk})} \tag{8.5}$$

6. Geometric mean minimum method

$$r_{ij} = \frac{\sum_{k=1}^{m} (x_{ik} \wedge x_{jk})}{\sum_{k=1}^{m} \sqrt{x_{ik} x_{jk}}} \tag{8.6}$$

7. Absolute value index

$$r_{ij} = e^{-\sum_{k=1}^{m} |x_{ik} - x_{jk}|} \tag{8.7}$$

8. Exponential similarity coefficient method

$$r_{ij} = \frac{1}{m} \sum_{k=1}^{m} e^{-(x_{ik} - x_{jk})^2 / s_k^2} \tag{8.8}$$

9. Absolute value reciprocal method

$$r_{ij} = \begin{cases} 1 & i = j \\ \dfrac{M}{\sum_{k=1}^{m} |x_{ik} - x_{jk}|} & i \neq j \end{cases} \tag{8.9}$$

where M is properly selected so that r_{ij} is in [0, 1] and separated.

10. Absolute value subtraction

$$r_{ij} = 1 - c \sum_{k=1}^{m} |x_{ik} - x_{jk}| \tag{8.10}$$

where c is appropriately selected so that r_{ij} is separated in [0, 1].

11. Non-parametric method

Let

$$x'_{ik} = x_{ik} - \bar{x}_i, \quad x'_{jk} = x_{jk} - \bar{x}_j$$

$$n_{ij}^+ = \{x'_{i1}x'_{j1}, x'_{i2}x'_{j2}, \cdots, x'_{im}x'_{jm}\}, \text{ positive number,}$$

$$n_{ij}^- = \{x'_{i1}x'_{j1}, x'_{i2}x'_{j2}, \cdots, x'_{im}x'_{jm}\}, \text{ negative number:}$$

$$r_{ij} = \frac{1}{2}(1 + \frac{n_{ij}^+ - n_{ij}^-}{n_{ij}^+ + n_{ij}^-}) \tag{8.11}$$

12. Closeness method

If the characteristics of x_i, x_j are normalized, let $x_{ik}, x_{jk} \in [0, 1]$ $(k = 1, 2, \ldots, m)$ the degree of similarity is taken as the closeness

$$r_{ij} = 1 - c(d(x_i, x_j))^\alpha \tag{8.12}$$

where c, α are properly selected the parameter values, $d(x_i, x_j)$ is various distances, you can take the Minkowski distance:

$$d(x_i, x_j) = \left(\sum_{k=1}^{m} |x_{ik} - x_{jk}|^p\right)^{1/p} \tag{8.13}$$

when $p = 1$ is the Hamming distance and $p = 2$ is the Euclidean distance.

13. Expert scoring

Experts are asked to directly rate the similarity between x_i and x_j, taking the average as the γ_{ij}. Generally, the percentage is used and then divided by 100 to obtain a decimal in the interval [0,1] as the similarity coefficient of the objects.

8.1.4 *Classifications*

The classification of transfer learning can be carried out from multiple dimensions, such as transfer learning definition, domain data, and transfer method:

1. Categorized by transfer learning definition

(1) Similarities and differences in feature space: i.e., whether Xs and Xt are the same;

(2) similarities and differences in feature distribution: i.e., whether $Ps(x)$ and $Pt(x)$ are the same;

(3) similarities and differences in marker spaces: i.e., whether Ys and Yt are the same;

(4) similarities and differences in conditional probability distributions: i.e., whether $Ps(y|x)$ and $Pt(y|x)$ are the same.

2. Categorized by domain data

Categorized according to the specifics of the domain data, it can be summarized into the architecture in Figure 8.1. For example, if there is a large amount of labeled data in the source domain and no or very little labeled data in the target domain, the technology of inductive transfer learning can be selected to solve the problems in the actual business.

3. Categorized by migration method

The method of classification according to transfer methods can be divided into the following four broad categories:

(1) **Instance-based transfer learning:** Instance-based migration is to directly assign different weights to different samples, for example, if there are similar samples, I will give them high weights so that the migration is completed.

(2) **Feature-based transfer learning:** Feature-based migration is the transformation of features. Assuming that the features of the source domain and the target domain are not in the same space, or that they are not similar in the original space, then we can find a way to transform them into a space and migrate them.

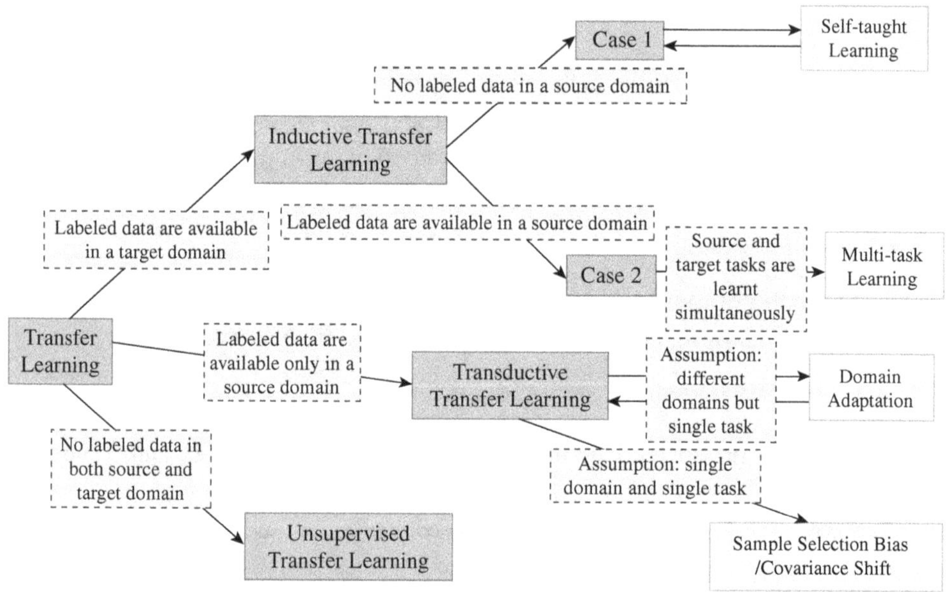

Fig. 8.1. Categories by domain data.

(3) **Model-based transfer learning:** Model-based migration is the reuse of parameters in the model. This type of method is especially used in neural networks because the structure of the neural network can be directly transferred. For example, the well-known fine-tuning is a good example of model parameter migration.

(4) **Relation-based transfer learning:** Relation-based migration is less used, which mainly means mining and using relationships for analogous migration. For example, the teacher attending the class and the students listening to the class can be compared to the scene of the company meeting, which is a kind of relationship transfer.

8.1.5 *Negative Transfer*

In transfer learning setting, labeled data is rare for a precise target task which often proposes an actual solution by using data from a linked source task. Though, when conveying knowledge from a less connected source, it may in reverse damage the target functioning; this occurrence is acknowledged as negative transfer. Regardless of its severity, negative transfer is typically defined in a casual manner, missing severe definition, cautious analysis, or methodical behavior. We have to achieve three main goals to avoid negative transfer: (a). Remove knowledge obtained in source domain which is destructive to target domain, (b). recollect knowledge that is useful in the target domain, and (c). recognize a cutting point at which the reskilling process is initiated for the finest knowledge transfer.

Transfer learning transfers knowledge from a source domain in which a lot of labeled training samples are available to target domain in which labeled training samples are absent. But data might go in another distribution in target domain supposing that the domains are interrelated. It becomes difficult to discover a strong degree of relationship between the source domain and target domain that gives rise to negative transfer. As per recent studies, few strategies can be used to prevent negative transfer: Maximum Mean Discrepancy (MMD) similarity metric and similar additional methods. Negative transfer is also seen in the application of sentence construction, vocabulary, culture, and coherence while converting from Chinese to English language.

8.2 Inductive Transfer Learning

The idea of inductive transfer learning algorithms is to increase the approximation of the target fT (•) in the target domain given target tasks are dissimilar from the source tasks. However, the source domains and the target domains may or may not be the same. For simplicity, we only consider the case where there is one source domain, D_S, and one target domain, D_T. More specifically, we denote the source domain data as $D_S = \{(x_{S1}, y_{S1}), \ldots, (x_{SnS}, y_{SnS})\}$, where $x_{Si} \in X_S$ is the data instance and $y_{Si} \in Y_S$ is the corresponding class label. In most cases, $0 \le n_T \ll n_S$.

Definition 8.2 (Transfer Learning). Given a source domain D_S and learning task T_S, a target domain D_T and learning task T_S, a transfer learning aims to help improve the learning of the target predictive function $f_T(\bullet)$ in D_T using the knowledge in D_S and T_S, where $D_S \ne D_T$, or $T_S \ne T_T$.

In the above definition, a domain is a pair $D = \{X, P(X)\}$. Thus, the condition $D_S \ne D_T$ implies that either $X_S \ne X_T$ or $P_S(X) \ne P_T(X)$. Similarly, a task is defined as a pair $T = \{Y, P(Y|X)\}$. Thus, the condition $T_S \ne T_T$ implies that either $Y_S \ne Y_T$ or $P(Y_S|X_S) \ne P(Y_T|X_T)$. When the target and source domains are the same, i.e., $D_S = D_T$, and their learning tasks are the same, the learning problem becomes a traditional machine learning problem.

In the case of self-taught learning (STL), labeled data is unavailable in source domain, whereas labeled data is available in the target domain. Self-taught learning is a deep learning methodology that contains two stages for the categorization. Foremost is a feature representation transfer which is learned from a huge gathering of the unlabeled data, whereas in the second stage, this learned representation is put on to labeled data to perform classification task. In STL, the label spaces among source domain along with the target domain might be dissimilar, which infers that side knowledge of source domain is inadequate to be used precisely.

Definition 8.3 (Inductive Transfer Learning). Given a source domain D_S and a learning task T_S, a target domain D_T and a learning task T_T, inductive transfer learning aims to help improve the learning of the target predictive function $f_T(\bullet)$ in D_T using the knowledge in D_S and T_S, where $T_S \ne T_T$.

Based on the above definition of the inductive transfer learning setting, a few labeled data in the target domain are required as the training data to induce the target predictive function. Inductive transfer learning comprises of source domain, i.e., D_S, and a learning task, i.e., T_{SL}, along with the target domain (D_T) plus a learning task (T_{TL}). Inductive transfer learning objectives to improve the learning of target $f_T P(\bullet)$ in target domain by utilizing the information in source domain and task, given T_{SL} is not equal to T_{TL}. The different approaches used in inductive transfer learning are instance transfer, feature representation transfer, parameter transfer, and relational knowledge transfer.

8.3 Transductive Transfer Learning

In 2007, Arnold *et al.* first proposed the term transductive transfer learning (Arnold *et al.*, 2007), where they required that the source and target tasks be the same, although the domains may be different. On top of these conditions, they further required that all unlabeled data in the target domain are available at training time, but we believe that this condition can be relaxed; instead, in our definition of the transductive transfer learning setting, we only require that part of the unlabeled target data be seen at training time in order to obtain the marginal probability for the target data.

Definition 8.4 (Transductive Transfer Learning). Given a source domain D_S and a corresponding learning task T_S, a target domain D_T and a corresponding learning task T_T, transductive transfer learning aims to improve the learning of the target predictive function $f_T(\bullet)$ in D_T using the knowledge in D_S and T_S, where $D_S \neq D_T$ and $T_S = T_T$. In addition, some unlabeled target-domain data must be available at training time.

In transductive transfer learning, a lot of labeled data is present in the source domain, whereas no labeled data is present in the target domain. In this setting, both source tasks and target tasks are similar, whereas there is a difference in domain only. Further two more cases arise in transductive transfer learning depending upon distinctive circumstances among the source and the target domain. In the first case, it is considered that feature spaces are different in the source and the target domain, i.e., $X_s \neq X_T$, and in case two, it is said that feature spaces among source domain and the target domain remain similar but they have different marginal, i.e., $P(X_S) \neq P(X_T)$.

The second case discussed above is associated with domain alteration for knowledge transfer. Transductive transfer learning is used in recognition of electroencephalogram signals and in spectrum optimization. Transductive transfer learning consists of a given source domain (D_S) which contains a learning task T_{SL} and the target domain (D_T) which contains a learning task T_{TL}. The aim of transductive transfer learning is to develop the learning of target $f_T P(\bullet)$ in target domain via knowledge in source domain and task, given $D_S \neq D_T$ and $T_{SL} = T_{TL}$. The different approaches used in transductive transfer learning are instance transfer as well as feature representation transfer.

Similar to the traditional transductive learning setting, which aims to make the best use of the unlabeled test data for learning, in our classification scheme under transductive transfer learning, we also assume that some target-domain unlabeled data be given. In the above definition of transductive transfer learning, the source and target tasks are the same, which implies that one can adapt the predictive function learned in the source domain for use in the target domain through some unlabeled target-domain data.

In general, we might want to learn the optimal parameters θ^* of the model by minimizing the expected risk:

$$\theta^* = \arg\min_{\theta \in \Theta} \; \mathrm{E}_{(x,y) \in P}[l(x, y, \theta)] \tag{8.14}$$

where $l(x, p, \theta)$ is a loss function that depends on the parameter θ. However, since it is hard to estimate the probability distribution P, we choose to minimize the empirical risk minimization (ERM) instead:

$$\theta^* = \arg\min_{\theta \in \Theta} \frac{1}{n} \sum_{i=1}^{n} [l(x_i, y_i, \theta)] \tag{8.15}$$

where n is the size of the training data.

In the transductive transfer learning setting, we want to learn an optimal model for the target domain by minimizing the expected risk:

$$\theta^* = \arg\min_{\theta \in \Theta} \sum_{(x,y) \in D_T} P(D_T) l(x, y, \theta) \tag{8.16}$$

However, since no labeled data in the target domain are observed in training data, we have to learn a model from the source domain data instead. If $P(D_S) = P(D_T)\mathrm{P}$, then we may simply learn the model by solving the following optimization problem for use in the target domain:

$$\theta^* = \arg\min_{\theta \in \Theta} \sum_{(x,y) \in D_S} P(D_S) l(x, y, \theta) \tag{8.17}$$

Otherwise, when $P(D_S) \neq P(D_T)$, we need to modify the above optimization problem to learn a model with high generalization ability for the target domain.

8.4 Model-Based Transfer Learning

Model-based transfer learning, also known as parameter-based transfer learning, assumes that the source task and target task share some common knowledge at the model level. In general, model-based transfer learning can be divided into transfer learning based on shared model components and transfer knowledge based on regularization. The former can be further divided into using Bayesian models and t using Gaussian processes.

8.4.1 *Bayesian Models*

Transfer learning based on probability models is mostly aimed at text data classification problems, which can be defined as Definition 8.5.

Definition 8.5 (Text data classification). Given one or more document sets D_l from the source domain for any document $di \in D_l$. It can be represented as an aggregation within the text but before, for each word $w_j, \in V$, where V represents the entire vocabulary. In the source domain, we have category labels y for each document d_i. Additionally, given the target domain D_u, only document data is included. The goal of this task is to achieve efficient and accurate label prediction on target domain D_u by jointly learning the data distribution of the source domain and the target domain itself.

Here $D = \{d_1, d_2, \cdots, d_m\}$ represent the data documents and $d = \{w_1, w_2, \cdots, w_N\}$ represents each document. Each w_i is a word in the document, we use V to represent the entire vocabulary, z to represent the topic of the document, and k is the number of topics.

1. Probabilistic latent semantic analysis (PLSA)

PLSA (Hoffman, 2001) is a statistical latent class model which introduces a hidden variable (latent aspect) z_k in the generative process of each element x_j in a document d_i. Given this unobservable variable z_k, each occurrence x_j is independent of the document it belongs to, which corresponds to the following joint probability: $p(x_j, z_k, d_i) = p(d_i)p(z_k|d_i)p(x_j|z_k)$. The joint probability of the observed variables is obtained by marginalization over the latent aspect z_k:

$$p(d_l, x_j) = p(d_l) \sum_{k=1}^{K} p(z_k|d_l)p(x_j|z_k) \tag{8.18}$$

A representation of the PLSA model is depicted in Figure 8.2.

The model parameters of PLSA are the two conditional distributions: $P(x_j|z_k)$ and $P(z_k|d_i)$. $P(x_j|z_k)$ characterizes each PLSA and remains valid for documents out of the training set. On the other hand, $P(z_k|d_i)$ is only relative to the specific documents and cannot carry any prior information to an unseen document.

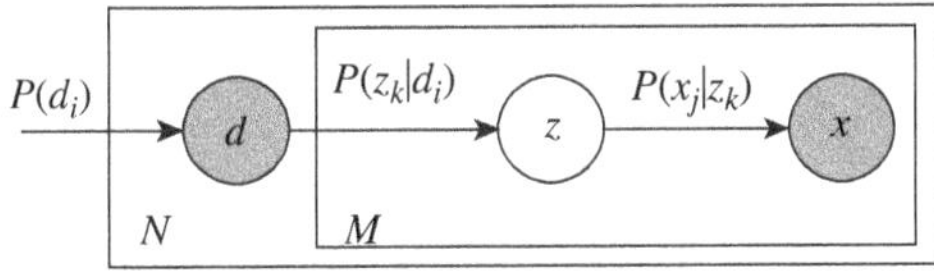

Fig. 8.2. PLSA model.

2. latent Dirichlet allocation (LDA)

In 2003, Blei *et al.* proposed latent Dirichlet allocation (LDA), which introduced a prior distribution of parameters based on the PLSA model (Blei *et al.*, 2003). A representation of the LDA model is depicted in Figure 8.3. The processes of document classification generation are as follows:

(1) Sample the topic distribution θ_i of document d_i generated from the Dirichlet distribution α.
(2) Sample from polynomial distribution θ_l to generate the z_j topic in d_i.
(3) Generate word distribution ϕ_{zj} corresponding to topic z_j by sampling from the Dirichlet distribution β.
(4) Generate word w_l by sampling from polynomial distribution ϕ_{zj}.

A representation of the LDA model is depicted in Figure 8.3.

8.4.2 *Gaussian Process (GP)*

Gaussian processes are a type of stochastic process in probability theory and mathematical statistics, which is an extension of multivariate Gaussian distributions and is applied in fields such as machine learning and signal processing. We can consider that the probability distribution of the source domain dataset D_l is D_l, and the probability distribution of the target domain dataset D_u is D_u. In naive Bayes transfer method, first, train based on D_l to obtain the initial model, then further use the maximum expectation algorithm to fit the local optimal model on D_u.

Based on the principle of local maximum posterior estimation, the model maximizes on D_l and D_u:

$$l(h|D_l, D_u) = \log p_{D_u}(h|D_l, D_u) \tag{8.19}$$

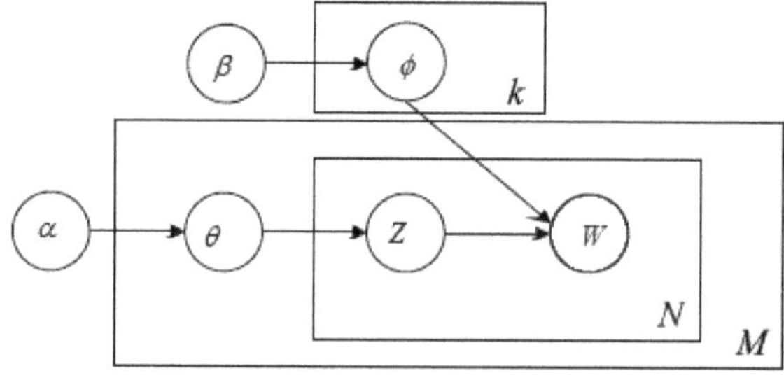

Fig. 8.3.   LDA model.

That is,

$$l(h|D_l, D_u) \propto \log p_{D_u}(h) + \sum_{d \in D_l} \log \sum_{c \in C} p_{D_u}(d|c, h) \cdot p_{D_u}(c|h)$$

$$+ \lambda \sum_{d \in D_u} \log \sum_{c \in C} p_{D_u}(d|c, h) \cdot p_{D_u}(c|h) \qquad (8.20)$$

Here, we use $\lambda \in (0, 1)$ for weakening the impact of unlabeled data D_u. Then use the following EM algorithm:

(1) Step E:

$$D_{D_u} = (c|h) \propto p_{D_u}(c) \prod_{w \in d} p_{D_u}(w|c) \qquad (8.21)$$

(2) Step M

$$p_{D_u}(c) \propto \sum_{i \in l, u} p_{D_u}(D_i) \cdot p_{D_u}(c|D_i)$$

$$p_{D_u}(w|c) \propto \sum_{i \in l, u} p_{D_u}(D_i) \cdot p_{D_u}(c|D_i) \cdot p_{D_u}(w|c, D_i) \qquad (8.22)$$

Algorithm 8.1. Naïve Bayes Transfer Classifier.

Input: A labeled training set D_l under the distribution D_l. An unlabeled test set D_u under the distribution D_u; A set C of all the class labels; A set W of all the word features; and the maximum number of iterations T.

Output: $p_{D_u}^{(T)}(c|d)$

1. Initialize $p_{D_u}^{(0)}(w|c)$, $p_{D_u}^{(0)}(c)$, *and* $p_{D_u}^{(0)}(w|c)$, via traditional naive Bayes classifier algorithm.
2. for t in 1 to T do
3. for each $c \in C$ and $d \in D_u$ do
4.   Calculate $p_{D_u}^{(t)}(c|d)$ via the Equation (8.21).
5. end for
6. for each $c \in C$ do
7.   Calculate $p_{D_u}^{(t)}(c)$ via the Equation (8.22).
8. for each $w \in W$ do
9.   Calculate $p_{D_u}^{(t)}(w|c)$ via the Equation (8.22).
10.   end for
11.   end for
12. end for

In recent years, a lot of researchers proposed several transfer learning methods which are applied in text mining applicated area and others.

8.5 Deep Transfer Learning

Deep transfer learning can be divided into four categories: instance-based deep transfer learning, mapping-based deep transfer learning, network-based deep transfer learning, and adversarial-based deep transfer learning.

1. Instance-based Deep Transfer Learning

Instance-based deep transfer learning is the use of a specific weight adjustment strategy to select a subset of instances from the source domain to complement the target domain training set by assigning appropriate weights to those selected instances. This type of scheme is based on the assumption that there is an intersection between the source and target domains.

This type of method generally assumes that the mapping relationship between x and y in the source domain and target domain is constant, the label space y is constant, and only the feature space x changes. The selection of samples must be simple (the computational complexity cannot be infinitely high), the diversity of the sampled data subset should be close to the distribution of the whole set and representative, and the sampled data itself should have typical characteristics.

2. Mapping-based Deep Transfer Learning

A map-based deep transfer learning is the mapping of source and destination domains to a new data space; in this new data space, instances from both domains are similar and suitable for federated deep neural networks. This class of scenarios is based on the assumption: Although two there are differences between the original domains, but they may be more similar in a well-designed new data space.

3. Network-based Deep Transfer Learning

Network-based deep transfer learning refers to the reuse of pre-trained parts of the network in the source domain, including its network structure and training parameters, which are migrated to the deep neural network used in the target as part of the new network structure. This type of scheme is based on the hypothesis that neural networks are similar to the processing mechanisms of the human brain; it is an iterative and continuous abstraction process. On the Internet, the front layer is considered a feature extractor, and the extracted features are generic.

4. Adversarial-based Deep Transfer Learning

The adversarial-based deep transfer learning introduces adversarial techniques in GAN to find migratory tables for both source and destination domains' attack.

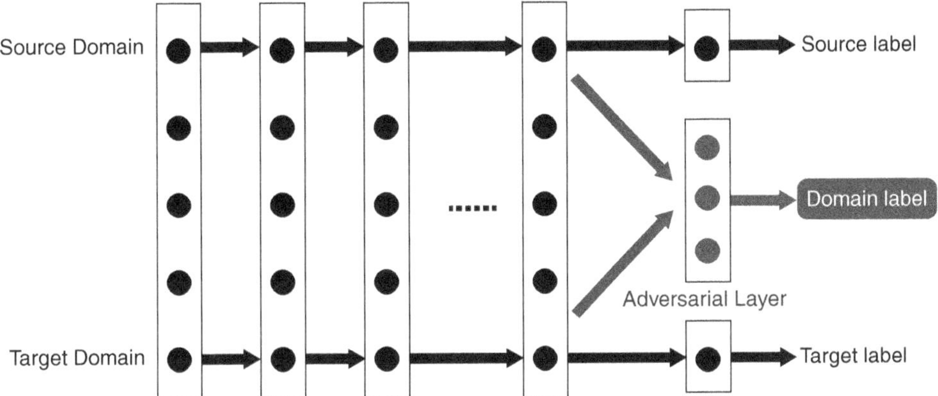

Fig. 8.4.   Sketch map of adversarial-based deep transfer learning.

Such schemes are based on the assumption that, in order to transfer effectively, a good representation should provide discriminating power for the primary transfer task and be indistinguishable between the source and target domains.

In Figure 8.4, which shows the training process on large-scale dataset in the source domain, the front layers of the network are regarded as a feature extractor. It extracts features from two domains and sends them to the adversarial layer. The adversarial layer tries to discriminate the origin of the features. If the adversarial network achieves worse performance, it means a small difference between the two types of features and better transferability, and vice versa. In the following training process, the performance of the adversarial layer will be considered to force the transfer network to discover general features with more transferability.

8.6 Heterogeneous Transfer Learning

Heterogeneous transfer learning refers to the knowledge transfer process in situations where the domains have different feature spaces. In addition to distribution adaptation, heterogeneous transfer learning requires feature space adaptation, which makes it more complicated than homogeneous transfer learning. In heterogeneous transfer learning, the feature spaces between the source and target are non-equivalent and are generally non-overlapping. In this case, $X_s \neq X_t$ and/or $Y_s \neq Y_t$ as the source and target domains may share no features and/or labels, while the dimensions of the feature spaces may differ as well. This method thus requires feature and/or label space transformations to bridge the gap for knowledge transfer, as well as handling the cross-domain data distribution differences.

Most heterogeneous transfer learning solutions fall into two categories when it comes to transforming the feature spaces: symmetric and asymmetric transformation. Symmetric transformation, illustrated in Figure 8.5(a), takes both the source feature space X_s and target feature space X_t and learns feature transformations to project each onto a common subspace X_C for adaptation purposes. This derived subspace becomes a domain-invariant feature subspace to associate cross-domain data and, in effect, reduces marginal distribution differences. Performing this brings the feature spaces for both domains together into a common feature representation where one can then apply traditional machine learning models, such as Support Vector Machines (SVM). Optimally, one can also apply models built for homogeneous transfer learning which consider the distribution differences and domain transfer ability observed in the subspace. Asymmetric transformation mapping, illustrated in Figure 8.5(b), transforms the source feature space to align with that of the target $(X_t \rightarrow TX_s)$ or the target to that of the source $(X_s \rightarrow TX_t)$. This, in effect, bridges the feature space gap and reduces the problem into a homogeneous transfer problem when further distribution differences need to be corrected. This approach is most appropriate when the source and target have the same class label space and one can transform X_s and X_t without context feature bias. Context feature bias occurs when there are conditional distribution differences between the domains as a feature in one domain may have a different meaning in another. In either category, once the issue of varied feature spaces is resolved, we may need to solve marginal and/or conditional distribution differences. This can be done through homogeneous adaption solutions which account for these distribution differences observed during cross-domain tasks.

Dai *et al.* (2017) provided a comprehensive survey of 38 methods which are designed to handle these heterogeneous transfer learning tasks. All methods have the commonality that they propose a methodology for knowledge transfer when

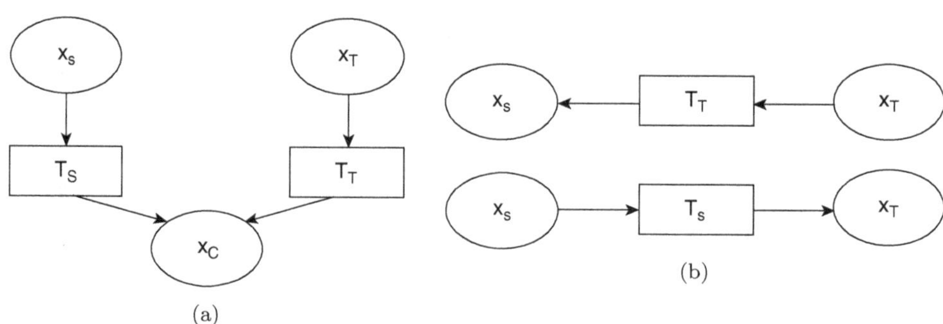

Fig. 8.5.   Heterogeneous transfer learning: (a) Symmetric and (b) asymmetric.

faced with differing feature spaces. Another pattern that can be noted is the use of Canonical Correlation Analysis (CCA) to solve the differing feature space issue.

Qi *et al.* (2011) proposed an HTL method for image classification. This method uses a Translator for Text to Images (TTI). TTI uses labeled source text data, text-image co-occurrence data, and limited labeled image target data. Image classification currently has two challenges: (1) labeled image data is relatively scarce and expensive to collect and (2) features of image data lack semantic meaning for class prediction as they represent visual features rather than conceptual ones. On the other hand, labeled text data is often more available than labeled image data and text features have more semantic meaning for predicting a class label. With this observation, this method proposes using transfer learning to exploit such text data to improve image classification. The issue becomes how to relate the text to the images for semantic knowledge transfer. To close this gap, this method uses a text-image co-occurrence matrix which contains images along with the text that occurred with them on the same webpage. Co-occurrence information is effective in this case because of the assumption that the text around an image is describing the concepts in such an image. This co-occurrence information is relatively inexpensive to collect and serves as a bridge to learn the correspondence for translating the semantic information between the features of the text source domain to that of the image target domain. This translation is done through a form of a feature transformation called a semantic translator function. This translator takes into account the source, target, and co-occurrence data and learns the correspondence between the text from the source and the images of the target through the co-occurrence bridge. Each translator for the source text contains a topic space which is a common subspace to associate the data for translation. As shown in Figure 8.6, each of the translators is

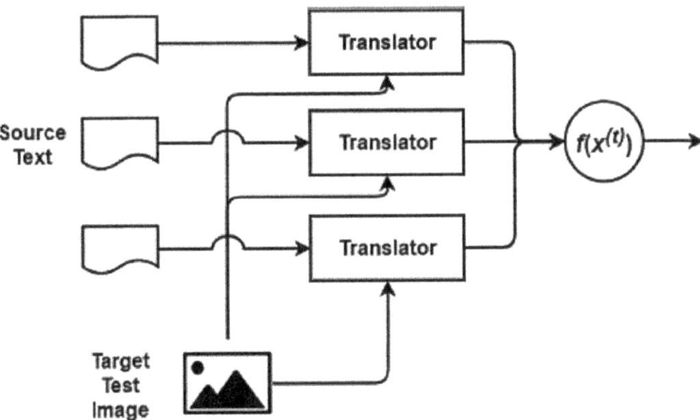

Fig. 8.6.   Illustration for TTI from text to image.

aggregated to create a final prediction function f(x(t)). The proposed TTI had the best performance in most cases, even when the number of labeled target samples was reduced.

8.7 Multi-task Transfer Learning

Multi-task learning is an approach to inductive transfer that improves generalization by using the domain information contained in the training signals of related tasks as an inductive bias. It does this by learning tasks in parallel while using a shared representation; what is learned for each task can help other tasks be learned better, as shown in Figure 8.7.

In the classification context, MTL aims to improve the performance of multiple classification tasks by learning them jointly. One example is a spam filter, which can be treated as distinct but related classification tasks across different users. To make this more concrete, consider that different people have different distributions of features which distinguish spam emails from legitimate ones, for example, an English speaker may find that all emails in Russian are spam, but not so for Russian speakers. Yet there is a definite commonality in this classification task across users, for example, one common feature might be text related to money transfer. Solving each user's spam classification problem jointly via MTL can let the solutions inform each other and improve performance. Further examples of settings for MTL include multi-class classification and multi-label classification.

Multi-task learning works because regularization induced by requiring an algorithm to perform well on a related task can be superior to regularization that prevents

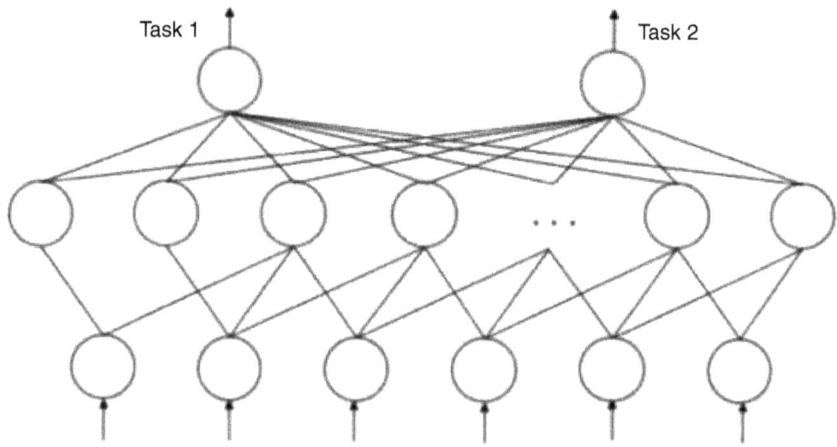

Fig. 8.7. Multi-task learning.

overfitting by penalizing all complexity uniformly. One situation where MTL may be particularly helpful is if the tasks share significant commonalities and are generally slightly under-sampled. However, as discussed in the following, MTL has also been shown to be beneficial for learning unrelated tasks.

Related to multi-task learning is the concept of knowledge transfer. Whereas traditional multi-task learning implies that a shared representation is developed concurrently across tasks, transfer of knowledge implies a sequentially shared representation. Large-scale machine learning projects such as the deep convolutional neural network GoogLeNet, an image-based object classifier (Christian *et al.*, 2015), can develop robust representations which may be useful to further algorithms learning-related tasks.

8.8 Domain Adaptation Transfer Learning

Domain adaptation is the ability to apply an algorithm trained in one or more "source domains" to a different (but related) "target domain". Domain adaptation is a subcategory of transfer learning. In domain adaptation, the source and target domains all have the same feature space (but different distributions); in contrast, transfer learning includes cases where the target domain's feature space is different from the source feature space or spaces.

The adaptation efficiency is directly correlated with two main terms that inevitably appear in almost all analyses: one term that depicts the divergence between the domains, and the other term that stands for the existence and the error achieved by the best hypothesis across the source and target domains. In 2010, Ben-David *et al.* proposed these two terms in the bounds by answering the following questions (Ben-David *et al.*, 2010):

(1) Is the presence of these two terms inevitable in the domain adaptation bounds?
(2) Is there a way to design a more intelligent domain adaptation algorithm that uses not only the labeled training sample but also the unlabeled sample of the target data distribution?

These two questions are very important, as answering them can help us obtain an exhaustive set of conditions that theoretically ensure efficient adaptation with respect to a given domain adaptation algorithm.

A domain adaptation leaner is a function

$$A: \bigcup_{m=1}^{\infty} \bigcup_{n=1}^{\infty} (X \times \{0, 1\}^m \times X^n \rightarrow \{0, 1\}^X \tag{8.23}$$

We say that $A(\varepsilon, \delta, m, n)$ learns T from S relative to H, if when given access to a labeled sample S of size m, generated by S, and an unlabeled sample T_u of size n, generated by T_X, with probability of at least $1 - \delta$, the learned classifier does not exceed the error of the best classifier in H by more than ε:

$$\Pr_{S\sim(S)^m, T_u\sim(T_x)^n} [R_T(A(S, T_u)) \leq cR_T(H) + \varepsilon] \geq 1 - \delta \qquad (8.24)$$

In other words, equation (8.24) says that the proper solving of the domain adaptation problem is achieved when the error of the returned hypothesis from a fixed hypothesis class w.r.t. the target distribution is bounded by c times the error of the best hypothesis on the target distribution plus a constant. Obviously, efficient solving of the proper domain adaptation is characterized by small δ and ε, and c close to 1. We also note that for both of the definitions given above, the inequality event can be reduced to $R_T(A(S; T_u)) \leq \varepsilon$ when the hypothesis class H contains a zero-error hypothesis, i.e., $R_T(H) = 0$.

Exercises

8.1 What is transfer learning? What is the difference between inductive and transductive transfer learning?

8.2 How to measure the similarity between target domain and source domain?

8.3 What is model-based transfer learning? Please list two types of model-based transfer learning.

8.4 Explain the principle of adversarial-based deep transfer learning.

8.5 Make an example to illustrate the heterogeneous transfer learning.

8.6 What is multi-task transfer learning?

8.7 Please develop an algorithm to implement the domain adaptation transfer learning.

Chapter 9

Federated Learning

Federated learning has emerged as an effective paradigm to achieve privacy-preserving collaborative learning among different parties. Compared to traditional centralized learning that requires collecting data from each party, in federated learning, only the locally trained models or computed gradients are exchanged, without exposing any data information. As a result, it is able to protect privacy to some extent.

9.1 Introduction

The traditional method of training AI models involves setting up servers where models are trained on data, often through the use of a cloud-based computing platform. However, over the past few years, an alternative form of model creation has arisen, called federated learning. Federated learning brings machine learning models to the data source rather than bringing the data to the model. Federated learning links together multiple computational devices into a decentralized system that allows the individual devices that collect data to assist in training the model.

In a federated learning system, the various devices that are part of the learning network each have a copy of the model on the device. The different devices/clients train their own copy of the model using the client's local data, and then the parameters/weights from the individual models are sent to a master device, or server, that aggregates the parameters and updates the global model. This training process can then be repeated until a desired level of accuracy is attained. In short, the idea behind federated learning is that none of the training data is ever transmitted between devices or between parties, only the updates related to the model are.

Federated learning can be broken down into three different steps or phases. Federated learning typically starts with a generic model that acts as a baseline and is trained on a central server:

(1) The generic model is sent out to the application's clients. These local copies are then trained on data generated by the client systems, learning and improving their performance.
(2) The clients all send their learned model parameters to the central server. This happens periodically, on a set schedule.
(3) The server aggregates the learned parameters when it receives them. After the parameters are aggregated, the central model is updated and shared once more with the clients. The entire process then repeats.

The benefit of having a copy of the model on the various devices is that network latencies are reduced or eliminated. The costs associated with sharing data with the server are eliminated as well. Other benefits of federated learning methods include the fact that federated learning models are privacy preserved, and model responses are personalized for the user of the device.

Figure 9.1 shows an example of the application of federated learning for the task of next-word prediction on mobile phones (Li *et al.*, 2019). To preserve the privacy of the text data and to reduce strain on the network, we seek to train a predictor in a distributed fashion rather than sending the raw data to a central server. In this setup, remote devices communicate with a central server periodically to learn a global model. At each communication round, a subset of selected phones performs local training on their non-identically distributed user data and sends these local updates to the server. After incorporating the updates, the server then sends back the new

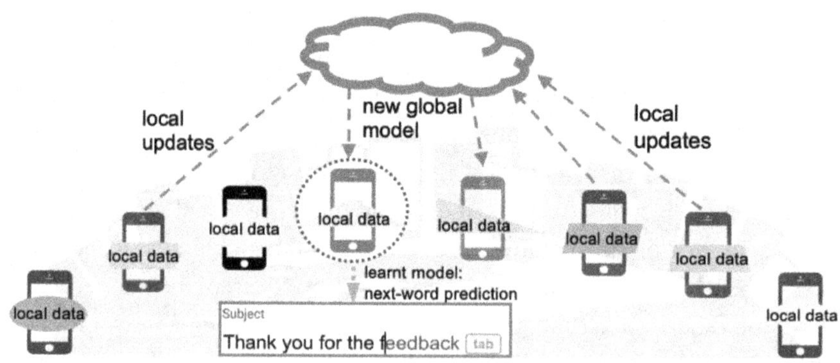

Fig. 9.1.   Next-word prediction on mobile phones.

global model to another subset of devices. This iterative training process continues across the network until convergence is reached or some stopping criterion is met.

Federated learning models include recommendation engines, fraud detection models, and medical models. Media recommendation engines, of the type used by Netflix or Amazon, could be trained on data gathered from thousands of users. The client devices would train their own separate models and the central model would learn to make better predictions, even though the individual data points would be unique to the different users. Similarly, fraud detection models used by banks can be trained on patterns of activity from many different devices, and a handful of different banks could collaborate to train a common model. In terms of a medical federated learning model, multiple hospitals could team up to train a common model that could recognize potential tumors through medical scans.

Four of the core challenges associated with solving the distributed optimization problem are as follows:

(1) **Expensive communication:** Communication is a critical bottleneck in federated networks, which, coupled with privacy concerns over sending raw data, necessitates that data generated on each device remain local. Indeed, federated networks are potentially comprised of a massive number of devices, e.g., millions of smartphones, and communication in the network can be slower than local computation by many orders of magnitude. In order to fit a model to data generated by the devices in the federated network, it is therefore necessary to develop communication-efficient methods that iteratively send small messages or model updates as part of the training process, as opposed to sending the entire dataset over the network. To further reduce communication in such a setting, two key aspects to consider are as follows: (a) reducing the total number of communication rounds or (b) reducing the size of transmitted messages at each round.

(2) **Systems heterogeneity:** The storage, computational, and communication capabilities of each device in federated networks may differ due to variability in hardware (CPU and memory), network connectivity, and power (battery level). Additionally, the network size and systems-related constraints on each device typically result in only a small fraction of the devices being active at once, e.g., hundreds of active devices in a million-device network. Each device may also be unreliable, and it is not uncommon for an active device to drop out at a given iteration due to connectivity or energy constraints. These system-level characteristics dramatically exacerbate challenges, such as straggler mitigation and fault tolerance. Federated learning methods that are developed and analyzed

must therefore (a) anticipate a low amount of participation, (b) tolerate hetero-geneous hardware, and (c) be robust to dropped devices in the network.

(3) **Statistical heterogeneity:** Devices frequently generate and collect data in a non-identically distributed manner across the network, e.g., mobile phone users have varied use of language in the context of a next-word prediction task. Moreover, the number of data points across devices may vary significantly, and there may be an underlying structure present that captures the relationship among devices and their associated distributions.

(4) **Privacy concerns:** Finally, privacy is often a major concern in federated learning applications. Federated learning makes a step toward protecting data generated on each device by sharing model updates, e.g., gradient information, instead of the raw data. However, communicating model updates throughout the training process can nonetheless reveal sensitive information, either to a third-party or to the central server. While recent methods aim to enhance the privacy of feder-ated learning using tools such as secure multi-party computation or differential privacy, these approaches often provide privacy at the cost of reduced model performance or system efficiency. Understanding and balancing these trade-offs, both theoretically and empirically, is a considerable challenge in realizing private federated learning systems.

9.2 Privacy-Preserving in Federated Learning

The federated setting poses novel challenges to existing privacy-preserving algo-rithms. Beyond providing rigorous privacy guarantees, it is necessary to develop methods that are computationally cheap, communication-efficient, and tolerant to dropped devices — all without overly compromising accuracy. Although there are a variety of privacy definitions in federated learning, typically they can be classified into two categories: global privacy and local privacy. Global privacy requires that the model updates generated at each round are private to all untrusted third parties other than the central server, while local privacy further requires that the updates are also private to the server.

Current works that aim to improve the privacy of federated learning typically build upon previous classical cryptographic protocols, such as SMC (Bonawitz *et al.*, 2016). Bonawitz *et al.* introduced an SMC protocol to protect individual model updates. The central server is not able to see any local updates but can still observe the exact aggregated results at each round. SMC is a lossless method and can retain the original accuracy with a very high privacy guarantee. However, the resulting method incurs significant extra communication cost.

9.3 Horizontal Federated Learning

Federated learning is a machine learning paradigm aimed at learning models from decentralized data, such as data located on users' smartphones, in hospitals, or banks, and ensuring data privacy. This is achieved by training the model locally in each node (e.g., on each smartphone, at each hospital, or at each bank), sharing the model-updated local parameters (not the data), and securely aggregating them to build a better global model.

Traditional machine learning requires all the data to be gathered in one single place. In practice, this is often forbidden by privacy regulations. For this reason, federated learning is introduced the goal being to learn from a large amount of data while preserving privacy. Horizontal federated learning is introduced in those scenarios, where datasets share the same feature space (same type of columns) but differ in samples (different rows).

Definition 9.1 (Federated learning). Define N data owners $\{O_1, \ldots O_N\}$, all of whom wish to train a machine learning model by consolidating their respective data $\{D_1, \ldots D_N\}$. A conventional method is to put all data together and use $D = D_1 \cup \ldots \cup D_N$ to train a model M_{SUM}. A federated learning system is a learning process in which the data owners collaboratively train a model M_{FED}, in which process any data owner O_i does not expose its data D_i to others. In addition, the accuracy of M_{FED}, denoted as V_{FED}, should be very close to the performance of M_{SUM}, V_{SUM}. Formally, let δ be a non-negative real number, if

$$|V_{FED} - V_{SUM}| < \delta \tag{9.1}$$

we say the federated learning algorithm has δ-accuracy loss.

Horizontal federated learning is a system in which all the parties share the same feature space. However, their user base may be significantly different. Such parties can collaboratively learn a model with the help of a server. Each party locally computes training gradients and masks them with some encryption or privacy-preservation technique. All parties send encrypted gradient to the server. The server aggregates them and sends the aggregated result to all parties. Parties update their model with the decrypted aggregated gradient and this way all parties share final model parameters. An example of horizontal federated learning is a set of banks located in the same city or region and thus sharing the same set of features, however very few common users. Horizontal federated learning can be used to learn a common model by agglomerating models learned in individual banks. Horizontal federal learning can be implemented with different machine learning algorithms without changing the main framework.

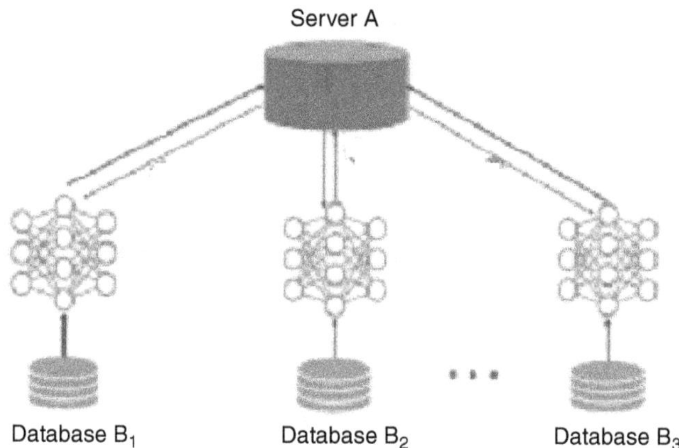

Fig. 9.2.   Client-server architecture.

Definition 9.2 (Horizontal federated learning). Assume owners i and j have datasets D_i *and* D_j. The conditions for horizontal federated learning are defined as

$$X_i = X_j, \quad Y_i = Y_j, \quad I_i \neq I_j, \quad \forall D_i, D_j, i \neq j \tag{9.2}$$

where we assume the owners' datasets have the same feature space and label space, that is, (X_i, Y_i) and (X_j, Y_j) are the same, I_i and I_j do not intersect or have very small intersections.

Two commonly used federated learning architectures client–server and peer-to-peer are introduced as follows.

1. Client–server architecture

Typical client–server architecture is shown in Figure 9.2 (Phong *et al.*, 2018).

The training process of client–server system usually contains the following four steps:

(1) Participants locally compute training gradients, mask a selection of gradients with encryption, differential privacy, or secret sharing techniques, and send masked results to the server.
(2) The server performs secure aggregation without learning information about any participant.
(3) The server sends back the aggregated results to participants.
(4) Participants update their respective models with the decrypted gradients.

2. Peer-to-peer architecture

Horizontal federated learning also can be implemented by peer-to-peer network, as shown in Figure 9.3 (Phong & Phuong, 2019). In this architecture, there is no central

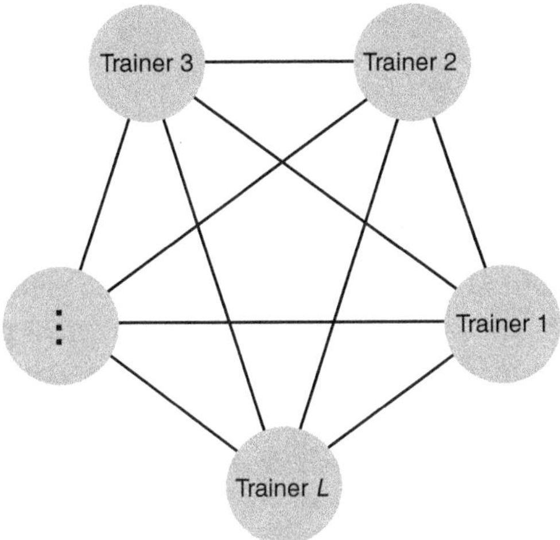

Fig. 9.3.   Peer-to-peer architecture.

server or coordinating server. In horizontal federated learning system, K participants are called as trainer or distributed trainer. Each participant only uses local data to train the same machine learning model, such as DNN. The training parties use secure links to transmit model parameter information to each other. To ensure communication security between any two parties, it is necessary to use security measures such as encryption methods based on public keys.

There are two communication methods: cyclic transfer and random transfer. In the cyclic transmission mode, the trainers are organized into a link, and the first trainer sends the current model parameters to its next trainer. After receiving model parameters from upstream, the training party will update the received model parameters with a small batch of data from the local dataset. Afterward, it will transfer the updated model parameters to the next training party, such as from training party 1 to training party 2, from training party 2 to training party 3, etc., from training party $(K\text{-}1)$ to training party K, and then from training party K back to training party 1. This process will be continuously repeated until the model parameters converge or reach the maximum allowed training time.

In random transmission mode, the kth training party selects i from $\{1, \ldots, L\}$ and sends the model parameters to the training party i. When the ith training party receives model parameters from the kth training party, it will use the mini-batch of data from the local dataset to update the received model parameters. Afterward, the ith training party also randomly selects a number j with moderate probability from $\{1, \ldots, L\}$ and sends its own model parameters to the training party j. This process will be repeated until K training parties agree that the model parameters converge or reach the maximum allowed training time. This method is also known as gossip learning.

9.4 Vertical Federated Learning

In vertical federated learning, participating parties do not expose users that do not overlap among the parties (Hardy *et al.*, 2017). Overlapping users are found by using an encryption-based user ID alignment. Since different parties have different features corresponding to the common users, vertical federated learning aggregates different features from different parties and computes the training loss and gradients in a privacy-preserving manner (Hardy *et al.*, 2017). Subsequently, computed gradients are used to train the model. Vertical federated learning assumes honest participants and there is no hard requirement of a third party. However, sometimes to secure computations between the participants, an additional party is introduced. An example of vertical federated learning is the case of cooperation between the online retailers and the insurers. They own their own feature space (and labels). However, they have a significant amount of common users. In such cases, vertical federated learning exploits the situation by merging the features together to create a larger feature space for machine learning tasks. The architecture of vertical federated learning system is shown in Figure 9.4 (Yang *et al.*, 2019).

Definition 9.3 (Vertical federated learning). Vertical federated learning is the process of aggregating these different features and computing the training loss and gradients in a privacy-preserving manner to build a model with data from both parties collaboratively. Under such a federal mechanism, the identity and the status of each participating party are the same, and the federal system helps everyone establish a common wealth strategy, which is why this system is called federated learning. For such systems, we can define

$$X_i \neq X_j, \quad Y_i \neq Y_j, \quad I_i = I_j \quad \forall D_i, D_j, i \neq j \tag{9.3}$$

Vertical federated learning is applicable to cooperative statistical analysis, association rule mining, secure linear regression, classification, etc.

From Figure 9.4, we can see the process of vertical federated learning system consists of two parts:

1. Encrypted entity alignment
Since the user groups of company A and company B are not the same, the system uses encryption-based user ID alignment technologies to confirm the common users

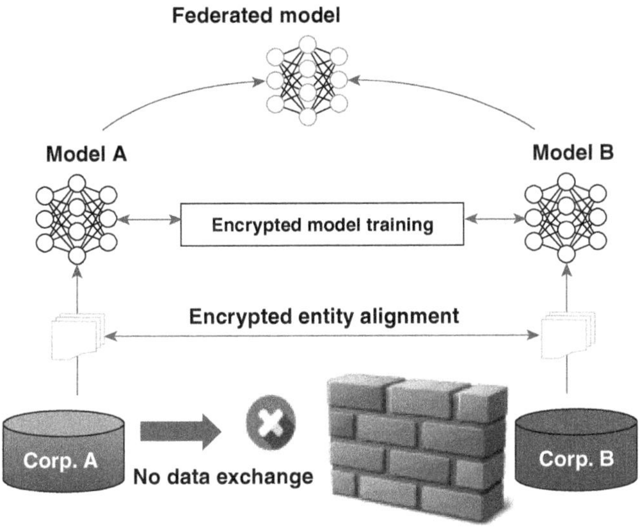

Fig. 9.4.   Architecture of vertical federated learning system.

of both parties without A and B exposing their respective data. During the entity alignment, the system does not expose users that do not overlap with each other.

2. Encrypted model training

After determining the common entities, we can use these common entities' data to train the model. The training process can be divided into the following four steps:

Step 1: Collaborators C create encryption pairs and send public key to A and B;

Step 2: A and B encrypt and exchange the intermediate results for gradient and loss calculations.

Step 3: A and B compute encrypted gradients and add additional masks, respectively, and also compute encrypted loss. A and B send encrypted values to C.

Step 4: C decrypts and sends the decrypted gradients and loss back to A and B; A and B unmask the gradients and update the model parameters accordingly.

9.5 Federated Transfer Learning

Federated transfer learning (FTL) system is a special case of federated learning and is different from both horizontal and vertical federated learning. In federated transfer learning, two datasets differ in the feature space. This applies to datasets collected from enterprises of different but similar nature. Due to the differences in the nature of business, such enterprises share only a small overlap in feature space. This is also

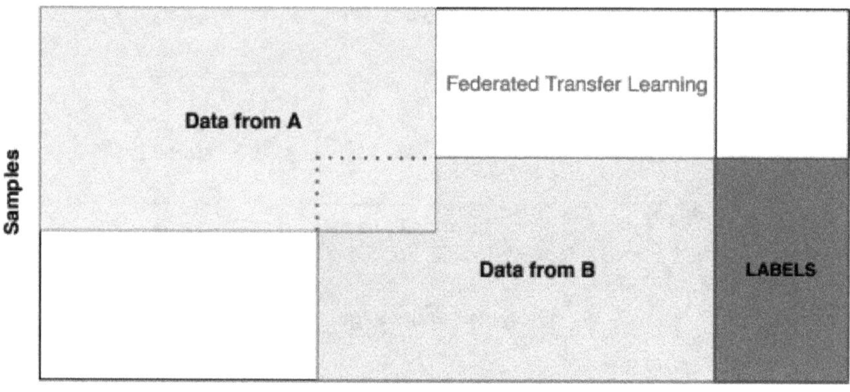

Fig. 9.5. Federated transfer learning.

applicable to the enterprises set up far in the globe. Thus in such scenarios, datasets differ both in samples and in feature space. Transfer learning techniques aim to build an effective model for the target domain while leveraging knowledge from the other (source) domains. A typical architecture of federated transfer learning is shown in Figure 9.5. Considering two parties A and B, where there is only a small overlap in feature space and sample space between A and B, a model learned on B is transferred to A by leveraging small overlapping data and features.

We recall that horizontal federated learning is used when there is a large overlap in the feature space between datasets and vertical federated learning is used when there is a large overlap in user/sample space between datasets. In contrast, FTL is used when there is a small overlap in both feature space and sample space, as shown by the dotted box in Figure 9.5. FTL ingests a model trained on source domain samples and feature space. Subsequently, FTL orients the model for reuse in target space such that the model is used for non-overlapping samples leveraging the knowledge acquired from source domain non-overlapping features. Thus, FTL covers the region in the right upper corner of Figure 9.5 by transferring knowledge from non-overlapping features from the source domain to the new samples in the target domain. The ability to use the transferred on non-overlapping data in A makes FTL different from vertical transfer learning.

9.6 Federated Reinforcement Learning

Supervised learning and unsupervised learning attempt to make the agent copy the dataset, i.e., learning from the pre-provided samples, Reinforcement learning (RL)

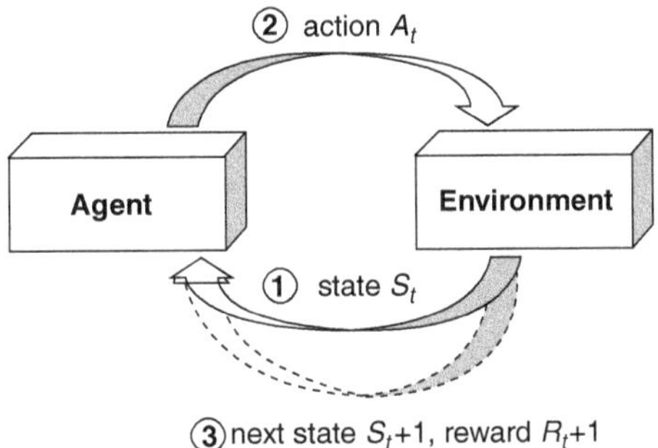

Fig. 9.6.   The basic reinforcement learning model.

is to make the agent gradually stronger in the interaction with the environment, i.e., generating samples to learn by itself. RL is a very hot research direction in the field of ML in recent years, which has made great progress in many applications. For example, the AlphaGo program developed by DeepMind is a good example to reflect the thinking of RL (Silver, 2016). The agent gradually accumulates the intelligent judgment on the sub-environment of each move by playing game by game with different opponents, so as to continuously improve its level.

The RL problem can be defined as a model of the agent–environment interaction, which is represented in Figure 9.6. The basic model of RL contains several important concepts, which are as follows:

(1) **Environment and agent:** Agents are a part of an RL model that exists in an external environment, such as the player in the environment of chess. Agents can improve their behavior by interacting with the environment. Specifically, they take a series of actions to the environment through a set of policies and expect to get a high payoff or achieve a certain goal.

(2) **Time step:** The whole process of RL can be discretized into different time steps. At every time step, the environment and the agent interact accordingly.

(3) **State:** The state reflects agents' observations of the environment. When agents take action, the state will also change. In other words, the environment will move to the next state.

(4) **Actions:** Agents can assess the environment, make decisions, and finally take certain actions. These actions are imposed on the environment.

(5) **Reward:** After receiving the action of the agent, the environment will give the agent the state of the current environment and the reward due to the previous action. Reward represents an assessment of the action taken by agents.

More formally, we assume that there are a series of time steps $t = 0, 1, 2, \ldots$ in a basic RL model. At a certain time step t, the agent will receive a state signal St of the environment. In each step, the agent will select one of the actions allowed by the state to take an action At. After the environment receives the action signal At, the environment will feed back to the agent the corresponding status signal $St + 1$ at the next step $t + 1$ and the immediate reward $Rt + 1$. The set of all possible states, i.e., the state space, is denoted as S. Similarly, the action space is denoted as A. Since our goal is to maximize the total reward, we can quantify this total reward, usually referred to as return with

$$G_t = R_{t+1} + R_{t+2} + \cdots + R_T \tag{9.4}$$

where T is the last step, i.e., S_T as the termination state. An episode is completed when the agent completes the termination action.

In addition to this type of episodic task, there is another type of task that does not have a termination state, in other words, it can in principle run forever. This type of task is called a continuing task. For continuous tasks, since there is no termination state, the above definition of return may be divergent. Thus, another way to calculate return is introduced, which is called discounted return, i.e.,

$$G_t = R_{t+1} + \gamma R_{t+2} + \gamma R_{t+3} + \cdots = \sum_{k=0}^{\infty} \gamma^k R_{t+k+1} \tag{9.5}$$

where the discount factor γ satisfies $0 \le \gamma \le 1$. When $\gamma = 1$, the agent can obtain the full value of all future steps, while when $\gamma = 0$, the agent can only see the current reward. As γ changes from 0 to 1, the agent will gradually become forward-looking, looking not only at current interests but also for its own future.

The results of RL are action decisions, called as the policy. The policy gives agents the action a that should be taken for each state s. It is noted as $\pi(A_t = a | S_t = s)$, which represents the probability of taking action $A_t = a$ in state $S_t = s$. The goal of RL is to learn the optimal policy that can maximize the value function by interacting with the environment. Our purpose is not to get the maximum reward after a single action in the short term but to get more reward in the long term. Therefore, the policy can be figured out as

$$\pi^* = \arg\max V_\pi(s), \quad \forall_s \in S \tag{9.6}$$

In RL, there are several categories of algorithms. One is value-based and the other is policy-based. In addition, there is also an actor-critic algorithm that can be obtained by combining the two, as shown in Figure 9.7.

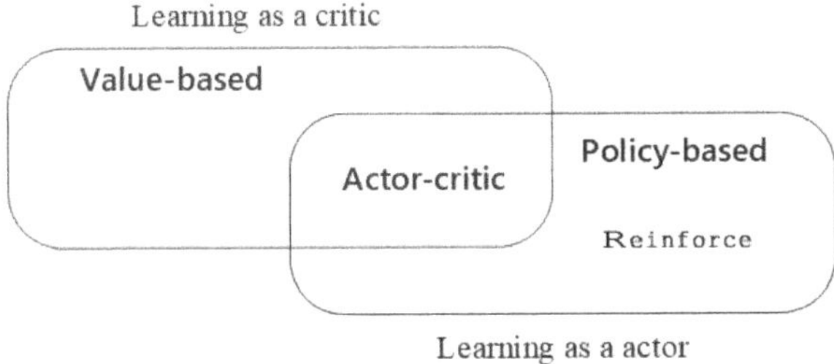

Fig. 9.7.   The categories of reinforcement learning.

1. Value-based Q-learning

Recursively expand the formulas of the action value function, and the corresponding Bellman equation is obtained, which describes the recursive relationship between the value function of the current state and subsequent state. The recursive expansion formula of the action value function $Q_\pi(s, a)$ is

$$Q_\pi(s, a) = \sum_{s',r} p(s', r|s, a) \left[r + \gamma \sum_{a'} \pi(a'|s') Q_\pi(s', a') \right] \tag{9.7}$$

where the function $p(s', r|s, a) = P_r\{S_t = s', R_t = r|S_{t-} = s, A_{t-1} = a\}$

defines the trajectory probability to characterize the environment's dynamics. $R_t = r$ indicates the reward obtained by the agent taking action $A_{t-1} = a$ in state $S_{t-1} = s$. Besides, $S_t = s'$ and $A_t = a'$ respectively represent the state and the action taken by the agent at the next moment t.

In the value-based algorithms, the above value function $Q_\pi(s, a)$ is calculated iteratively, and the strategy is then improved based on this value function. If the value of every action in a given state is known, the agent can select an action to perform. In this way, if the optimal $Q_\pi(s, a)$ can be figured out, the best action a^* will be found. There are many classical value-based algorithms, including Q-learning and State–Action–Reward–State–Action (SARSA) (Thorpe, 1997).

Q-learning is a typical widely used value-based RL algorithm. It is also a model-free algorithm, which means that it does not need to know the model of the environment but directly estimates the Q-value of each executed action in each encountered state by interacting with the environment. Then, the optimal strategy is formulated by selecting the action with the highest Q-value in each state. This strategy maximizes the expected return for all subsequent actions from the current state. The most important part of Q-learning is the update of Q-value. It uses a table, i.e., Q-table, to store all Q-value functions. Q-table uses state as row and action as column.

Each (s; a) pair corresponds to a Q-value, i.e., Q(s; a), in the Q-table, which is updated as follows:

$$Q(s, a) \leftarrow Q(s, a) + \alpha \left[r + \gamma \max_{a'} Q(s', a') - Q(s, a) \right] \tag{9.8}$$

where r is the reward given by taking action a under state s at the current time step. s' and a' indicate the state and the action taken by the agent at the next time step, respectively. α is the learning rate to determine how much error needs to be learned, and Υ is the attenuation of future reward. If the agent continuously accesses all state-action pairs, the Q-learning algorithm will converge to the optimal Q-function. Q-learning is suitable for simple problems, i.e., small state space, or a small number of actions. It has high data utilization and stable convergence.

9.7 Deep Reinforcement Learning

With the continuous expansion of the application of deep learning, its wave is also swept into the RL field, resulting in Deep Reinforcement Learning (DRL), i.e., using a multi-layer deep neural network to approximate value function or policy function in the RL algorithm. DRL mainly solves the curse-of-dimensionality problem in real-world RL applications with large or continuous state and/or action space, where the traditional tabular RL algorithms cannot store and extract a large amount of feature information.

Q-learning, as a very classical algorithm in RL, is a good example to understand the purpose of DRL. The big issue with Q-learning falls into the tabular method, which means that when state and action spaces are very large, it cannot build a very large Q-table to store a large number of Q-values. Besides, it counts and iterates Q-values based on past states. Therefore, on the one hand, the applicable state and action space of Q-learning are very small. On the other hand, if a state never appears, Q-learning cannot deal with it. In other words, Q-learning has no prediction ability and generalization ability at this point.

In order to make Q-learning with prediction ability, considering that neural network can extract feature information well, Deep Q Network (DQN) is proposed by applying deep neural network to simulate Q-value function. In specific, DQN is the continuation of Q-learning algorithm in continuous or large state space to approximate Q-value function by replacing Q-table with neural networks.

In addition to the value-based DRL algorithm such as DQN, we summarize a variety of classical DRL algorithms according to algorithm types by referring to some DRL-related surveys, including not only the policy-based and actor-critic DRL algorithms but also the advanced DRL algorithms of Partially Observable Markov Decision Process (POMDP) and multi-agents.

9.8 Federated Learning Applications

Federated learning typically applies when individual actors need to train models on larger datasets than their own but cannot afford to share the data in itself with others (e.g., for legal, strategic, or economic reasons). The technology yet requires good connections between local servers and minimum computational power for each node.

In smart manufacturing, there is a widespread adoption of machine learning techniques to improve the efficiency and effectiveness of industrial processes while guaranteeing a high level of safety. Nevertheless, privacy of sensitive data for industries and manufacturing companies is of paramount importance. Federated learning algorithms can be applied to these problems as they do not disclose any sensitive data.

Federated learning seeks to address the problem of data governance and privacy by training algorithms collaboratively without exchanging the data itself. Today's standard approach of centralizing data from multiple centers comes at the cost of critical concerns regarding patient privacy and data protection. To solve this problem, the ability to train machine learning models at scale across multiple medical institutions without moving the data is a critical technology. *Nature Digital Medicine* published the paper "The Future of Digital Health with Federated Learning" in September 2020, in which the authors explore how federated learning may provide a solution for the future of digital health and highlight the challenges and considerations that need to be addressed.

Wang & Feng (2024) applied federated learning in insurance precision marketing scenarios. The paper proposed the FedPV-FS method, which is the federated feature selection based on public verifiable covert (PVC) and verifiable secret sharing (VSS). Feature selection is a key aspect of federated learning tasks. The relevant algorithms are broadly divided into three categories: filter, wrapper, and embedded methods:

(1) **Filter:** Filter methods use variable ranking techniques as the principle criteria for variable selection by ordering.
(2) **Wrapper:** Wrapper methods use the predictor as a black box and the predictor performance as the objective function to evaluate the variable subset.

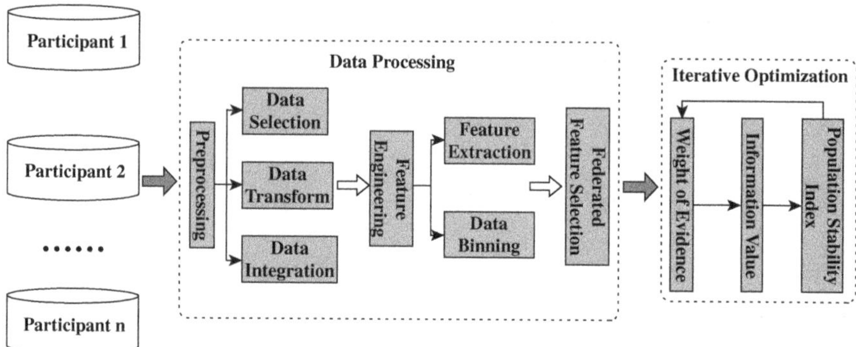

Fig. 9.8.   Federal feature selection.

(3) **Embedded:** Embedded methods want to reduce the computation time taken up for reclassifying different subsets which is done in wrapper methods. Embedded methods that remove unimportant features during training seem to be a good fit for federated learning, however they must be adapted to support distributed training and keep communication overhead low.

Federated feature selection for multiple participants requires the use of principles of secure multi-party computation. Figure 9.8 shows the federal feature selection. First, data pre-processing such as data selection, transformation, and integration needs to be performed on each participant. Feature extraction and data binning are also carried out as part of feature engineering. Then, during the data binning process, equal-width binning using unsupervised learning is applied for continuous data types, while chi-squared binning using supervised learning is used for discrete data types to enhance the ability to handle missing attribute values. Next, according to the number of participants required, two-party federated feature selection based on publicly verifiable secrets and multi-party federated feature selection based on verifiable secret sharing are designed. Finally, feature selection rules are established using metrics such as weight of evidence, information values, and population stability index for iterative optimization.

Data binning is the method used to reduce the impact of minor observation errors, which can group multiple continuous values into fewer "bins". Generally, Data binning is necessary to discretize feature variables. After feature discretization, the machine learning models become more stable and reduce the fitting risk.

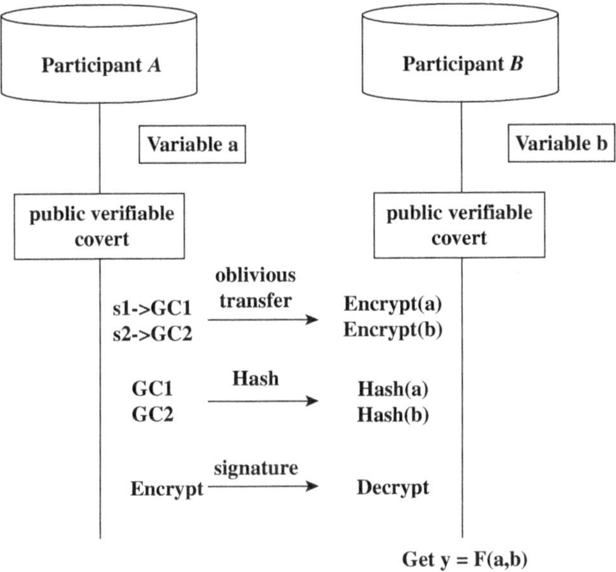

Fig. 9.9.   Two-party feature selection.

Data binning is generally divided into two methods: equal-width binning and chi-square binning.

An encryption scheme based on public verifiable covert (PVC) is used for federated feature selection between two participants. The detailed process is shown in Figure 9.9.

Assuming that participants A and B respectively provide data a and b, we hope to securely calculate the function F(a, b). Using a deterrence factor of 50% as an example, two-party feature selection based on PVC scheme is shown in Algorithm 9.1.

Wang & Feng (2024) use an encryption scheme based on Verifiable Secret Sharing (VSS) for federated feature selection between multiple participants. The detailed process is shown in Figure 9.10.

The coordinator is responsible for splitting and sharing keys. Each participant is responsible for sharing verification and recover secretly between them. The specific process refers to Algorithm 9.2.

According to Algorithm 9.2, Verifiable Secret Sharing scheme can resist $(n - 1)/2$ malicious participants. Since $gs = ga0 \bmod p$ is publicly available, this scheme is computationally secure.

Algorithm 9.1. Two-party feature selection.

Input: data a from participant A;

data b from participant B;

Output: prediction function F(a,b);

1: *A* selects two random seeds *s1* and *s2*, while *B* and *A* randomly select one of them by using 1-out–2 (let's assume *B* obtains *s1*);

2: *A* encrypt *s1* and *s2* to generate *GC1* and *GC2* by using Garbled Circuit;

3: *B* and *A* get the cipher text of wire inputted by *B*;

4: *A* hashes *GC1* and sends *hash*(*GC1*) to *B*;

5: *A* hashes *GC2* and sends *hash*(*GC2*) to *B*;

6: *A* signs above processes and sends the signature to *B*;

7: *B* can generate *GC1* based on *s1* and simulate step 3 to 5;

8: If $F(a,b)$ is inconsistent with what *A* has sent, the signatures will be published as evidence of *A*'s wrongdoing. If consistent, *B* can use *GC2* for actual calculation.

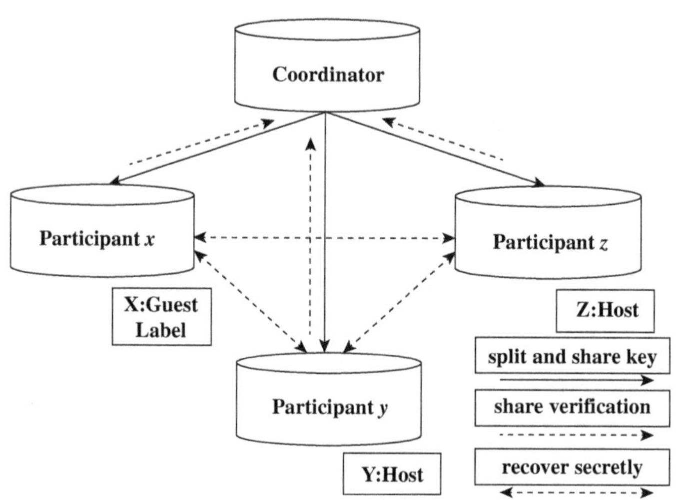

Fig. 9.10. Multi-party feature selection.

Algorithm 9.2.   Multi-party feature selection.

Input: p: *a large prime number with a length of no less than 512 bits*;

        q: a large prime factor with a length of no less than 160 bits;

        g: a qth order element in Z_p*;

        k: the threshold value;

        n: the number of sharers;

Output: *s: the secret that satisfies VSS scheme*;

1. The coordinator randomly selects $f(x) = a_0 x_0 + a_1 x_1 + \cdots + a_{k-1} x_{k-1}$ in Z_p and lets $a_0 = s$, which is the secret to share;
2. The coordinator share $s_j = f(j)\ mod\ q$ to $P_j (j = 1, 2, \ldots, n)$ secretly and broadcast $a_i = g^{a_i}\ mod\ p\ (i = 0, 1, 2, \ldots, k-1)$;
3. Each participant P_i verify $g^{s_j} = a_0^j + a_1^j + \cdots + a_{k-1}^j)\ mod\,p\ (j = 1, 2, \ldots, n)$, if the equation does not hold, the shared s_j received by P_j is invalid;
4. When k or more participants cooperate to recover secrets, each participant P_j broadcasts the shard s_j to other participants;
5. When the sharing of all collaborators is effectively verified, the collaborator can calculate the secret s using Lagrange Approximation.

Exercises

9.1 What is federated learning? What are its main characteristics?

9.2 Through examples, explain the difference between horizontal and vertical federated learning.

9.3 How to construct mobile phones' federated learning platform?

9.4 Develop an algorithm to implement federal feature selection.

9.5 Draw a framework on how to apply the federated learning in bank union.

Chapter 10

Artificial General Intelligence

Artificial General Intelligence (AGI) could learn to accomplish any intellectual task that human beings or animals can perform. AGI is similar or related to notions like strong AI, human-level AI, complete AI, and some others.

10.1 Overview

Artificial Intelligence (AI) started with "thinking machine" or "human-comparable intelligence" as the ultimate goal. After several decades, "general-purpose system", "integrated AI", and "human-level AI" become less taboo topics, as shown by several related meetings. Since 2008, several research communities have emerged, with similar focuses and overlapping participants from Artificial General Intelligence, Biologically Inspired Cognitive Architectures, Advances in Cognitive Systems, and IEEE Task Force on Towards Human-like Intelligence.

In mainstream AI, deep learning has made impressive progress in recent years, which raises many people's hopes on "human-level" AI once again. The claim "The Turing Test has been passed" and the success of AlphaGo in the board game Go renewed the discussion on what "artificial intelligence" is really about and how to reach it (Wang & Goertzel, 2008), (Wang & Goertzel, 2012).

The general purpose nature of AGI has also obtained different interpretations over the years, as meaning. AGI projects attempt to duplicate human intelligence at different aspects of abstraction, such as structure, behavior, capability, function, and principle. Intelligence is produced by the human brain. Therefore, to build an intelligent computer means to simulate the brain structure as faithfully as possible. About the behavior, intelligence is displayed in how human beings behave. Therefore, the goal should be to make a computer behave exactly like a human. Intelligence is evaluated by problem-solving capability. Therefore, an intelligent system should be able

253

to solve certain practical problems that are currently solvable by humans only. As you know, intelligence is associated with a collection of cognitive functions, such as perceiving, reasoning, learning, acting, communicating, and problem-solving. The goal is to reproduce these functions on a computer. Background: Psychology, linguistics, etc. Intelligence is evaluated by problem-solving capability. Therefore, an intelligent system should be able to solve certain practical problems that are currently solvable by humans only. From a function point of view, intelligence is associated with a collection of cognitive functionality, such as perceiving, reasoning, learning, acting, communicating, and problem-solving. Therefore, the goal is to reproduce these functions in computers. According to certain general principles, an intelligent system should always do the right thing.

The ultimate goal of AGI is to reproduce intelligence as a whole, while on the other hand, engineering practice must be step-by-step. To resolve this dilemma, we can use three overall strategies to develop, that is, hybrid, integrated, and unified. Obviously, the selection of development strategy partially depends on the selection of the research objective. Current research on artificial general intelligence is an active area. Mainly research projects contain brain-like intelligence, large language model, agent-based AGI, logic-based AGI, brain–computer integration, etc. Next, we are going to introduce these projects, respectively.

10.2 Brain-like Intelligence

Intelligence science studies the basic theory and implementation technology of intelligence, which is a frontier interdisciplinary subject created by brain science, cognitive science, artificial intelligence, etc. (Shi, 2021). Intelligence science is an important way to realize human-level artificial intelligence.

The goal of AI research is to make computers have human-like behaviors, such as listening, talking, reading, writing, thinking, learning, and adapt to ever-changing environments, etc. In 1977, E. Feigenbaum, a young scholar of Stanford University and graduate student of Simon put forward the concept of knowledge engineering in the 5th International Joint Conference on Artificial Intelligence (IJCAI, 1977), which marked the transition from the traditional reasoning to the knowledge-centered research in artificial intelligence research.

Knowledge is the national wealth and information industry is vital for a country's development. The fifth-generation computer-intelligent computer symposium was held in Tokyo Japan in October of 1981. Professor Moto-Oka Tohru from Tokyo University proposed "the Fifth-Generation Computer System: FGCS". After that,

Japan made an ambitious plan to develop the fifth-generation computers in 10 years. In the summer of 1982, Japan established "the new generation of computer technology institute" (ICOT) headed by Fuchi Kazuhiro. Japan's Ministry of International Trade and Industry fully supported the plan and the total investment budget reached $430 million. Eight large enterprises including Fujitsu, NEC, Hitachi, Toshiba, Panasonic, and Sharp were invited to this project.

It took almost ten years on the project for ICOT colleagues, who even had no time for normal lives and spent all time between the lab and their apartments. However, the outcome of the FGCS was somehow tragic. Its failure in 1992 might come from the bottleneck of key technologies, such as human–machine dialogue and program automatic proving. After that, Professor Fuchi Kazuhiro had to return to his university. Also, some thought that the FGCS was not a total failure, in that it achieved some expected goals in the first two phases. In June 1992, ICOT demonstrated the prototype of FGCS with 64 processors for parallel processing, which had similar functions to human left brain and could perform advanced precision analysis on proteins.

The failure of FGCS pushed people to find a new way to research on intelligence. Intelligence requires not only function simulation but also mechanism simulation. Intelligence requires not only top-down reasoning but also bottom-up learning as well, which may be finally combined to achieve human-level Intelligence. The perceived components of the brain including various feelings such as vision and auditory, movements, and language cortex regions play not only the role of input/output channel but also contribute to thinking activities directly.

In December 2001, the National Science Foundation and the Department of Commerce of the United States in Washington organized experts and scholars from government departments, scientific research institutions, universities, and industry to discuss the issue of converging technologies to improve human performance. Based on the papers and conclusions submitted by the conference, in June 2002, the National Science Foundation and the U.S. Department of Commerce jointly put forward a meeting technical report. According to the report, cognitive science, biology, informatics, and nanotechnology are developing rapidly at present. The organic combination and integration of these four sciences and related technologies form convergence technology, which is abbreviated as NBIC.

In the 21st century, with the development of convergence technology, the combination of life science and information technology has given birth to interdisciplinary Intelligence Science. In 2002, the Intelligence Science Laboratory of the Institute of Computing Technology, Chinese Academy of Sciences established the world's first Intelligence Science Website: http://www.intsci.ac.cn/.

10.2.1 *Blue Brain Project*

The Blue Brain Project initiated by Markram of the Federal University of Technology in Lausanne, Switzerland, has been implemented since 2005 (Markram, 2006). After 10 years of effort, the computation and simulation of cortical functional columns in specific brain regions have been completed completely. But overall, there is a big gap to be crossed in order to realize the simulation of cognitive function.

10.2.2 *Human Brain Project*

In 2013, H. Markram conceived and led the planning of the Human Brain Project (HBP) of the European Union, which was selected as the flagship technology project in the future of the European Union, and won the financial support of 1 billion euros, becoming the most important human brain research project in the world. The goal of this project is to use supercomputers to simulate human brain, to study the working mechanism of human brain and the treatment of brain diseases in the future, and to promote the development of brain-like artificial intelligence. The scientists involved come from 87 research institutions in EU Member States.

10.2.3 *BRAIN Program in the United States*

The United States proposed "Brain Research Through Advancing Innovative Neurologies (BRAIN)". The U.S. brain program focuses on the research and development of new brain research technologies, to reveal the working principle of brain and the mechanism of major brain diseases. Its goal is to not only lead the frontier scientific development but also promote the development of related high-tech industries, like the human genome program. An additional $4.5 billion will be invested over the next 10 years. The brain program puts forward nine priority areas and goals, which are as follows: identifying neural cell types and reaching consensus; mapping brain structure; developing new large-scale neural network electrical activity recording technology; developing a set of tools for regulating neural circuit electrical activity; establishing the relationship between neural electrical activity and behavior; integrating theory, model, and statistical method. This chapter analyzes the basic mechanism of human brain imaging technology and establishes the mechanism of human brain data collection, the dissemination of brain science knowledge, and personnel training.

10.2.4 *Brain Simulation System SPAUN*

The human brain is a highly complex organ, and if scientists want to build an artificial brain model, they must first understand how our brain works, specifically

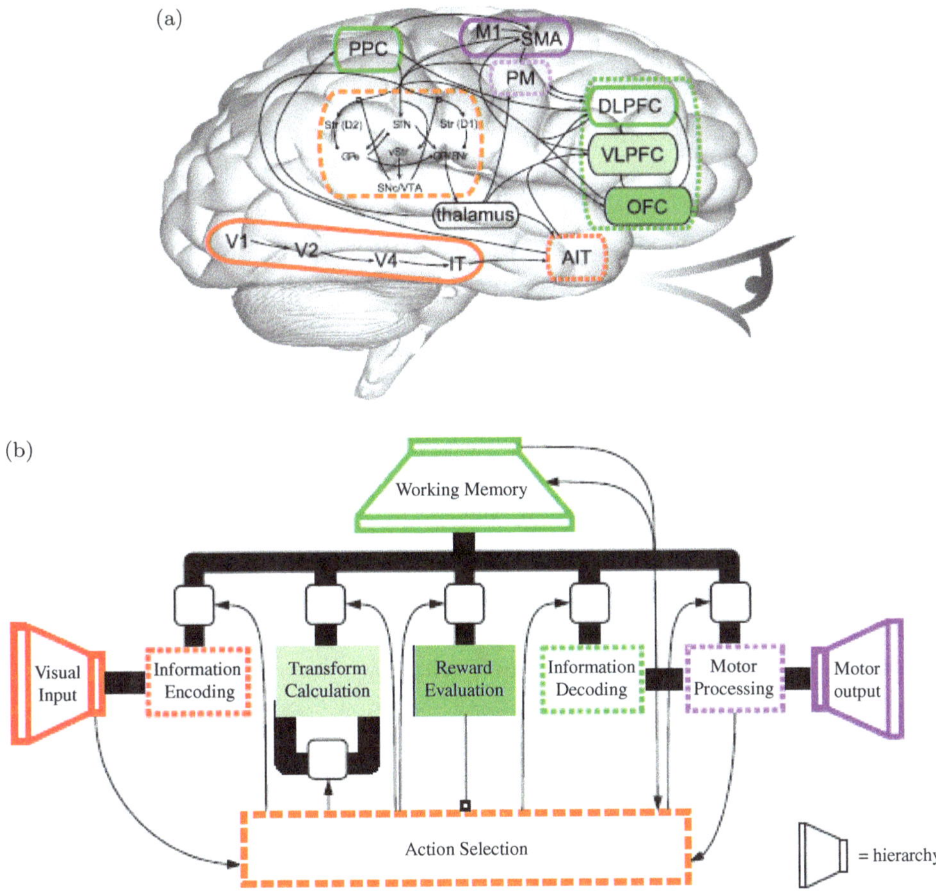

Fig. 10.1.   SPAUN's brain simulation model: (a) The anatomical architecture of SPAUN. (b) The functional architecture of SPAUN.

the computational tasks that each part of the brain is responsible for, and how these computational functions are implemented on a neural network system. In November 2012, *Science* published an article by Eliasmith *et al.* (Eliasmith *et al.*, 2012) describing a large-scale model of the human brain, as shown in Figure 10.1. This brain model is capable of simulating a variety of complex human behaviors, a development that marks another big step forward for scientists in the field of artificial intelligence.

The visual image information seen by the artificial brain is first "compressed" to remove irrelevant or redundant information. Elijah Smith's group uses a multi-level restricted Boltzmann machine algorithm when compressing the graphics information, which belongs to the operating mechanism of a feed-forward neural network system, and each layer of finite Boltzmann machine processing can obtain a kind of graph feature information, and after multiple rounds (layer) of finite Boltzmann machine processing, all the relevant information of the whole graph can be obtained. Then, this graphic information is distributed to the sub-systems of the artificial brain that correspond to the visual center of the real human brain, corresponding to the primary and secondary visual cortex, the striatal cortex, and the inferior temporal cortex, respectively. In terms of motor function, the SPAUN system takes a similar approach, breaking down simple action commands, such as drawing the number "6", into many simple actions, and then combining these actions to draw a "complex" 6. All relevant calculations are based on the theory of optimal control, which also includes the operation of auxiliary and primary motor centers. This artificial brain model, which compresses signals and accompanies movements, solves the "breadth problem" that the brain needs to process when interacting with the environment, and the previous artificial models have always been unsure of how to deal with the large amount of sensory information, and at the same time, they have not known which choice to make when faced with many action alternatives.

The cognitive apparatus of the SPAUN system actually consists of two intersecting components: a working memory system equivalent to the prefrontal cortex region of the human brain and an action selection system equivalent to the basal ganglia and thalamus of the human brain. This action selection system controls the current state of the artificial brain and is also partly inspired by the theory of the current popular basal ganglia model. The SPAUN system's memory work system uses a completely new set of algorithms that draw on neural system algorithms from the field of computational neuroscience and convolutional memory theories from the field of mathematical psychology. This set of nervous system algorithms enables the SPAUN system to have a networked information storage mechanism, and the convolutional memory theory allows the SPAUN system to effectively combine the past information with the latest received information. So the SPAUN system can effectively perform repetitive behaviors, such as writing the first and last digit in a column of numbers, which is unmatched by other artificial brain models.

Eliasmith *et al.* also used another memory system to automatically infer relationships between past and current signals. This automatic reasoning function implies the most basic syntactic function, and the number recognition and reproduction function presented by the SPAUN system indicates that this syntactic function will be realized one day in the future. This reproducibility is directly related to the symbolic computing function, which is very common in computer science and connectionist theoretical writings. The SPAUN system uses these calculations to pass the most basic IQ tests.

In the SPAUN system, the sub-systems corresponding to the prefrontal cortex region serve as a link between abstract operations, symbolic operations, and the activity of individual neuronal cells. With regard to convolutional memory function, Elijah Smith *et al.* made a very interesting prediction by estimating that the rate of neuronal cell activation (the average number of action potentials present per unit of time) would gradually accelerate as memory work was completed in succession.

In each module of the SPAUN system, the actual information is processed by a large number of activated nerve cells. This connection between the high-level computation performed by neural networks under physiological conditions and the low-level computation performed by relying on individual neuronal cells is reconstructed in the SPAUN system by relying on the so-called "neural engineering framework" (Eliasmith & Anderson, 2003), which is particularly adept at arbitrary mathematical operations in activated neural networks' vector operation. The system assumes that information is read linearly at the rate of neural activation and then converts the information into neural activity in a nonlinear manner. In this way, the information processed in each sub-system is assigned to each neuronal cell, and this pattern is also very consistent with the conclusions of brain electrophysiology research, such as the fact that our brains respond differently to different sensory stimuli (input) signals or motor output signals.

Based on the development ideas of Eliasmith, etc., it is not surprising that the SPAUN system does not mimic the real brain in some way, for example, the responsiveness of several parts of the system in several aspects (including the most basic statistical areas). It's all very different from the actual situation of the brain. We don't yet know how far these problems can be improved in the future or to what extent these biases are a true reflection of the inconsistencies in the level of basic response within our brains. The biggest problem with the SPAUN system is its hard-wired nature and inability to learn new tasks (functions). However, the structure of the SPAUN system is very flexible and does not stick to a single task, and there are many parts of the SPAUN system that have the potential to learn, such as the multi-layer processing system of graphic information and the action selection system. As for the ability to learn more generally, such as learning a completely new task, this may be deliberately left by Eliasmith, etc. In fact, what the SPAUN system lacks is precisely what we lack in terms of our understanding of our own brains. Elijah Smith *et al.* have incorporated a large number of brain research results into the SPAUN system, and the work itself has shown us a theory of brain working, which of course does not include the mechanisms related. In addition, Eliasmith, etc. offer the possibility of developing AI systems on a large scale, top-down. The advent of the SPAUN system sets a new benchmark for this work and provides a new way to focus on reproducing as many brain functions as possible and performing more complex walks rather than just thinking about how to put together as many nerve cells or amounts of information as possible.

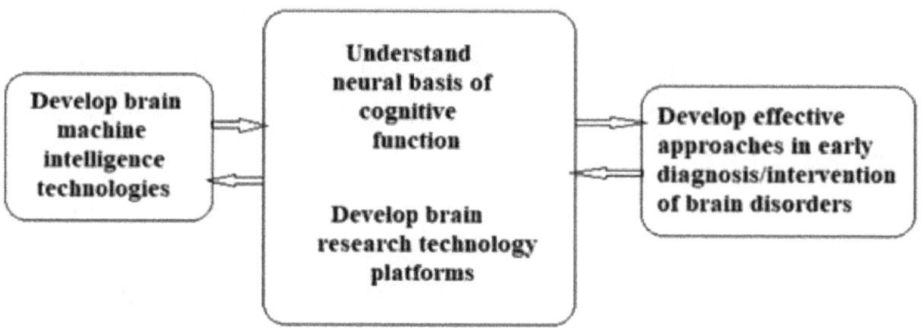

Fig. 10.2. China Brain Project.

10.2.5 *China Brain Project*

China Brain Project takes the neural basis of understanding the cognitive function of brain as the research subject and takes the brain-like intelligence technology and the diagnosis and treatment methods for major brain diseases as the two wings shown in Figure 10.2 (Poo *et al.*, 2016).

10.3 Large Language Model

Intelligence is displayed in how human beings behave. Therefore, the goal should be to make a computer behave exactly like a human. Large Language Models (LLMs) typically refer to language models that contain hundreds of billions (or more) of parameters and are trained with a large amount of textual data to generate natural language texts or understand the meaning of language texts. These parameters are trained on a large amount of text data, representing significant progress in the field of AI, and their performance is constantly evolving as the complexity and scale of these models increase.

From the 2019 Google T5 to the OpenAI GPT series, large models with explosive parameter numbers continue to emerge. It can be said that the research on LLMs has been greatly promoted in both academia and industry. In September 2020, OpenAI authorized Microsoft to use the GPT-3 model, making Microsoft the first company in the world to enjoy GPT-3 capabilities. On November 30, 2022, OpenAI in the United States released Chat Generative Pretrained Transformer, a chat robot program. The emergence of ChatGPT has attracted widespread attention from all sectors of society. Within just two months of its release, 100 million users participated, making it the fastest growing user product in history. On March 15, 2023, OpenAI released the multi-modal pretrained large model GPT4.0.

In February 2023, Google released the chat robot Bard, which is driven by Google's big language model LaMDA. On March 22, 2023, Google opened the public beta for Bard, first targeting the United States and the United Kingdom and gradually launching it in other regions in the future.

On February 7, 2023, Baidu officially announced that it would launch ERNIE Bot, which was officially launched on March 16. The underlying technical foundation of ERNIE Bot is the Wenxin Big Model. The underlying logic is to provide services through Baidu AI Cloud, attract enterprise and institutional customers to use APIs and infrastructure, jointly build AI models, develop applications, and achieve industrial AI benefits.

The functions of a large language model include scaling, training, ability stimulation, alignment tuning, and tool utilization:

(1) **Zoom:** Scaling is a key factor in increasing the capacity of LLM models. Initially, GPT-3 increased the model parameters to 175 billion, followed by PaLM further increasing the model parameters to 540 billion. Large-scale parameters are crucial for emergence ability. Scaling is related not only to model size but also to data size and total computation.

(2) **Training:** Due to its huge scale, successfully training a powerful LLM is very challenging. Therefore, distributed training algorithms are needed to learn the network parameters of LLMs, often using various parallel strategies in conjunction. In order to support distributed training, optimization frameworks such as DeepSpeed and Megatron LM have been used to promote the implementation and deployment of parallel algorithms. In addition, optimization techniques are also important for training stability and model performance, such as restarting training loss spikes and mixed accuracy training. Recently, GPT-4 has developed special infrastructure and optimization methods to utilize much smaller models to predict the performance of larger models.

(3) **Ability stimulation:** After pretraining on a large-scale corpus, LLMs are endowed with the potential to solve general tasks. However, when LLMs perform a specific task, these abilities may not be explicitly demonstrated. Therefore, designing suitable task instructions or specific contextual strategies to stimulate these abilities is very useful, such as thinking chain prompts that help solve complex reasoning tasks through intermediate reasoning steps. In addition, instruction tuning can be further carried out on LLMs with natural language task descriptions to improve their generalization ability for unseen tasks.

(4) **Alignment tuning:** Due to the training of LLMs to capture the data features of pretrained corpora (including high-quality and low-quality data), they are likely to generate toxic, biased, and harmful text content. In order to align LLMs with human values' consistent values, Instrument GPT has designed an efficient tuning method that utilizes reinforcement learning and human feedback to enable

LLMs to follow expected instructions. ChatGPT was developed on a technology similar to Instrument GPT, demonstrating strong alignment capabilities in generating high-quality, harmless responses.

(5) **Tool utilization:** LLMs are essentially text generators trained on large-scale pure text corpora, so they do not perform as well in tasks with poor text expression, such as numerical calculations. In addition, the ability of LLMs is limited by pretrained data and cannot capture the latest information. In response to these issues, people have proposed using external tools to compensate for the shortcomings of LLMs, such as using calculators for precise calculations and using search engines to retrieve unknown information. ChatGPT utilizes external plugins to learn new knowledge online, which can widely expand the capabilities of LLMs.

The technological progress of LLMs has had a significant impact on the entire AI field and will completely change the way people develop and use AI algorithms. ChatGPT is a natural language processing tool driven by artificial intelligence technology. It can engage in conversations by understanding and learning human language, interact based on the context of the conversation, and even complete tasks, such as writing emails, video scripts, copywriting, translation, writing codes, and writing papers. Following is an introduction to the key technologies used in ChatGPT.

10.3.1 *Transformer Network*

In 2017, Vaswani A and others released the Transformer Network (Vaswani *et al.*, 2017), which greatly changed the methods used in various sub-domains of artificial intelligence and developed into the basic model for almost all artificial intelligence tasks today. The transformer network is based on self-attention mechanism and supports parallel training of models, laying a solid foundation for large-scale pretraining models. Since then, natural language processing has opened up a new paradigm and greatly promoted language modeling and semantic understanding, achieving the explosive ChatGPT.

As shown in Figure 10.3, the converter network consists of two parts, with the encoding part on the left and N encoders; on the right is the decoding section, consisting of N decoders. The encoding part encodes the input sequence (text), and the decoding part continuously decodes the next word element using autoregressive method, ultimately completing the transformation from sequence to sequence and outputting. Due to its excellent parallelism and capacity, converter networks have become the main architecture for developing various LLMs, making it possible to extend language models to hundreds of billions of parameters. Generally speaking,

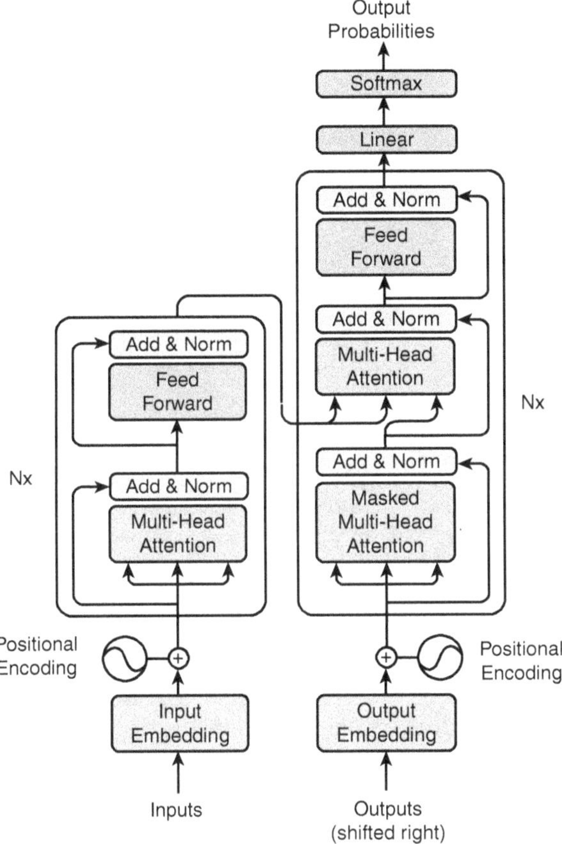

Fig. 10.3. The transformer-model architecture.

the mainstream architecture of existing LLMs can be roughly divided into three categories, namely encoder–decoder, temporary decoder, and prefix decoder. Since the emergence of converter networks, various improvements have been proposed to improve their training stability, performance, and computational efficiency. Researchers discuss the corresponding configurations of the converter network, including normalization, position encoding, activation function, attention mechanism, and bias. Pretraining plays a crucial role in encoding general knowledge from large-scale corpora into large-scale model parameters. For training LLMs, there are two commonly used pretraining tasks: language modeling and denoising self-coding.

10.3.2 *Multi-Head Self-Attention Mechanism*

The biggest innovation of the instance transformer network using the converter network for Chinese to English translation is the complete use of the multi-head

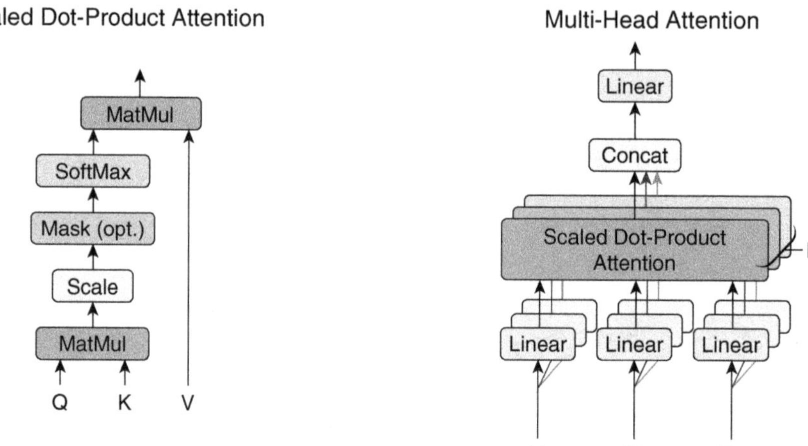

Fig. 10.4. Multi-head self-attention mechanism.

self-attention mechanism, the architecture of which is shown in Figure 10.4 (Vaswani *et al.*, 2017). Self-attention is when the model processes the semantic correlation between a certain token of the current input sequence and other tokens in the sequence, and different "heads" focus on different aspects of semantics. The self-attention mechanism is to directly calculate the attention weight of each position in the sentence in the coding process through some operation and then calculate the implicit vector representation of the whole sentence in the form of weight sum. The defect of the self-attention mechanism is that the model will over-focus on its own position when encoding the information of the current position, and the multi-head attention mechanism is used to solve this problem.

The transformer network relies entirely on the attention mechanism, supporting great parallelization. The encoder and decoder of the converter network both use the same multi-head attention structure, except that in the encoder, the attention is bidirectional, while in the decoder, the attention is only allowed to focus on the earlier position in the output, as shown in Figure 10.7; V, K, and Q are fixed single values, the linear layer has three, and the scaled dot-product attention has three, i.e., three heads, similar to stacking three single-head attentions. Then concat or sum, and finally linear. In fact, long attention is to perform single attention multiple times, and then combine the results.

10.3.3 *Human Feedback Reinforcement Learning*

Reinforcement Learning with Human Feedback (RLHF) is a type of artificial intelligence model that uses human feedback to achieve model learning in the process of prediction (inference) so that the model output is consistent with human intentions and preferences, and continuously optimized in a continuous feedback loop to produce better results (Christiano *et al.*, 2017). In fact, in the process of AI development, there has always been human interaction in the model training stage, which is also known as human-in-the-loop (HITL), but in the prediction stage, there is more human participation, that is, human-out-of-the-loop (HOOTL). In these five years of endeavor, reinforcement learning through human feedback has enabled natural language processing to learn from human feedback in the inference stage. This is a new innovation in the field of natural language processing, which can be described as a person and a model working hand in hand to build a beautiful new AI. From a technical point of view, reinforcement learning with human feedback is a type of reinforcement learning, which is suitable for those scenarios where it is difficult to define a clear loss function for optimizing the model, but it is easy to judge how well the model predicts the performance, that is, it is easier to evaluate the behavior than to generate the behavior. In the idea of reinforcement learning, an agent learns through interaction with its environment, which is commonly found in various game AIs. In 2016, AlphaGo beat the core skill of the world champion of chess, Lee Seok of South Korea, which is reinforcement learning. Human feedback reinforcement learning does not start with natural language processing, for example, in 2017, OpenAI and DeepMind collaborated to explore whether human feedback reinforcement learning systems can effectively interact with the real world. The experimental scenarios were Atari games, simulated robot movements, and so on. These achievements were subsequently applied to large language models by OpenAI and DeepMind, optimizing the language model through human feedback, and aligning the output of the model with expected goals, such as InstructionGPT and FLAN. These results indicate that incorporating human feedback reinforcement learning significantly improves the quality of generated text compared to baselines without using human feedback reinforcement learning, while also enabling better generalization to new fields. Figure 10.5 shows the framework of human feedback

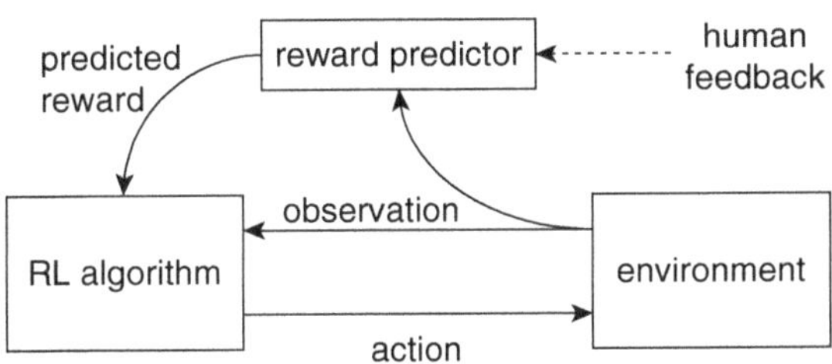

Fig. 10.5. Human feedback reinforcement learning framework.

reinforcement learning, where the reward predictor is learned, which is different from traditional reinforcement learning. In traditional reinforcement learning, the reward function is manually set.

10.3.4 *Language Model*

The GPT series implemented the autoregressive language model through the decoder of the transformer network (Radford *et al.*, 2018), using a multi-tasking training method to train the model. Autoregression is very common in time series analysis, such as ARMA and GARCH, which are typical autoregression models. In language models, the autoregressive model predicts the next word from a set of words based on a given context each time and defines a direction (usually positive, meaning guessing the next word/word from front to back in a sentence). Taking "A Red Apricot Comes Out of the Wall" as an example, in the autoregressive language model, given the context of "A Red Apricot" to predict the next "apricot" character, followed by a given "A Red Apricot" to predict the next "out" character, and then based on the given "A Red Apricot Comes Out" to predict the "wall" character, the cycle continues until the entire sequence is predicted and output. There are various options to choose the output marker sequence predicted by the model, such as greedy decoding, beam search, Top-K sampling, nucleus sampling, and temperature sampling.

BERT used a multi-task learning method to train models from large-scale corpus and fine-tune them in specific tasks. ERNIE adopts a model architecture similar to BERT and incorporates a knowledge graph, enabling the model to better understand semantics using prior knowledge.

10.3.5 *In-Context Learning*

In-context learning has become popular with GPT-3. In GPT-3, a method that can effectively complete many natural language processing tasks by providing only a few examples is called situational learning. Intuitively speaking, situational learning is giving a model some prompts containing task inputs and outputs and attaching an input for prediction at the end of the prompts. The model predicts the results of the task and outputs them based on the prompts and predicted inputs. Therefore, situational learning is sometimes referred to as prompt-based learning.

The predictive results of situational learning perform very well in the case of large models but perform poorly in the case of small models. Simply put, big models make situational learning useful. This is because situational learning relies on the conceptual semantics learned by language models and implicit Bayesian reasoning, which relies on large-scale pretrained models learning potential concepts, learning long-distance dependencies from document-level corpus, and maintaining long-distance coherence, thought chains, and complex reasoning, among others.

Situational learning can effectively adapt the model to new tasks with significant differences in input distribution and training distribution, which is equivalent to learning specific tasks through "learning" examples during inference, allowing users to quickly build the model through new use cases without the need for fine-tuning training for each task. Situational learning built on top of large language models usually only requires a few prompt examples to work properly, which is very intuitive and useful for experts in non-natural language processing and artificial intelligence fields.

In the next 5 to 10 years, a super large model that integrates multiple modalities such as language, vision, and speech will greatly enhance the ability of reasoning and generation. At the same time, by integrating human prior knowledge into the super large knowledge graph and knowledge computing engine, the accuracy of artificial intelligence reasoning and decision-making will be greatly improved. Such artificial intelligence systems can not only adapt to the vast majority of tasks in different modalities of the real world like humans and complete tasks at a level that even surpasses the vast majority of ordinary people but also effectively assist humans in various imaginative and creative tasks.

The AI big model accelerates the implementation of universal artificial intelligence. Microsoft researchers stated in their paper "Sparks of Universal Artificial Intelligence: Early Trials of GPT-4" that, in addition to mastering language proficiency, GPT-4 can also solve novel and difficult tasks in fields such as mathematics, programming, vision, medicine, law, and psychology without the need for any special prompts. In addition, in all of these tasks, the performance of GPT-4 is very close to the human level and often significantly exceeds previous models, such as ChatGPT. The field of large models in China is also vigorously advancing. Beijing Zhiyuan Artificial Intelligence Research Institute has launched the 1.75 trillion parameter Wudao 2.0, Huawei has collaborated with Pengcheng Laboratory to launch the Pangu large model with 100 billion parameters, Baidu has collaborated with Pengcheng Laboratory to launch the Wenxin large model with 260 billion parameters, and Alibaba Damo Institute has launched the 10 trillion parameter multi-modal M6 large model.

10.3.6 *OpenAI Video Generation Model*

The Sora released by OpenAI on February 16th 2024 is another heavyweight release in the field of text-to-video models, following GPT's text-to-text and DALL · E's text-to-image. It is another innovative leap toward AGI.

Sora initially abandoned task-specific visual models, such as recurrent networks, generative adversarial networks, autoregressive transformers, and diffusion models, and firmly chose to follow the path of universal and scalable visual models. Since only universal and scalable models can be applied to scaling law and achieve intelligent breakthroughs in large computing power and massive data training.

Sora's breakthrough in video modeling is of great significance for Artificial General Intelligence (AGI). Language is the interaction between people, while vision is the interaction between people and the world — not just the physical world but also the digital world. The product of AGI era includes various media forms, such as text, images, audio, video, and spatial computing.

The visual big model represented by Sora will greatly innovate the production method of videos, the way people interact with the surrounding environment, and bring new development momentum to the field of embodied intelligence robots.

Sora authors take inspiration from large language models which acquire generalist capabilities by training on Internet-scale data. The success of the LLM paradigm is enabled in part by the use of tokens that elegantly unify diverse modalities of text — code, math, and various natural languages. In this work, they consider how generative models of visual data can inherit such benefits. Whereas LLMs have text tokens, Sora has visual patches. Patches have previously been shown to be an effective representation of models of visual data. They find that patches are a highly scalable and effective representation for training generative models on diverse types of videos and images. OpenAI trains a network that reduces the dimensionality of visual data and extracts a sequence of spacetime patches which act as transformer tokens. The patch-based representation enables Sora to train on videos and images of variable resolutions, durations, and aspect ratios.

10.4 Agent-Based AGI

10.4.1 *OpenCog*

OpenCog as a software framework aims to provide research scientists and software developers with a common platform to build and share artificial intelligence programs. The long-term goal of OpenCog is acceleration of the development of beneficial AGI. OpenCogPrime is a specific AGI design being constructed within the OpenCog framework. Figure 10.6 shows the high level of OpenCogPrime (Goertzel, 2009). It comes with a fairly detailed, comprehensive design covering all aspects of intelligence. The hypothesis is that if this design is fully implemented and tested on a reasonably sized distributed network, the result will be an AGI system with general intelligence at the human level and ultimately beyond.

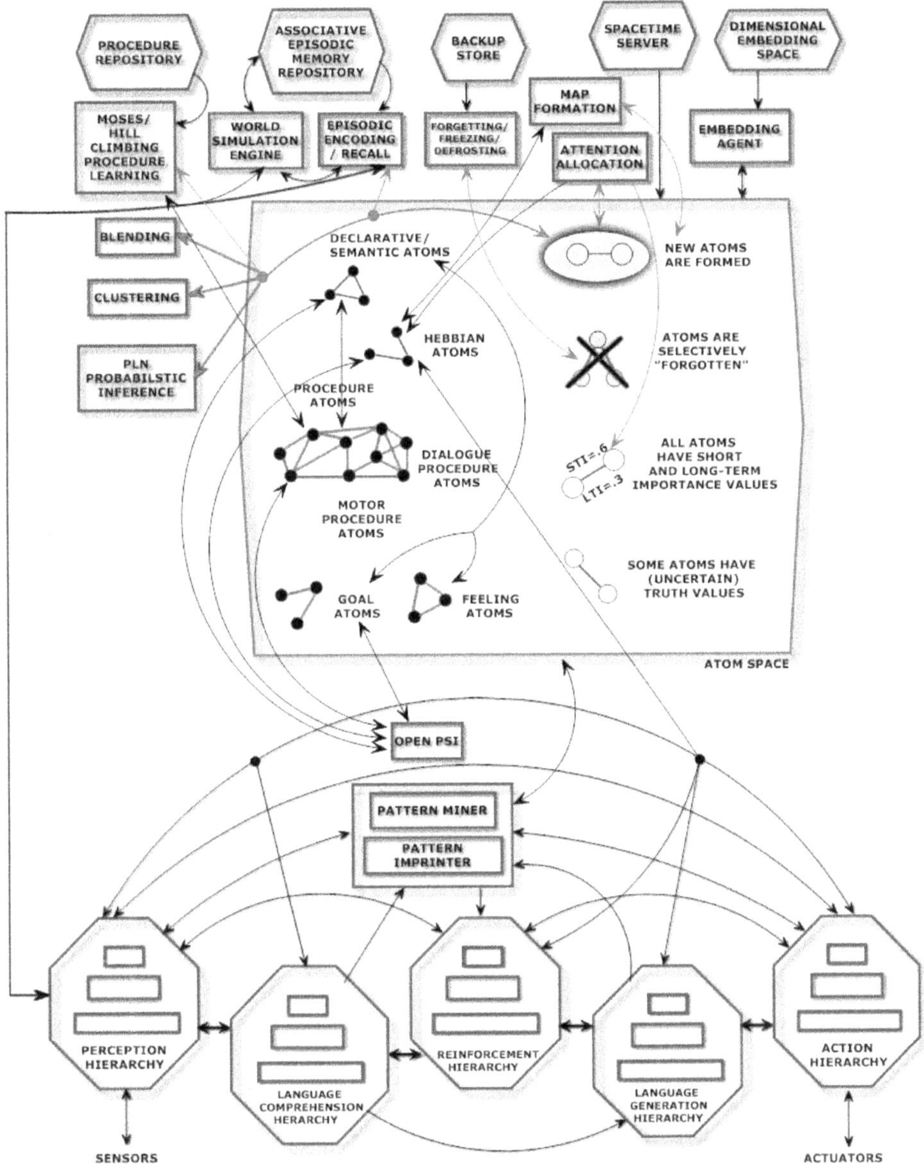

Fig. 10.6.   High-level architecture of OpenCogPrime.

Current OpenCogPrime involves extending the virtual dog work via using OpenCog to control virtual agents in a game world inspired by the game Minecraft. These agents are initially specifically concerned with achieving goals in a game world via constructing structures with blocks and carrying out simple English communications. Representative example tasks would be as follows:

(1) learning to build steps or ladders to get desired objects that are high up,
(2) learning to build a shelter to protect itself from aggressors,
(3) learning to build structures resembling structures that are shown,
(4) learning how to build bridges to cross chasms.

While an OpenCogPrime-based AGI system could do a lot of things, they are initially focusing on using OpenCogPrime to control simple virtual agents in virtual worlds. They are also experimenting with using it to control a Nao humanoid robot.

10.4.2 *MicroPsi*

The MicroPsi agent architecture describes the interaction of emotion, motivation, and cognition of situated agents, mainly based on the Psi theory of Dietrich Dörner (Bach, 2012). The Psi theory addresses emotion, perception, representation, and bounded rationality but being formulated within psychology has had relatively little impact on the discussion of agents within computer science. MicroPsi is a formulation of the original theory in a more abstract and formal way, at the same time enhancing it with additional concepts for memory, the building of ontological categories, and attention. The architecture of MicroPsi is shown in Figure 10.7.

The agent framework uses semantic networks, called node nets, that are a unified representation for control structures, plans, sensory and action schemas, Bayesian networks, and neural nets. Thus, it is possible to set up different kinds of agents on the same framework. The agent possesses a number of innate desires (urges) that are the source of its motives. Events that raise these desires are interpreted as negative reinforcement signals, whereas satisfaction of a desire creates a positive signal. Currently, there are urges for intactness, energy (food and water), affiliation, competence, and reduction of uncertainty. The levels of energy and social satisfaction (affiliation) are self-depleting and need to be raised through interaction with the environment. The cognitive urges (competence and reduction of uncertainty) lead the agent into exploration strategies but limit these to directions, where the interaction with the environment proves to be successful. The agent may establish and pursue sub-goals that are not directly connected to its urges, but these are parts of plans that ultimately end in the satisfaction of its urges. The execution of internal behaviors and the evaluation of the uncertainty of externally perceivable events create a feedback on the modulators and the cognitive urges of the agent.

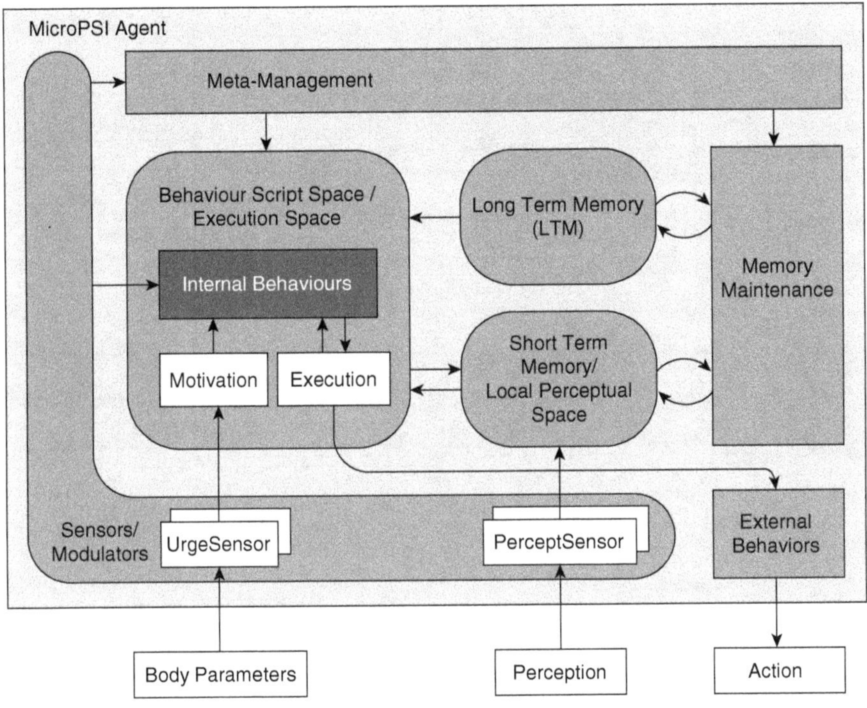

Fig. 10.7. The architecture of MicroPsi.

10.4.3 *LIDA*

LIDA was developed by Franklin (Franklin *et al.*, 2014) and Baars (Baars & Franklin 2009). LIDA is based on a comprehensive, conceptual computational model that covers most of human cognition, with modules or processes for perception, working memory, episodic memories, consciousness, procedural memory, action selection, perceptual learning, episodic learning, deliberation, volition, and non-routine problem-solving. LIDA technology is based on the LIDA cognitive cycle, a sort of cognitive atom. LIDA model is ideally suited to provide a working ontology that would allow for the discussion, design, and comparison of AGI systems. This model is mainly based on the global workspace theory of consciousness, which implements and enriches the theory of psychology and neuropsychology.

The LIDA cognitive cycle can be divided into three phases: comprehension, (awareness), understanding, and action selection (see Figure 10.8) (Franklin *et al.*, 2014) (Snaider *et al.*, 2012). It starts in the upper left corner and flows roughly clockwise. The comprehension phase begins with stimulus input, activating the primary

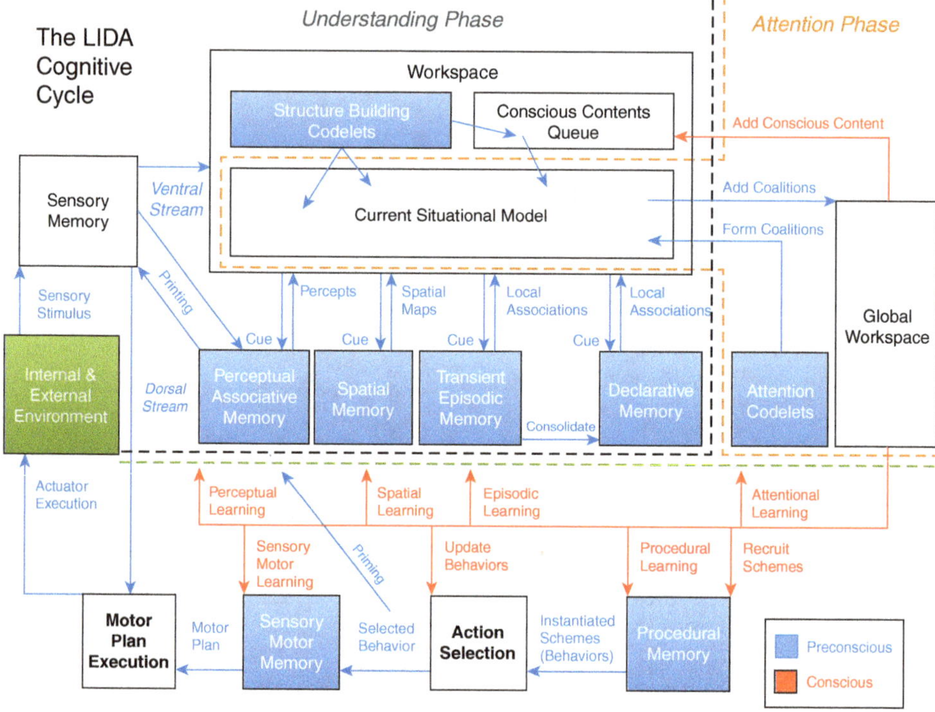

Fig. 10.8. LIDA cognitive cycle.

feature detector in sensory memory. The output signal is sent to sensory associative memory, where higher-level functional detectors are used for the detection of more abstract entities, such as objects, categories, actions, and events. The competition for global workspaces then selects the most prominent, relevant, important, and urgent connections, and their content becomes the content of awareness. The content of this consciousness is then broadcast throughout the space, initiating the selection phase of action.

10.5 Logic Based AGI

In 2007, Wang published *The Logic of Intelligence* (Wang, 2007) and pointed out that intelligence is the capacity of a system to adapt to its environment while operating with insouciant knowledge and resources. NARS (Non-Axiomatic Reasoning System) is an intelligent reasoning system. It answers questions according to the knowledge originally provided by its user. NARS is fully based on a working definition of intelligence as the ability of an information processing system to adapt to its environment while working with insufficient knowledge and resources (Wang, 2000). NARS does not use first-order predicate logic. Instead, each piece of knowledge in NARS has the form "$SrP < f; c >$". Here S is the subject term and P is the predicate term. In the simplest situation, both of them are words. r is an inheritance relation. In the NARS system, three types of inheritance relations can be used:

(1) "$S \subset P$" means that "S is a special type of P".
(2) "$S \in P$" means that "S is an instance of P".
(3) "$S = P$" means that "S and P are similar to each other".

"$< f, c >$" is the truth value of the sentence, where f is the "frequency", a real number in $[0, 1]$, indicating the ratio of positive evidence among all evidence of the relation, and c is the "confidence", a real number in $(0, 1)$, indicating the amount of evidence the system has on the relation.

An intelligence usually requests a reasoning function. A reasoning system in general has the following components:

(1) a formal language for knowledge representation, as well as communication between the system and its environment,
(2) a semantics that determines the meanings of the words and the truth values of the sentences in the language,
(3) a set of inference rules that match questions with knowledge, infer conclusions from promises, and so on,
(4) a memory that systematically stores both questions and knowledge and provides a working place for inferences,
(5) a control mechanism that is responsible for choosing premises and inference rules for each step of inference.

The development of NARS takes an incremental approach consisting of four major stages. At each stage, the logic is extended to give the system a more expressive language, a richer semantics, and a larger set of inference rules; the memory and control mechanism are then adjusted accordingly to support the new logic.

In NARS, the notion of reasoning is extended to represent a system's ability to predict the future according to the past and to satisfy the unlimited resource demands using the limited resources supply, by flexibly combining justifiable micro steps into macro behaviors in a domain-independent manner.

10.6 Brain–Computer Integration

Brain–computer integration is a new intelligent system based on brain–computer interface technology, which is integrated with biological intelligence and machine intelligence. Brain–computer integration is an inevitable trend in the development of brain–computer interface technology. In the brain–computer integration system, the brain and the brain, the brain and the machine not only are the signal level of the brain–machine interoperability but also need to integrate the brain's cognitive ability with the computer's computing ability. But the cognitive unit of the brain has a different relationship with the intelligent unit of the machine. Therefore, one of the key scientific issues of brain–computer integration is how to establish the cognitive computing model of brain–computer integration.

Brain–computer integration system has three remarkable characteristics: (a) more comprehensive perception of organisms, including behavior understanding and decoding of neural signals; (b) organisms also as a system of sensing, computation body and executive body, and information bidirectional exchange channel with the rest of the system; (c) comprehensive utilization of organism and machine in the multi-level and multi-granularity will achieve system intelligence greatly enhanced.

Brain–computer integration adopts a hierarchical architecture. Human beings analyze and perceive the external environment through the acquired perfect cognitive ability. The cognitive process can be divided into the perception and behavior layer, decision-making layer, and memory and intention layer, forming mental thinking. The machine perceives and analyzes the external environment by detecting data, and the cognitive process can be divided into the awareness and actuator layer, planning layer, and belief and motivation layer, forming formal thinking. The same architecture indicates that humans and machines can merge at the same level, and cause-and-effect relationships can also be generated at different levels. Figure 10.9 is the cognitive model of brain–computer integration (Shi & Huang, 2019). In this model, the left part is simulated human brain in terms of CAM mind model; the right part is computer-based on ABGP agent.

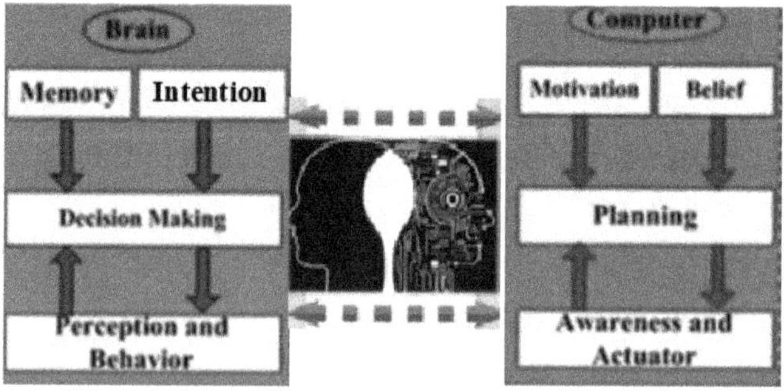

Fig. 10.9. Cognitive model of brain–computer integration.

Exercises

10.1 How to understand artificial general intelligence?

10.2 Explain what brain-like intelligence is.

10.3 Would you please list all brain-like projects in the worldwide and compare their research focuses?

10.4 What is large language model? Give the key technical issues.

10.5 What is transformer network ? Why has it got a lot of applications?

10.6 OpenAI announced Sora system. Why is it an important step to AGI?

10.7 Describe OpenCog architecture and explain the main function of each component.

10.8 How did Baars embed his conciseness into the LIDA system and how to improve the functions of the LEDA system?

10.9 What is NARS principle?

10.10 How to implement brain–computer integration? What are the key issues?

Chapter 11

Crowd Computing

Crowd computing is a new collaborative computing paradigm for heterogeneous agents in human–machine–physical integration environments. It is the ability to effectively utilize and manage various resources, achieve goals, and create value through group cooperation in the world. Multi-agent systems mainly solve the problem by physically or logically distributed systems; it is an important approach for crowd computing.

11.1 Introduction

The computing paradigm has mainly gone through the host era, PC era, and network computing era (distributed computing, grid computing, cloud computing, etc.). Network computing mainly emphasizes computer interconnection and computing power sharing. With the development of smart terminals and wireless communication technology, many devices currently have sensing capabilities, and their intelligent processing capabilities have also been improved. Everything is intelligent. How to effectively use the intelligence of all things to achieve high efficiency in sensing capabilities, computing capabilities, and control capabilities? Collaboration ultimately forms stronger intelligence, which is the fundamental starting point of crowd computing.

A deep analysis of general types and substantive characteristics of future Web-based industrial operation systems can easily identify a smart interconnected network system that takes on the features of large scale, openness, self-organization, and ecologicalization, such a system is called "Crowd Cyber System" in short. In terms of components, this kind of network system takes the form of smart entities composed of a great many intelligently interconnected individuals, enterprises, governmental agencies, and articles, with a sort of supply–demand relationship between them. In terms of major behaviors on the network, the interaction between any smart entities boils down to some sort of transaction or behavior. From the perspective of systems, the network system is a ternary system of information, physics, and society. Openness, interconnection, cooperation, and sharing constitute the basic logic of the network system, while stability, efficiency, innovation, and development are the targets the system aims to fulfill.

Crowd computing is based on the organic integration of heterogeneous intelligence, with the goal of improving the multi-dimensional capabilities of individuals and groups, and aims to take advantage of the complementarity of human–machine–physical heterogeneous intelligence resources, differences in capabilities, collaboration, and competitiveness, explore and develop basic mechanisms, computing models, core algorithms, and system platforms, such as swarm intelligence perception and cognition, swarm intelligence collaboration and aggregation, and swarm intelligence learning and evolution, in order to build a system with self-organization, self-learning, self-adaptation, and self-performance, and Provide support for the group intelligence space with globalization capabilities.

Crowd computing is an enabling technology in the era of intelligent interconnection of humans, machines, and objects. Human groups, information space, and all entities are realizing extensive and deep interconnection. Humanity has entered the era of intelligent interconnection of all things with the three-dimensional integration of humans, machines, and things. In particular, the integrated development of the Internet of Things and artificial intelligence has enabled intelligent connections among mobile devices carried by human groups, the widely deployed Internet of Things, and various forms of the Internet, breaking the information isolation in the three-dimensional space. Humans, machines, and things are integrated into the intelligent interconnection of all things become possible. Among them, how to realize mutual stimulation and cross-domain collaboration between human, machine, and physical elements (i.e., intelligent agents) is one of the key issues. Crowd intelligence computing is group intelligence computing for the fusion environment of humans, machines, and objects. It aims to build a new collaborative computing paradigm for heterogeneous agents across space. Its models, methods, and platforms are the key to supporting the three-dimensional integration and intelligent interconnection of humans, machines, and objects.

As a new collaborative computing paradigm for heterogeneous agents in a human–machine–physical fusion environment, crowd intelligence computing aims to take advantage of the complementarity of human–machine–physical heterogeneous intelligence resources, the difference in capabilities, and the collaboration and competition between them, and explore and develop basic mechanisms, computing models, core algorithms and system platforms, such as swarm intelligence perception and cognition, swarm intelligence collaboration and aggregation, and swarm intelligence learning and evolution, in order to build a system with self-organization, self-learning, self-adaptation, and self-evolution capabilities. The group intelligence space provides support. The core of crowd intelligence computing lies in "swarm" (swarm of drones, unmanned vehicles, robots, or a mixed group thereof) and "intelligence" (perception ability and autonomous intelligence). Different from traditional computing equipment, swarm intelligence has the ability to move the environment, and tasks are more complex and changeable due to their nature, action, groupness, and interactivity. It is necessary to propose a full-stack solution from the system level that combines hardware architecture, software support, algorithm framework, etc. Building a ubiquitous group intelligence computing system with "group intelligence chips, group intelligence operating systems, and group intelligence algorithm framework" as the core is a key issue that needs to be solved in future computing.

Crowd computing is based on the organic integration of heterogeneous agents, with the goal of improving the multi-dimensional capabilities of individuals and groups, and enhances the advantages of group intelligence through collaboration. Research in this field not only involves computing theory, artificial intelligence, and network science but is also closely related to cognition, biology, sociology, etc. and has distinct multi-disciplinary integration characteristics. Specifically, crowd intelligence computing has three main characteristics:

(1) **Cross-domain interweaving:** The intelligent connection of all things makes the three-dimensional space of society, information, and physics not only connected in an orderly manner but also deeply intertwined with each other, forming a smart space of cross-domain integration.
(2) **Behavioral diversity:** Heterogeneous agents coexisting in cross-domain space have diverse behaviors, such as movement, action, cooperation, competition, and games.
(3) **Scenario adaptability:** Crowd intelligence computing models and their system implementation need to adapt to different application fields and scenarios and have the capabilities of autonomous evolution and dynamic adaptation.

Crowd computing faces new challenges at the levels of mechanisms, models, methods, and platforms, which mainly include the following three aspects:

(1) Non-deterministic heterogeneous intelligence emerges from crowd intelligence and is intertwined across domains. The interactions, relationships, and crowd intelligence aggregation results of group intelligence are uncertain. It is difficult

to meet the deterministic needs of group intelligence computing. It is necessary to explore the emergence of crowd intelligence and mechanism and construct deterministic models or rules to achieve effective control of the results of crowd intelligence emergence.

(2) **The evaluability of crowd computing:** Heterogeneous crowd entities have different abilities and complex computing scenarios. It is necessary to build an adaptive evolutionary group intelligence computing model and its formal verification method to realize the complementary resources/capabilities of the group intelligence entities, collaborative scheduling and performance enhancement.

(3) **Guarantee of the quality of crowd intelligence:** Heterogeneous crowd entities gather dynamically and the demand for crowd services is diverse, resulting in uneven crowd intelligence collaboration cognition and decision-making accuracy and completeness. A suitable quantitative evaluation system needs to be designed and methods to effectively support high-quality crowd computing services.

Core scientific issues and key technologies for crowd computing are as follows:

(1) **Mechanism of crowd emergence:** Explore the internal factors of crowd emergence, discover the mechanism of crowd emergence, and lay a theoretical foundation for the construction of crowd intelligence computing paradigm and model; on this basis, further explore the interaction between heterogeneous crowd entities, collaboration, competition, and game processes, forming a new evolutionary learning guidance theory for swarm intelligence.

(2) **Crowd computing model:** Construct an abstract model of the core elements, key roles, and typical relationships of group intelligence computing, study self-organization, self-adaptation, and self-evolution representation methods of crowd behavior, and establish a new paradigm for heterogeneous crowd intelligence fusion computing and its collaborative operators and mechanisms to complete the formal verification of the crowd intelligence computing model.

(3) **Crowd computing algorithms and their architecture:** Research and design core algorithms for crowd intelligence computing such as distributed collaborative cognition, heterogeneous crowd intelligence collaborative computing, and multi-role collaborative decision-making optimization; design and adapt to different ubiquitous computing scenarios A new group intelligence computing architecture for crowd intelligence algorithms and its dynamic adaptation strategies and mechanisms.

(4) **Crowd computing operating system:** Explore the structure of the crowd intelligence computing operating system, focusing on unified representation and

management of heterogeneous resources, self-organizing scheduling, and collaboration of group intelligence tasks, scene-driven adaptive services, and quality assurance, to provide crowd intelligence. Intelligent computing applications provide a unified management and interaction platform. For example, the CrowdOS operating system developed by Northwestern Polytechnical University of China is a typical crowd computing operating system.

The numerous intelligent agents (enterprises, institutions, individuals, and articles) in the crowd cyber system can give full play to their wisdom and potentiality, interact with each other, and produce various kinds of unexpected nonlinear behaviors and effects, which cannot be explained by an existing discipline but rather need the basic theories and frontier achievements of multiple disciplines. In the following section, we introduce the agent.

11.2 The Essence of Agent

Agent refers to a software or hardware entity that can act autonomously. The concept of agent was proposed by Minsky and he introduced the concept of society and social behavior into computing systems in his book *Society of Mind* (Minsky, 1985). Minsky calls such individuals as agents.

11.2.1 *The Concept of Agent*

In English, for the term "agent", there are three main meanings: first, referring to the people accountable for their actions; second, agent is defined as an article to produce a certain effect, the physical, chemical, or biological significance of the active things; third, the delegate that a person receiving the commission and do actions on behalf of other persons. In computer science and artificial intelligence fields, an agent can be viewed as an entity which perceives its environment through sensors and acts upon that environment through effectors. If the agent is a person, the sensors have eyes, ears, and other organs, hands, legs, mouth, and other parts of the body are the effectors. If it is the robot, cameras are sensors and all moving parts are the effectors. Generally, an agent has a structure as shown in Figure 11.1 (Russell & Norvig, 1995). It has part or all of the following characteristics:

(1) **Autonomy:** This is a basic characteristic of the agent, that is, it can control its own behavior. The autonomy of the agent is reflected in the following: the acts of the agent should be active and spontaneous; the agent should have its own goals or intentions. According to the requirements of the goal and environment, the agents should plan for their own short-term acts.

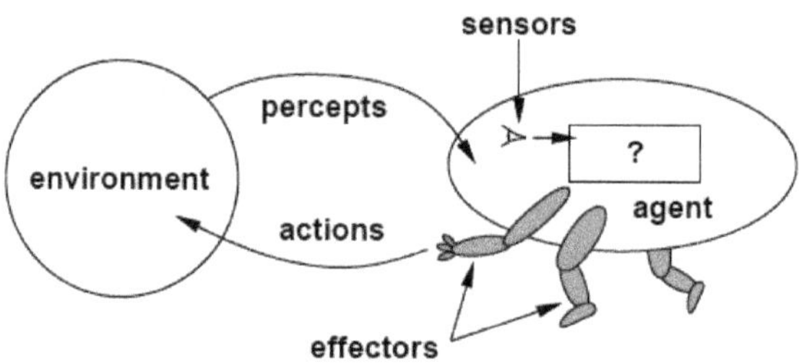

Fig. 11.1. The agent structure.

(2) **Interactive:** That is the awareness and influence of the environment, whether the agent survives in the real world (such as robots and Internet services on the agent) or the virtual world (such as the agents in virtual shopping malls). They are able to sense the environment and can change the environment through behaviors. An agent is not called "Agent" if it cannot affect the environment.

(3) **Collaborative:** Usually, the agent is not alone there but survives in a lot of the agent world. Good and effective cooperation among agents can greatly enhance the multi-agent system performance.

(4) **Communication:** This is also the fundamental characteristic of an agent. The so-called communications are the means of information exchange between the agents. Furthermore, the agents and people should be able to carry out a certain sense of "conversation". To undertake the task, agents' collaboration and negotiation are based on communication.

(5) **Longevity (or coherence time):** Traditional procedures are activated when needed and stopped when the operator ends. The agent is different. It should at least be "fairly long" for the time to run. Although not the indispensable characteristics of agents, it is generally believed that these are the important characteristics of agents.

In addition, some scholars also raised that the agent should have adaptive, personality, and other characteristics. In practical applications, the agent needs to take action under certain restrictions of time and resources. Therefore, the agents in the real world should have real-time characteristics besides the general natures.

At present, the research on multi-agent system is very active. Multi-agent systems are trying to use the agents to simulate the human behaviors. Agents are used in the real world and the community simulation, robots and intelligent machinery, and other fields. In the real-world survival, the agents need to face a changing environment. In such an environment, agents not only maintain a timely response to emergency situations but also plan for the use of certain short-term strategies. And then they are through the world and the other main modeling to predict the future of the state, as well as through communication language to realize cooperation or negotiation with other agents (Shi, 2000).

11.2.2 *Rational Agent*

Bratman's research on intention from philosophy generates a broad impact on artificial intelligence. He believes the balance between belief, desire, and intention can solve the problem effectively. In the open world, rational agents cannot be driven by intention, desire, and their combination since there is belief-based intention that exists between desire and planning. The reason is as follows:

(1) Agent actions are restricted to limited resources. Once agents decide what to do, they build a limited format commitment.
(2) In multi-agent environment, agents need commitment to coordinate the actions of agents. If there is no commitment, no actions are existent. Intention is the choice of commitment.

In an open and distributed environment, the actions of a rational agent are restricted by intention. The intention is represented:

(1) If an agent wants to change its intention, it needs reasons to do so.
(2) An agent cannot insist on its unimportant intention without considering the change of environment.

The main purpose of rational balance is to make a rational act in conformity with the characteristics of the environment. The so-called characteristics of the environment not only refer to the objective conditions but also include environmental factors in the social groups, such as social groups on the rational judgment of the law. Bratman presented the intention-action principle: If A takes action B under the current intention is reasonable, then Agent A changes intention to action. Action B is reasonable.

At the specific time, agent is shown as follows:

(1) Performance test provides a performance measure of the degree of success.
(2) Agents can percept all things; we will call this sense of history as the perception sequence.
(3) Agents know the environment.
(4) Agents can perform the actions.

An ideal agent can be defined: for each sequence may be the perception, the ideal rational agents, on the basis of the evidence provided by the perception sequence and the agent's inner knowledge, should do the desired actions to make its performance the largest.

11.2.3 *BDI Model*

BDI agent model can be described by the following elements:

(1) a group of belief about the world,
(2) a group of goals that agents want to achieve,
(3) a plan base to describe how to reach the goal and change the belief,
(4) an intention structure to describe currently how the agent achieves the goal and changes the beliefs.

11.3 Agent Architecture

Agent can be seen as a black box. It perceives the environment through the sensor and affects the environment with the actor.

11.3.1 *Agent Basic Architecture*

Human being, if we see an agent, perceives the environment with eyes, ears, nose, etc. It affects the environment using hands, legs, etc. Robot agent usually holds some cameras for getting environment information and the motor is their actor. Software agent uses codes as their sensor and actor (Figure 11.2).

Besides communication with the environment, most agents need to deal with the received information to achieve their goals. Figure 11.3 shows the work process of an intelligent agent. After receiving the information, the first step for agents is

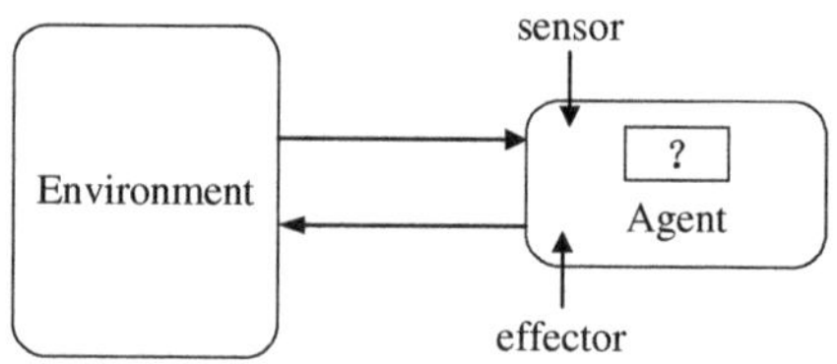

Fig. 11.2.   Agents interact with environments.

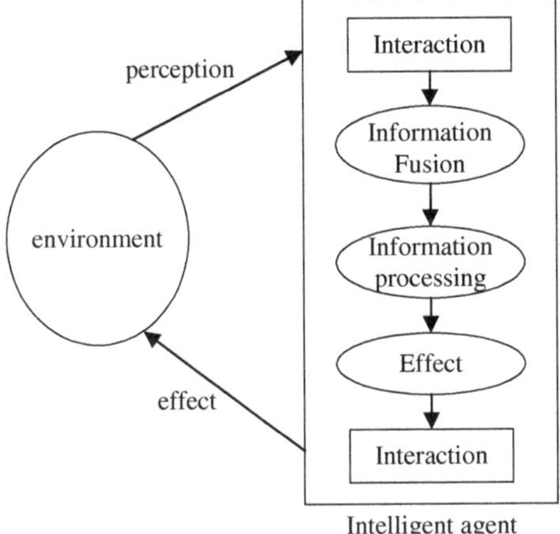

Fig. 11.3.   The work process of intelligent agent.

syncretizing the information to make it acceptable by knowledge base of agents. Information syncretizing is very important because the results from different communication components are often heterogeneous and the representation is different. For example, for the same thing, the information provided by human beings and received by agents may be different. So, information syncretizing must identify and differentiate correctly this kind of variance.

Once the agents receive the outer information, the information process becomes the key task of agents since it reflects the real function of every agent. The purpose of information processing is to explain useful data and form a concrete plan. Since every agent has its own goal, the impact of the inner goal must be considered as a part of the whole impact. If agents know the impact, they need to adopt actions to realize the goal. To form a plan, agents may prescribe knowledge, including new conditions to reflect the concrete actions. However, this is not obligatory since sometimes agents need not plan when taking action. If communicating with environmental objects, action model will adopt the proper communication component. The control task is also part of the action component.

From the discussion above, the agents can be defined as a map from perceptive knowledge to actions. Suppose O is the perceptive set and A is the possible action which can be achieved by agents. The agent function $f: O^* \rightarrow A$ defines the actions of all agents. The task of artificial intelligence is to design agent programs which can realize the map from perceptive knowledge to actions. The skeleton of an agent program is as follows:

Algorithm 11.1. Skeleton-Agent(percept) return action.

Input static: memory

Method:

1. *memory* ← Update-Memory(*memory, percept*);
2. *action* ← Choose-Best-Action(*memory*);
3. *memory* ← Update-Memory(*memory, action*);
4. return *action.*

The agents will change their memory to reflect the new percept after every function call. An ideal rational agent hopes to achieve the best performance for every percept.

Not every agent's action is the reflection of new conditions. Agents can also create their own new plans. In this condition, the knowledge of the information provider is useful only at a specific time. This is the main difference between reactive agents and deliberative agents.

11.3.2 *Deliberative Agent*

Deliberative agent (or cognitive agent) is a distinct symbol model which includes the reasoning ability about the environment and intelligent actions. It keeps the tradition of classical artificial intelligence. It is a knowledge-based system. Environmental model is implemented in advance to form the main knowledge base. There are two problems with this architecture:

Conversion problem: How to translate the real world to correct symbol description?

Representation/reasoning problem: How to represent the real entity and process? How to let the agents in possession of the ability to make decisions according to reasoning the information at a limited time?

The first problem results in the research of computer vision and natural language processing. The second problem induces the research on knowledge representation, autonomic reasoning, and planning. In the real world, these problems are a bit difficult since the representation is very complex. So, the deliberative agents have their own limitations in dynamic environment. Without the necessary knowledge and resource, it is difficult to add the new information and knowledge about environment into the existing model.

Deliberative agent is a kind of active software. Different from concrete domain knowledge, it has knowledge representation, problem-solution representation, environment representation, and communication protocol. According to the way of

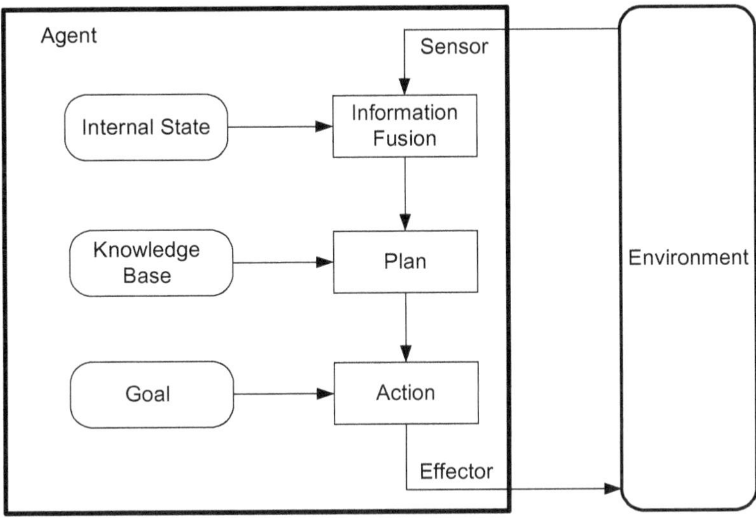

Fig. 11.4.   The architecture of deliberative agent.

thinking, deliberative agents can be divided into abstract thinking agents and visual thinking agents. Abstract thinking agents have abstract concepts. They think through the symbol information processing. However, visual thinking agents think using visual material. It adapts the connection theory of nerve mechanism.

Figure 11.4 gives the architecture of deliberative agent. Agent gets outer information through the sensors. After then, the agents syncretize the information according to the inner states and generate the description of the current state. They create the plan with the support of knowledge base then. After then, agents can form a series of actions which is taken by actors to affect the environment.

Algorithm 11.2.   Deliberate-Agent algorithm.

Input: static: *environment,* //describing the current world environment

  Kb,   //knowledge base

  plan   // Plan

Method:

1. *environment* ← Update-World-Model(*environment,percept*)
2. *state* ← Update-Mental-State(*environment,state*)
3. *plan* ← Decision-Making(*state,kb,action*)
4. *environment* ← Update-World-Model(*environment,action*)
5. return *action*

In the program above, the Update-World-Model function generates the abstract description of the current world environment from percept. The Update-Mental-State function revises the agent's inner mental state according to the percept environment. The knowledge base includes general knowledge and real knowledge. Agents apply the knowledge to make decisions with the Decision-Making function.

The BDI model which is the core of a typical deliberative agent can be described with the following elements:

1. a group of beliefs about the world,
2. current goals,
3. a plan base which describes how to achieve the goal and revise the belief,
4. an intention structure to describe how to achieve the current goals and revise the current beliefs.

According to BDI Architecture, Rao and Georgeff proposed a simple BDI interpreter (Rao & Georgeff, 1992):

Algorithm 11.3. BDI-Interpreter.

```
 1. BDI-Interpreter ( ) {
 2. initialize-state();
 3. do
 4.     options := option-generator(event-queue, B, G, I);
 5.     selected-options := deliberate(options, B, G, I);
 6.     update-intentions(selected-options, I);
 7.     execute(I);
 8.     get-new-external-events();
 9.     drop-successful-attitudes(B,G,I);
10.     drop-impossible-attitudes(B,G,I);
11.     until quit
```

11.3.3 *Reactive Agent*

The problems in traditional artificial intelligence are reflected in the deliberative agent without any change. The main criticism focuses on the rock-bound architecture. The agents work in a dynamic environment. So, they need to make a decision according to the current conditions. However, their intention and plan were developed on the symbol model of past specific times. There is little change about that.

Rule-based rock-bound extends the disadvantage since the conversion between planer, scheduler, and executor is time cost. The implementation of condition of scheduler changes more or less. The symbol algorithms of deliberative agents are often ideal and decidable which cause the high complication. In dynamic environment, it is more important to meet the requirement than the plan optimization. On the other way around, deliberative agents are good at mathematics proof of plans.

Different from the deliberative agents, reactive agents include no world model. They have no complex symbol reasoning (Wooldridge & Jennings, 1995). Figure 11.5 gives the architecture of reactive agents. The condition-action rules connect the cognition and actions. The rectangle in the figure represents the current inner states of the decision process. The ellipse shows the background knowledge which is used in the process.

Algorithm 11.4.   Reactive-agent algorithm.

Reactive-agent (percept) returns action

      static: state, describe the current world state

            rules, a group of condition-actions rule

1. state ← Interpret-Input(percept)
2. rule ← Rule-Match(state, rules)
3. action ← Rule-Action[rule]
4. return action

In the program, the Interpret-Input function generates the abstract description of current states from the percept. The Rule-Match function returns a rule which is matched with the description of current states.

Reactive agents are related to Professor Brooks at MIT who proposed behavior-based artificial intelligence (Brooks, 1991a). He thinks intelligent behaviors are the result of communication between agents and the environment. Brooks is not only a critic but also a practice expert. He implemented some of Robots without symbol reasoning. These Robots is a subsumption architecture system. Subsumption architecture is a hierarchical behavior which can finish the tasks. Every behavior tries to control the robot which causes competition between them. The behaviors in the base layer represent the relative original behaviors which have high priority. As a result, the implementation of this kind of architecture is relatively simple. Brooks alleged the agent with this kind of architecture can finish the tasks by symbol-based artificial intelligence.

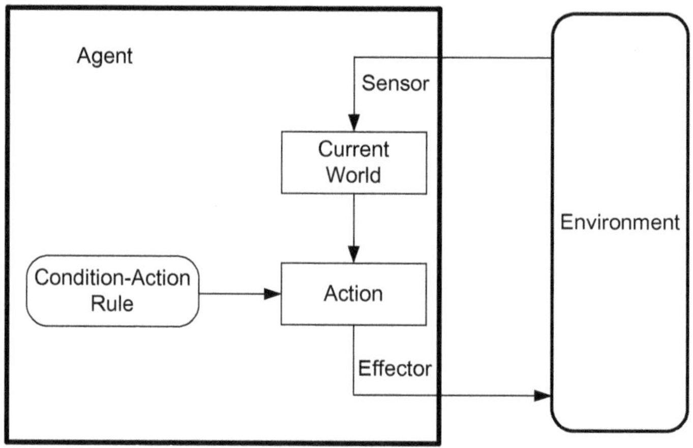

Fig. 11.5.　The architecture of reactive agent.

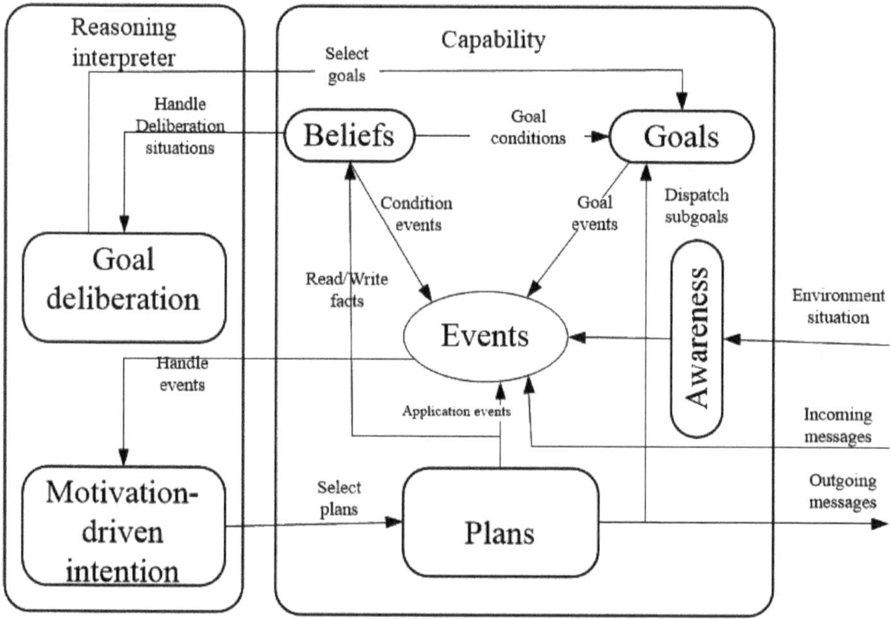

Fig. 11.6.　Agent ABGP model.

11.3.4 *Agent ABGP Model*

A cognitive model for multi-agent collaboration should consider external perception and internal mental state of agents. A 4-tuple framework ⟨Awareness; Belief; Goal; Plan⟩ is proposed for the agent, whose architecture is shown in Figure 11.6.

Awareness is an information pathway connecting to the world (including natural scenes and other agents in a multi-agent system). M. Endsley pointed out awareness has four basic characteristics (Endsley, 1995):

(a) Awareness is knowledge about the state of a particular environment.
(b) Environments change over time, so awareness must be kept up to date.
(c) People maintain their awareness by interacting with the environment.
(d) Awareness is usually a secondary goal, that is, the overall goal Endsley's opinion.

Beliefs can be viewed as the agent's knowledge about its setting and itself. Goals make up the agent's wishes and drive the course of its actions. Plans represent agent's means to achieve its goals.

ABGP agent considers not only the internal mental state of the agent but also the cognition and interaction of the external scene, which plays an important role in the decision-making of the agent (Shi & Huang, 2019). ABGP agent is the core component of brain–computer collaborative simulation environment.

11.4 Collaborative Crowd Action

In computer sciences, one of the most challenging goals is to build a computer system which can cooperate with each other (Kraus,1997). Computer system becomes more and more complex. To integrate the intelligent agents is more challenge. The cooperation between agents is the key point for cooperative work. Besides, the multi-agent cooperation is an important concept for distinguishing multi-agent systems from other distributed computing, object-oriented systems, and expert systems (Doran *et al.*, 1997). Coordination and cooperation are the core problems of multi-agent research since based on the intelligent agent. Coordinating the knowledge, desire, intention, plan, and action to implement the cooperation is the main goal of a multi-agent system.

Coordination is a kind of character that multi-agents interact to finish a group of activities. Coordination is the adaptation to the environment. Coordination is to change the intention of agents. The reason for multi-agent to coordinate is that the agents have intentions. Cooperation is a special example of multi-agent coordination. Multi-agent research is based on the human being society. In human being society, communication is indispensable. The communication is between the conflict and non-conflict. At the same time, in an open, dynamic multi-agent environment, agents with different goals and resources must coordinate with each other. For example, there will generate dead lock when resource conflict without coordination. On the other side, if one agent cannot finish the goal independently, it needs the cooperation of other agents.

In a multi-agent system, cooperation improves not only the ability of a single agent but also the performance of the multi-agent system and increases the agent's ability to solve the problem which makes the system more agile. Through cooperation, multi-agents can solve more real problems and extend applications. Although for the single agent, it only cares about its own requirements and goals. Its design and implementation can be independent of other agents. However, in the multi-agent system, agents do not exist alone. The agents' actions must meet the society rules although they are complied agents. The relationship between the multi-agent systems makes the interaction and cooperation have great conditionality to the agent's implementation and design.

In the current phase, the research on multi-agent cooperation has two species. The one is borrowing the approach from the research on multi-entities' behaviors (Kraus, 1997). The other is from the view of intention, goal, and plan to research the multi-agent cooperation, e.g., FA/C model (Lesser, 1991), the Joint Intention Framework (Levesque, 1990), and share plan (Grosz & Sarit, 1996). The later method has a bigger application scope. Each theory is only adapted to some specific coordination environments. Once the environment changes, for example, the number of agents, types, and interaction relationships disagree with theory, the cooperation based on this theory loses its superiority. The later approach focuses on the planning and solution of problems and assumes the process of coordination is different. Some agents find the cooperative partners first and then generate the plans. Some may generate a plan for the problem first and then take cooperative actions. Some may generate partial global planning (PGP) to regulate its actions for the cooperative goals (Lesser, 1991). The latter two cooperative methods seem relaxed and lack the necessary disposal mechanism. They require the shared cooperative plan between agents. The first approach has bigger uncertainty and the cooperative plans are influenced by the cooperative team.

In a multi-agent system, the agent is autonomic. The knowledge, desire, intention, and behavior of agents are different. To coordinate the cooperative work is the necessary condition for problem solution and keeping the efficiency. Many researches on organization theory, politics, sociology, social psychology, anthropology, jurisprudence, and economics are applied in multi-agent systems. Multi-agent coordination is the process in that multi-agents interact with each other for consistent work way. Coordination can avoid the deadlock and livelock of multi-agents. Deadlock means a multi-agent cannot take the next action; the livelock is the state of that multi-agent consecutive work but there is no result. There are many coordination methods:

- organizational structuring,
- contracting,
- multi-agent planning,
- negotiation.

From the point of social psychology, multi-agent cooperation has the following types:

(1) Cooperation: put its own interest in the second place.
(2) Selfish: put the cooperation in the second place.
(3) Completely selfish: do not consider any cooperation.
(4) Completely cooperation: do not consider its own interest.
(5) The hybrid of cooperation and selfishness.

The interaction between agents has two kinds of relationships: negative and positive relationship. Negative relationship causes conflict. The resolution of conflicts constructs coordination. Positive relationship represents the plan of multi-agent has overlap or some agent has the excluded ability. Every agent can get help through the cooperation.

Before the mid-1980s, research on coordination and cooperation in distributed intelligence mainly focused on helping each other to realize the goal without conflicts. This kind of research is used in distributed problem solutions. In the mid-1980s, Rosenschein did a deep research in his doctoral thesis on multi-agent interaction when agents had the goals conflicts. He applied the game theory to build the static model for rational agent interaction (Rosenschein,1986). It then becomes the theory foundation for the multi-agent coordination and cooperation. After then, many researchers used the game theory to formalize the multi-agent cooperation. All of these researches are to build a model to coordinate the actions or through the cooperation to realize the goal with the inconsistent goals condition. Some of the research considered time preference. Some of the research is open environment oriented.

MIT's research adopted the meta-communication to coordinate multi-agent computing under the foundation of intensifying FA/C and PGP approaches. Macintosh applied the heuristic approach to introduce cooperation to machine theorem proof. Sycara researched the negotiation based on the labor and capital problem (Sycara *et al.*, 1996). The approach applied the heuristic and constraint satisfaction technology to solve the distributed search problems. It adopted the asynchrony trace to resume the inconsistent search policies. The disadvantage is that this approach needs an arbitrage machine to solve the conflicts. Conry researched multi-step negotiation under the multi-goal and resources. Hewitt proposed the distributed artificial intelligence approach which challenges Rosenschein's static interaction model. The real world is open and dynamic. The coordination and cooperation is also open and dynamic. Computing bionomics argues that the agents need not have strong reasoning ability in the open and dynamic environment. They can gradually coordinate the relationship with the environment through consecutive interaction to make the whole system have the evolution ability. That is similar to an ecosystem. In the BDI model, it emphasizes the agent's intention, desire, and intention's rational balance in the interaction process.

Shohan proposed the artificial agent society needs a law to regulate the agents' actions. Every agent must follow the law and believe that other agents also abide by the law. This law can regulate the agents' behaviors on one side. On the other side, it can insure other agents' action style. This guarantees the realization of agents' behaviors. Decker proposed a dynamic coordination algorithm in distributed sensor network (Decker, 1995). The agent can revise the sensor domain according to the unified standard which can automatically reach the coordination. This dynamic reconstruction domain can realize the system load balance in the whole system. Its performance is better than static sub-area algorithm. Especially, this approach can decrease the fluctuation of performance which is more adapted to dynamic real conditions.

In multi-agent planning, agents can regulate the pertinence of goals to realize autonomous behavior coordination. German Wei proposes an approach to form action and goal's correlation value for specific problems through distributed learning. The other autonomic coordination approach is the Markov process proposed by Kosoresow. Markov process is a quick probability coordination approach when the agent's goal and preference are consistent. If Markov process is applied as the agent's reasoning mechanism, it can analyze the astringency of multi-agent interaction and average astringency time. When the agent's goal and preference are not consistent, it can check the inconsistency at a specific time point and solve the conflict by submitting it to a higher coordination protocol. In order to describe a group of agent actions, we need a common intention to connect the member's behavior. Jennings adopts common duty concept to emphasize the function of intention as the action controller. It prescribes how the agents act in the cooperative problem solution. This

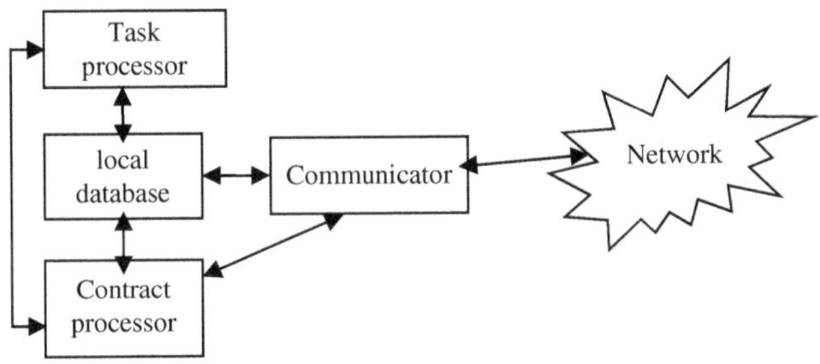

Fig. 11.7.   Architecture of node in contract net protocol.

common duty can provide function instruction for architecture design and provide a standard for inspecting problem solutions and exception disposal. They realized this common duty in the GRATE system.

11.4.1 *Contract Net Protocol*

In 1980, Smith proposed the contract net protocol in distributed problem solution (Smith, 1980). This protocol was widely applied in multi-agent coordination. The communication among agents has the same message format. The real contract net system provides a contract protocol based on the contract net protocol. It prescribes the task assignment and the roles of agents. Figure 11.7 shows the architecture of a node in the contract net protocol.

The locale database includes the knowledge base related to the node which is used to negotiate the information in the current state and problem-solution process. The other three components execute their tasks according to local knowledge base. The communicator is in charge of communicating with other nodes. The node can only connect the network through the communicator. In particular, the communicator should understand the message sent and received.

The contract processor judges the tasks, sends the application, and finishes the contract. It can also analyze the reached information. Finally, the contract processor executes the coordination for whole nodes. The task of task processor is dealing with the tasks. It receives the tasks from the contract processor, applies the local database for a solution, and returns the result to the contract processor.

Contract net protocol partitions the task into sub-tasks. It designates a node as the manager node. The manager node is familiar with the sub-tasks (Figure 11.8).

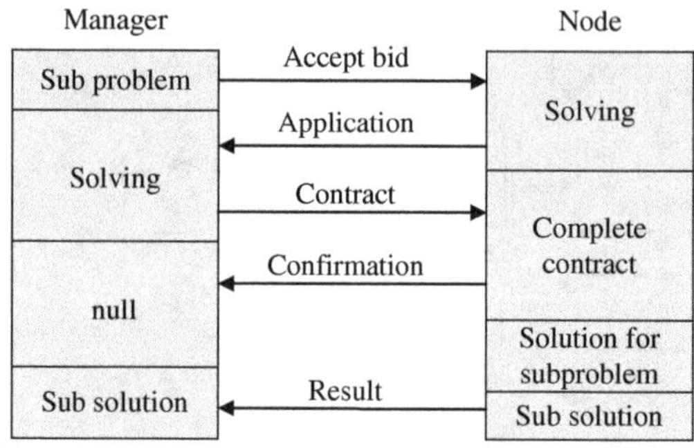

Fig. 11.8.   The negotiation process in contract net protocol.

The manager node provides the bids. Those are unsolved sub-problem contracts. The message format in contract net protocol is as follows:

For example:

TO:	All nodes
FROM:	Manager
TYPE:	Task bid announcement
ContractID:	xx-yy-zz
Task Abstraction:	<description of the problem>
Eligibility Specification:	<list of the minimum requirements>
Bid Specification:	<description of the required application information>
Expiration time	<latest possible application time>

The bids are open for all agents. They solve the bids through the contract processor. They apply the local database to solve the current resources and agent knowledge. The contract processor decides which tasks are worth doing. If they want to do, they need to inform the manager using the following message format:

TO:	Manager
FROM:	Node X
TYPE:	Application
ContractID:	xx-yy-zz
Node Abstraction:	<description of the node's capabilities>

The manager must select nodes to give the contract to the most proper node in the whole application. It visits the concrete knowledge solution and approach to select the best performance and assigns the sub-problem to the node. According to the message, the assignment contract is as follows:

TO:	Node X
FROM:	Manager
TYPE:	Contract
ContractID:	xx-yy-xx
Task Specification:	<description of the subproblem>

The communicator sends the confirmation message for accepting the contract to the manager. Once the problem is finished, the solved problem is sent to the manager. The promising node is in charge of the solution of the problem. The contract net protocol is for task allocation. The node cannot accept other agents' current state information. If the node cannot finish the task because of the resource or ability limitation, it can divide the task into some sub-tasks and allocate the sub-tasks to other nodes. In this condition, the node is a manager role and provides the bids. Every node can be a manager, bid applicant, or contract member.

The traditional contract net protocol is extended to infer the negotiation process. One of these extensions is publicizing the bid file. All the nodes can join the bid process. This requires communication and abundant resources. The manager must evaluate many lot of bid files and use lot of resources. The big load of the manager is caused by the publicizing the bid files. First, the manager needs the ability to inform part of the nodes. You can imagine that if the manager knows the knowledge of every node, it can estimate the candidate nodes for dealing with the sub-problems. Second, the common bid request can be canceled. If the unsolved problems can use the previous methods, the manager can contact the previous solution nodes. Once the resource can be used, the manager and the node write down the contract. Besides, the node can self-bid. In this condition, many open bids only investigate the new tasks. The bid request is only needed when it cannot find the proper bids.

The second side of contract net system extension is influencing the real contract assignment. In the traditional protocol, the manager needs to wait for the node's information after the assignment. Before the coming of the information, the manager does not know whether the node accepts the contract. The node cannot form a contract and build the constraint of contract after a bid. One suggestion for the extension is to build the contract constraint at an early time. For example, when a node is bid, it can provide the possible clause of accepting a promise. The acceptance is not easy acceptant or refuse. It can hold some reference and conditions. The biggest time limit for contract confirmation is extending further. If there is no confirmation for the contract in the limited time, the manager will stop the contract. The contract process can send the information and avoid the manager for waiting a long time. The managers can assign the contract again before the longest time interval.

11.4.2 *Partial Global Planning*

The most important trait of PGP approach is the ability of each agent is determined in multi-agent system (Durfee,1991). That is collecting the current states to reach the goals. The agents can optimize their tasks with knowledge. PGP provides a flexible concept and coordinates the distributed problem solution component.

The basic condition to apply the PGP is there are several agents for the whole problem to work. An agent is part of PGP and considers the other agents' action and relation to reach their own conclusion. This knowledge is seen as PGP. Figure 11.9 gives an example to explain the basic work principle. Two agents work for two sub-problems (A or B). Each agent sends information to their cooperative agent. Agent 1 informs agent 2 its current sub-problem A. At the same time, agent 2 informs its sub-problem B. Every agent can know the cooperative agent's condition according to the information. For example, agent 1 knows its sub-problem A2 is determined by agent 2's sub-problem B. It can inform agent 2. PGP's process can be divided into four steps:

(1) Each agent creates partial planning.
(2) Communicate and exchange rules between agents.
(3) Create partial planning.
(4) Revise and optimize the PGP.

At the beginning of the coordination, each agent must create partial planning and solve the assigned tasks. Every partial planning at least has two different layers. The whole structure includes the important steps for solving the problem which reflects the long planning for solving the problem. It includes detailed information about every sub-problem.

Once the partial plan is finished, agents exchange the knowledge with each other. Each agent needs to have some specific organizational knowledge. It can

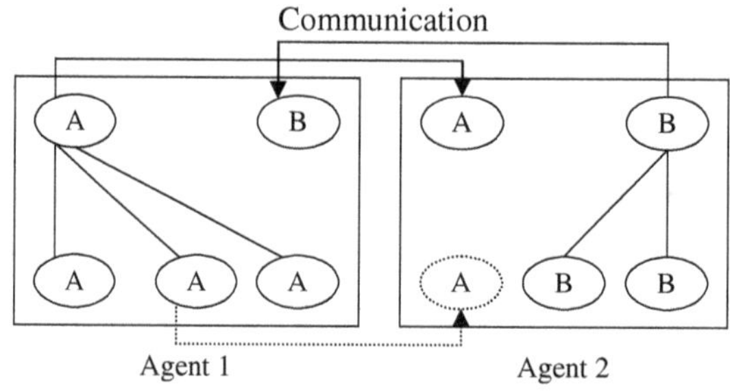

Fig. 11.9. Partial global planning.

determine other agents' roles and which information is interesting. Meta organization determines the layers of agents. Once agents receive the plan information from other agents, it must organize the information with the PGP format. It also needs to check if there is new information including the dependence on the inner plan. After that, they form logic components with the sub-plans. PGP consists of the following components:

(1) **Goal:** Goal includes the basic information, the reason for existent, long goal and the comparable priority.
(2) **Plan activity diagram:** It includes other agents' tasks and their current states, detailed plan, expected results, and some experiments.
(3) **Solution construction diagram:** This diagram includes how the agents communicate and cooperate among agents. The size and time of the sent plan are very important.
(4) **State:** State includes reporting PGP's whole important information, e.g., other agents' plan and time mark.

PGP core planer is to analyze the information sent by other agents and determine if they work for the same goal. PGP planer integrates the knowledge through planning the activity diagram. It can forecast the behavior and result of other agents. Plan activity diagram forms the foundation of future work. Plan activity compares the partial plan and new knowledge to create a revised partial plan. At this time, they create the solution construction diagram. PGP planner generates a revised partial plan as the final result. After the concrete plan is sent to the specific agents, the planer uses the current knowledge about the whole system to optimize and enrich the minutia.

There are many advantages to use the partial planner. The most important advantage is high dynamic system behavior. All plans can adapt to the environment solution at any time and the whole system has high ability and effect. PGP can identify the change in the old plan. The system must send the revision to the node as soon as possible since revision can inference their work. If any small revision needs to be sent to the node, the whole system will be high load. In this condition, PGP agrees on small inconsistencies and only informs the big revisions. The system developers must develop the fault tolerance system and designate a threshold to indicate the revision is important. The second advantage is effective and avoiding redundancy. Two or more agents have similar problems and register their PGP, and reassign and construction. The work assignment is very effective.

The old PGP model has many constraints, especially in the heterogeneous agents' different problem solutions. How to deal with dynamic agents change solution policy again? How to finish the concrete tasks in the limited time? How to negotiation between agents? There are some methods to extend the PGP model. For example, using the general partial global planning (GPGP) provides new components to meet the need (Decker,1995).

Osawa proposed an approach to construct a cooperative plan in an open environment based on robots' collaboration (Osawa,1993). In the approach, rational agents use the "multi-world model" to construct an incomplete single-agent plan and then according to the effect-based model to balance the cost. This process can be described as follows:

(1) Requestor sends RFP to call-board.
(2) Leisure agent applies an RFP from call-board.
(3) Call-board sends RFP to leisure agents.
(4) Leisure agent generates the plan.
(5) Leisure agent sends it to requestor.
(6) Requestor investigates the possibility of cooperation.
(7) Requestor sends cooperation rewards.
(8) Applicants form cooperative plan.

The utility is computed using the following formula:

$$\text{utility}(a, g) = \text{worth}(a, g) - \text{cost}(\text{plan}(a, g))$$

The average utility is the principal of cooperation. Although Osawa solved the cooperative problems in an open environment, it is not a good way to add the utility to be divided for average value since the utility is only the order relation of agent to goal. We can not compare the utilities between different kinds of agents.

11.4.3 *The Planning Based on Constraint Propagation*

In multi-agent systems, the coordination work of agents can be realized by situated action and planning. The method-based situated action emphasized the interaction between agents and the environment, and the effects of perception–action looping upon agent actions. The planning decides the sequence of agent actions which will be executed, and the sequence will be decided by some searching way. Compared with situated action, planning can make the agent's actions more correct and more rational, therefore planning is applied generally in deliberative agent.

Due to the lack of considering coordination, cooperation among agents in a multi-agent system, generally planning algorithms, such as UCPOP, GraphPlan, and SNLP, cannot be applied directly. In a multi-agent environment, the actions among agents are concurrent, which are not controlled by a single agent. An agent can make a system more coordinated by communication and protocol to try to change the actions of other agents and to complete the work which cannot be complicated itself by requesting the cooperation of other agents. Classical planning algorithms assure that an agent is the only one who can change the environment, therefore the planning result cannot ensure that there is no conflict with other agents, and the solution capability of an agent is limited. The distributed planning in dynamic environment hypothesizes that the paroxysm in environment is not foreseeable, and the planning and executing are handled alternately. This method cannot ensure the reliability and completeness of planning.

A planning, especially partial order planning, can be assured by the constraints in planning steps as follows: using temporal constrains to designate the time sequence of executing steps and using code signation constraints to designate the executing steps. Planning solving is a process which adds and refines constraints gradually. The conflict of coordination and cooperation among agents can be accomplished by designating the temporal constraints and code signation constraints. The resource confliction of agent actions and the services which are offered by one agent for other agents can be presented by causal link and can be solved by adding temporal constraints to agent actions. At the same time, the cooperation actions among agents, as two agents cooperating to take up a desk, must be described by a single action description. The correctness of an agent planning is assured by truth criterion in multi-agent system. We propose a multi-agent concurrent partial order planning algorithm (Shi *et al.*, 2002). Each agent is planning for the target to be accomplished in parallel. When an agent decides to add a new action, it accomplishes the coordination of multi-agent system by the constraint propagation with other agents.

To describe the planning action, there are too many constraints in the accomplishment of practical schedulers using STRIPS, and it is too difficult to use situated checking computations.

In 1989, Pednault proposed a kind of action described as language ADL (Pednault *et al.*, 1989). We expand ADL to describe the planning actions in multi-agent system (Shi *et al.*, 2002). The represent capability of it is more powerful than STRIPS but less powerful than first-order logic.

The semantic of ADL is based on the mathematics structure which describes the world statements. An action a is represented by a statement couple $<s, t>$ in ADL, and action a produces a statement "t" when it is executed at statement s. The relations between a statement s and a statement description ϕ are represented by a symbol $\models$, denoted as $s \models \phi$.

The action template in ADL represents a set of possible actions, which is described by four selectable clauses: Precond, prerequisite; Add and Delete, the set of formulas added or deleted in relation R on statement t; Update, a set of relations which describe the functions how to transform from s to t.

Multi-agent planning means that one or more concurrent processes of agent search their respective targets in their respective planning space simultaneously. In order to coordinate the actions of different agents, we have to add partial constraints between actions which may conflict with each other. The partial constraint distributes in different agents, so there should be a problem in judging consistency in the partial constraint.

Algorithm 11.5. Consistency Checking of distributed partial order constraint.

Calculate_transition_closure(α, A_i)

  Input: planning $\langle S, B, O, L \rangle$, constraint $A_i \prec A_j$,

  Output: transitive closure $T_<(A_i)$.

1. send information to all agents to set temporal constraint O in its planning to read only mode;
2. compute of local transitive closure $T_{\le}^{a}(A_i)$ of action A_i in agent α;
3. $T_<(A_i) = T_{\le}^{a}(A_i)$;
4. let SET(agentn) be actions of the agent in $T_<(A_i)$.
5. start a new thread for every nonempty SET(agentn)
6. {
7. $T_<(A_i) = T_<(A_i)+$ request (calculate_transition_closure(agentn, SET(agentn));
8.   return.
9. }
10. after all subthreads return, return $T_<(A_i)$;
11. demand of all agents to disable the read-only mode of constraint.

In algorithm 11.6, the cooperation between agents involves two phases: task assignment for every agent and sub-goal planning of agent. Task assignment means that the global goal of multi-agent system is decomposed and is assigned to one or a group of agents in the system so as to accomplish the target. Here we emphasize the later problem, or say, in the beginning of planning, every agent has its own goal, which may be assigned to the agent or be formed to meet its own interest. Furthermore, for multi-agent system and agents in the system, we assume the following:

(1) Any effect of all actions is a deterministic function of system status when the job and action are being performed.
(2) Agent possesses all the knowledge of its own action, the actions of other agents, and the initial states of the system.

(3) The change of system can only be caused by actions of agents.

(4) No matter whether an agent could make planning successfully, it will keep the promise of actions with other agents in the planning phase so as to enable other agents to accomplish their planning.

Communication between agents is in accord with message transmission protocol in communication language ACL specification. In particular, the communication should be without delay, and the message sent to the same target will arrive according to the sequence of the message sent.

The architecture of agents is given in Figure 11.10. Due to requirements of negotiation between agents and judging consistency in constraint, in every agent, there are three independent threads, which are as follows:

1. **Planning thread:** Responsible for executing algorithm 11.6 for solving its own planning problems.
2. **Constraints maintenance thread:** Responsible for calculating the local transitive closure of the action on its own actions set S and judging consistency in the distributed partial order planning.
3. **Communication and negotiation thread:** responsible for sending action constraints and accomplishing cooperation and coordination of actions between agents.

Every agent could solve its own sub-goal planning according to algorithm 11.6. In algorithm 11.6, agents construct and maintain planning according to the UCPOP framework; when needed for requesting services from and cooperating with other agents to judge consistency in the distributed partial order constraint, agents communicate with each other.

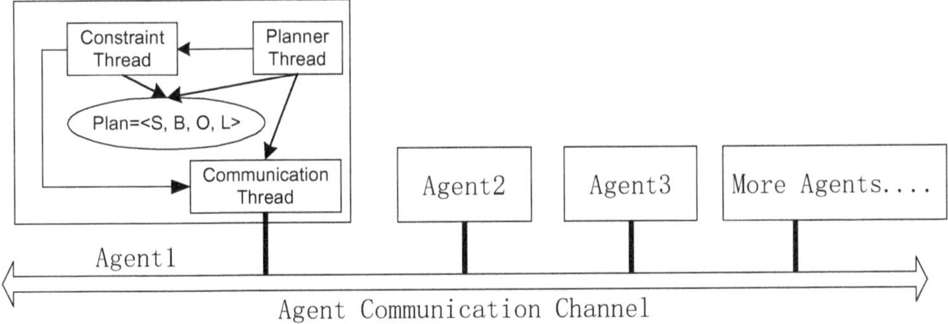

Fig. 11.10.   The architecture of planning agents.

Algorithm 11.6. Planning algorithm.

---

planning $(\langle S, B, O, L \rangle, G, \Lambda)$

1. Termination conditions: if G is empty, return $\langle S, B, O, L \rangle$;
2. Targets resolution: get a target $\langle Q, A_c \rangle$ from G,

    (i) If Q is a conjunction of Q_i, add every $\langle Q_i, A_c \rangle$ into G, and go to step 2;

    (ii) If Q is a disjunction of Q_i, randomly choose a Q_k, and add it into G, and go to step 2;

    (iii) If Q is a character and $A_p \xrightarrow{\neg Q} A_c$ exists in L, return failure.

3. Select operator: randomly choose an action existing in S, or from Λ choose a new action A_p, which has effection e and an universal clause p, where $p \in T(\theta_\ell)$ and $MGU(Q, p) \neq \perp$ (Q and p have a general resolution); if there is not any action satisfying conditions in Λ and there is a agent β whose action $A_{\beta j}$ satisfies conditions, send a request to β; if there are joint-actions satisfying conditions, send requests to all executors of joint-actions; the agent which receives a request executes the step 4 and 5 in its planning P_b, if it succeeds, β returns $\mathrm{Comm}(A_{\beta j})$ and add joint-actions or $\mathrm{Comm}(A_{\beta j})$ into S, and go to 2; otherwise, β returns failure.

4. Enabling new action: Let $S' = S$, $G' = G$. If $A_p \notin S$, add A_p into A', and add $\langle preconds(A_p) \backslash MGU(Q, R, B), A_p \rangle$ into G', and add non-cd-constraints(A_p) into B'.

5. Protection of Causal Link: for every causal link $l = A_i \xrightarrow{p} A_j$ and every action A_t which may be a threat to l, choose one from following three solutions (if do not choose any solution, return failure)

    (i) Upgrade: for O judge consistency in the distributed partial order constraint, if constraint is consistent, $O' = O' \cup \{A_j < A_t\}$

    (ii) Degrade: for O judge consistency in the distributed partial order constraint, if constraint is consistent, $O' = O' \cup \{A_t < A_j\}$

    (iii) Facing: if l which A_t becomes a threat to is a conditional conclusion, let condition be S and conclusion be R, and add $\langle \neg S \backslash MGU(P, R), A_t \rangle$ into G'.

6. Recursive call: if B is inconsistent, return failure; otherwise, call $(\langle S', B', O', L' \rangle, G, \Lambda)$.

Algorithm 11.6 is sound but not complete. Soundness means that for a planning problem if an agent finds a planning using a planning algorithm, the planning is the solution for the problem. Algorithm 11.6 is not complete because there is a temporal and equivalent constraint relationship when an agent and other agents search in planning space, which means that the planning space of agents has been limited. In particular, in order to not cause parallel planning to be too complex and prolix, we do not provide a mechanism that when an agent cannot find a solution after the backtracking process, the agent could ask other agents to adopt the backtracking process or to relax constraints, which will expand its current solution space and the agent could continue searching.

11.4.4 *Ecological-Based Cooperation*

At the end of the 1980s, a new subject — the ecology of computation — is presented. It is a subject to research the behavior and resource application. It spurns the traditional closed, static algorithm for solving the problem. It thinks the world is an open, evolved, concurrent ecosystem which solves the problem with collaboration. Its development is related to the research of open information system.

The distributed system is similar to the society and biologic organization. This kind of open system is different from the current computer system. It computes the tasks asynchronously. Its nodes can generate the process in other machines. These nodes can make decisions according to incomplete and late knowledge. There is no center control node. It solves the problem according to the communication and cooperation of many nodes. These characteristics consist of a concurrent combination. Its communication, policy, and competition are similar to the ecosystem. Hewitt proposed an open information system concept (Hewitt, 1991). He argued that incomplete knowledge, asynchronous computing, and inconsistent data are inevitable in open computing system. Human society, especially the problems in sciences, can be solved by cooperation.

Computing bionomics considers the computing system as an ecosystem. It introduces make biologic mechanism, e.g., mutation in computing system. This revision causes the change of life gene which forms the diversity to improve the ability to adapt to the environment. This mutation policy becomes a method to improve its own ability in artificial intelligence system. Miller and Drexler discussed a series of evolution models, for example, the ecosystem and commerce market, and pointed out the differences with the ecosystem. They think a direct computing market is the most ideal system model.

Incomplete knowledge and late information are the inner characteristics of computing ecosystem. The dynamic activities research under these constraints is important. Huberman and Hogg proposed and analyzed the process of dynamic games (Huberman & Tad, 1987). They pointed out that if there are many choices when the processes finish the tasks, dynamic graduate process may become nonlinear fluctuation and chaos. This indicates the stable policy for computing biologic system is not existent. They also discussed possible common rules and the importance of cooperation and compared them with the biologic ecosystem and human organization. Similar to dynamic theory, Rosenschein and Geneserth proposed static policy theory to solve the conflicts in nodes with different goals.

Famous ecosystem models include biologic ecosystem model, species evolution model, economic model, and sciences group's society model. Large ecosystem's intelligence surpasses any single intelligence.

1. Biologic ecosystem model

It is the most famous ecosystem which has typical evolution characteristic and hierarchical traits. This trait is reflected in the food chain. For complex biologic ecosystem, all the species consist of a close network-food chain. This system's main roles are catchers and preys. Life is dependant on life. The large ecosystem consists of a small ecosystem.

2. Species evolution model

Species evolution depends on genes. From the plant heredity to model genetics, researchers indicate the combination of genes' importance in the species' evolution. The gene pool consists of a group of genes from a species. The biological organization is the carrier of genes. If the environment changes, the selection mechanism will change. This change inevitably causes the change of gene pool. The change of specific species is named gene stream. A species always experiences isolation, gene flow, and change circle. In the beginning, a geographically isolated group develops lonely and the gene quickly flows into the inner structure. Due to the open, the system can communicate and compete to realize the survival of the fittest.

3. Economics model

Economic system is similar to biologic ecosystem. In commerce market and ideal market, evolution determines economic entities' decisions. The choice mechanism is market encouragement mechanism. The evolution is rapid. The relationship between the enterprise and consumer and the interenterprise is dependent on each other. Decision-maker can adopt effective approaches in order to purse long-term interest, especially sustaining losses in business in a short time.

11.4.5 *Game Theory Based Negotiation*

In multi-agent system, negotiation has many understandings. A point of view is that negotiation is about the allocation of resource and sub-problem. Another view thinks the negotiation is peer-to-peer direct negotiation. The goal of negotiation is to build cooperation among a group of agents. Agents have their own goals. Negotiation protocol provides the possible basic rules and the negotiation process and the foundation of communication. The policy of negotiation is decided by the concrete agents. Although the agent developers provide different extent ability of negotiation, they must guarantee the match between protocol and policy. That is to say, the policy can run in the protocol.

From single agent, the goal of negotiation is to improve their states and support other agents without affecting themselves. The agents must make a trade-off to maintain the ability of the whole system. From this view, the negotiation communication can be divided into some types:

(1) **Symmetry cooperation:** The result of the negotiation is better than the past result and the inference by other agents is positive.

(2) **Symmetry trade-off:** Agents would like to achieve their goals. Negotiation means the trade-off of participants which will decrease the effects. However, the negotiation can not ignore the existence of other agents. They can only take the policy of trade-off and make the participants accept the result.

(3) **Asymmetry cooperation/trade-off:** The effect on one agent's cooperation is positive, but to the other agent, the trade-off is required.

(4) **Conflict:** Due to the conflicts of goals of multi-agents, they can not reach an acceptable result. The negotiation must stop before the result.

11.4.6 *Intention-Based Negotiation*

Grosz applied the belief, desire, and intention theory to the agent negotiation (Grosz & Sarit, 1996). This method does not use the sub-plan but uses the intention for negotiation to decrease the communication. BDI theory thinks that the agents' actions were not the desire and plan but the result of the combination of belief and desire. The sub-plan to realize the intention is generated by this intention. One intention is correspondence with some sub-plans. Agents' communication need not exchange the sub-plans and only exchange the intentions. However, Grosz's approach assumes the agents are completely cooperative.

Zlotkin's work greatly improves the static negotiation theory of two agents. However, to open multi-agent system, his approach is not applicable. Current BDI team approaches lack tools for quantitative performance analysis under uncertainty. Distributed partially observable Markov decision problems (POMDPs) are well suited for such analysis, but the complexity of finding optimal policies in such models is highly intractable. Nair and Tambe have proposed a hybrid BDI-POMDP approach, where BDI team plans are exploited to improve POMDP tractability and POMDP analysis improves BDI team plan performance (Nair & Milind, 2005).

11.4.7 *Team-Oriented Collaboration*

Coordination between large teams of highly heterogeneous entities will change the way complex goals are pursued in real-world environments. Scerri and his colleagues proposed a Machinetta approach which combines team-oriented programming and proxy architecture to overcome the limitations of effective coordination between very large teams of highly heterogeneous agents (Scerri, 2003). The main advantages to Machinetta approach display one or a combination of the characteristics: large scale, dynamic environment, and integration of humans. By connecting the Machinetta proxies with the graphical development tool for constructing team plans, the team-oriented programming programmer gains a good idea of what is going on in the plan and how to make effective changes to it.

Token-based coordination is a process by which agents attempt to maximize the overall team reward by moving tokens around the team. If an agent were to know the exact state of the team, it could use a Markov decision process (MDP) to determine the expected utility-maximizing way to move tokens. Unfortunately, it is infeasible for an agent to know the complete state, however, it is illustrative to look at how tokens would be passed if it were feasible. Then, by dividing the monolithic joint activity into a set of actions that can be taken by individual agents, we can decentralize the token routing process where distributed agents, in parallel, make independent decisions of where to pass the tokens they currently hold. Thus, we effectively break a large coordination problem into many small ones.

Algorithm 11.7. Token pass local model which makes decision process for agent α to pass incoming tokens (Xu *et al.*, 2005).

Method:

1. while true do
2.   *Tokens*$(\alpha) \leftarrow$ *getToken*(*sender*);
3.   for all $\Delta \in$ *Tokens*$(\in)$ do
4.   if *Acceptable*(Δ, α) then
5.   if Δ.*type* $==$ *Res* then
6.   *Increase*$(\Delta, threshold)$;
7.   end if
8.   else
9.   *Append*(*self,* Δ.*path*);
10.   for all $\Delta_i \in H_\alpha$ do
11.   *Update*$(P_\alpha[\Delta], \Delta_i)$;
12.   end if
13.   if $(\Delta$.*type* $==$ *Res*$)||(\Delta.$ *type* $==$ *Role*$)$ then
14.   *Decrease*$(\Delta$.*threshold*);
15.   end if
16.   *acquaintance* $\leftarrow$ *Choose*$(P_\alpha [\Delta])$
17.   *Send*(*acquaintance,* Δ);
18.   *AddtoHistory*(Δ);
19.   end if
20.   end for
21. end while

In the above model, $P\alpha$ is the decision matrix agent α uses to decide where to move tokens. Initially, agents do not know where to send tokens, but as tokens are received, a model can be developed and better routing decisions made. That is, the model, $P\alpha$, is based on the accumulated information provided by the receipt of previous tokens. Algorithm 11.8 shows the reasoning of agent α when it receives incoming tokens from its acquaintances via function getToken(sender) (line 2). For each incoming token Δ, function Acceptable(α, Δ) determines whether the token will be kept by α (line 4). When a resource is kept, its threshold is raised (line 6). If α decides to pass Δ, it will add itself to the path of Δ (Line 9) and Update($P\alpha[\Delta]$, Δ_i) will update how to send Δ according to each previously received token Δi in α's history (line 11). If Δ is a resource or role token, its threshold will be decreased (line 14). Then α will choose the best acquaintance to pass the token to according to $P\alpha[\Delta]$ (line 16) and record Δ in its history, Hα (line 18).

11.5 Crowd Perception

Mobile Crowd-Sensing (MCS) is a new sensing paradigm that takes advantage of the extensive use of mobile phones that collect data efficiently and enable several significant applications. In a traditional sensing framework, the sensing level relies on a network of dedicated and fixed sensing nodes. The framework introduces many drawbacks, such as higher cost, inefficient sensing coverage, maintenance issues, and lack of scalability. The new paradigm shift toward MCS addresses the aforementioned drawbacks by replacing the dedicated sensing nodes with the MCS level. The remaining MCS framework levels resemble the corresponding levels in a typical wireless sensor network. Figure 11.11 shows an overview of the MCS framework, which is divided into several levels: crowd-sensing, data transmission, data collection, and applications on the server side.

Mobile crowd-sensing can be considered as crowdsourcing where the resource provided by the crowd is their sensing capabilities. Crowdsourcing is a group of outsourcing techniques that utilize independent, volunteer, and paid human resources to complete a specific task (Abualsaud *et al.*, 2018). It is also a process in which a task, a project, or a problem is performed by a group of private and geographically isolated participants. The participating members are compensated or provided with recognition once the problem is solved or the task is completed. Smartphone-based crowd-sensing takes advantage of the tremendous growth in network-monitoring applications.

The mobile crowd-sensing framework mandates an infrastructure composed of four main components, which are data collection, communication media, data

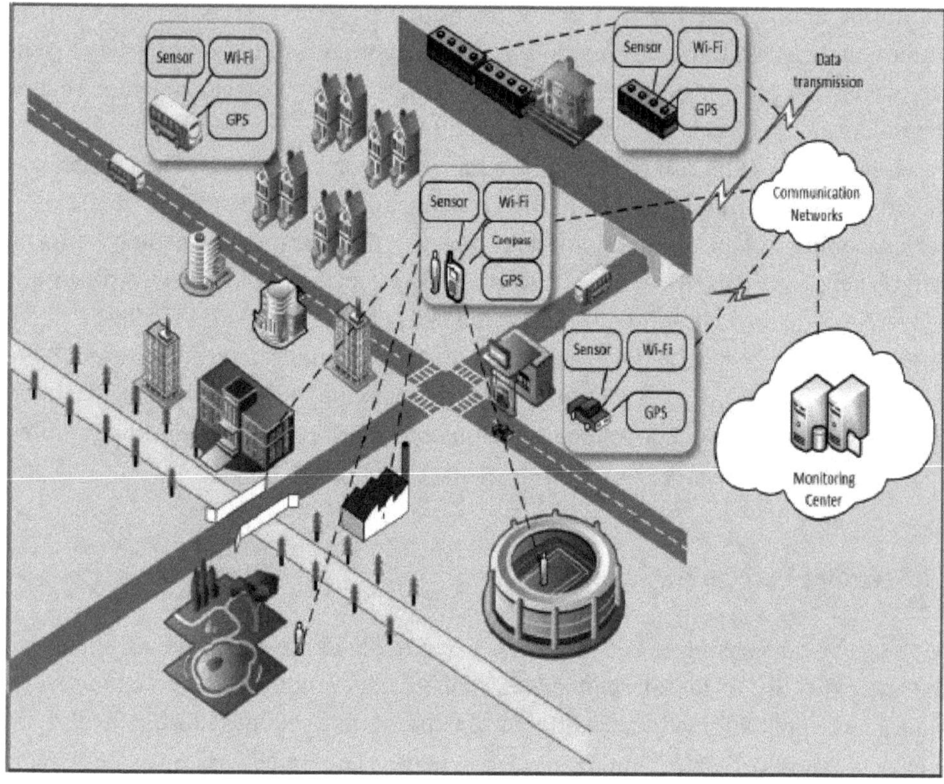

Fig. 11.11. The MCS framework.

source aggregation/fusion, and data storage and classification. The explanation of each component can be introduced as follows:

(1) data collection, with the help of an expert group to maintain a larger participant base and deliver the required verification of the usability of the collected sensing data,

(2) communication medium, which could be 3G/4G/GSM, WiFi, or Bluetooth depending on the required range of transmission,

(3) data aggregation/fusion, which is the process that collects the data from different sensor nodes and, based on a decision criterion, pre-processes the data and transfers it to another node/base station,

(4) data storage and classification, which is mandated by the combination of human and machine intelligence resulting from the participation of human in sensing data. Based on the wealth of data, both human and machine intelligence can be used for data processing.

The mobile crowd-sensing framework mandates an infrastructure composed of four main components, which are data collection, communication media, data source aggregation/fusion, and data storage and classification. The explanation of each component can be introduced as the following:

(1) Data collection, with the help of an expert group to maintain larger participant base and deliver the required verification of the usability of the collected sensing data.
(2) Communication medium, which could be 3G/4G/GSM, WiFi or Bluetooth depending on the required range of transmission.
(3) Data aggregation/fusion, which is the process that collects the data from different sensor nodes, and based on a decision criterion pre-processes the data and transfers it to another node/base station.
(4) Data storage and classification, which is mandated by combination of human and machine intelligence resulting from the participation of human in sensing data. Based on the wealth of data, both human and machine intelligence can be used for data processing.

MCS also explores the data fusion from different sensor nodes acting as spatially distributed data sources. Therefore, using MCS, both online and offline-denoted data can be leveraged by participants exploring through-space data fusion to develop modern applications. Several distinctive research challenges grow from the mobile crowd-sensing paradigm, such as proper incentive mechanisms, data collection, and through-space data fusion. Moreover, MCS represents a mixture of human and machine intelligence that is not explored so far. A data fusion node collects the data from numerous nodes, and based on a decision criterion, it fuses the data with its own and transfers it to another node. The advantages are that it reduces the traffic load and conserves the battery of the smartphone. The data comes from different sensors, databases, or more accurate datasets. The data fusion algorithm is very important for any mobile monitoring system. Finally, data fusion takes place closer to the sensors, especially for raw sensor data, to reduce the network load resulting from many sensors collecting data from different locations.

Data fusion significantly improves the accuracy of the statistical model by combining specific information of several heterogeneous sensing systems, where the datasets are represented in different feature spaces. This makes it hard to investigate relationships between the heterogeneous data, even in case the datasets are related to each other. A statistical approach to data fusion combines sensor datasets in a robust way as they benefit from variances between devices and their complementary features. Castrignanò *et al.* (Castrignanò *et al.*, 2017) combined multiple sources of information in a statistical framework. The main advantage of the statistical framework is that straightforward probability models are used to describe the different relationships between sensors, taking into account the uncertainty behind it and the change of support. However, there are several statistical methods that analyze heterogeneous data based on Bayesian and machine-learning methods. Other research work introduces a novel model using heterogeneous data fusion through fully convolutional neural networks to achieve semantic labeling. The authors presented the residual correction as a way of learning how to fuse predictions from a dual-stream structure.

11.6 The Emergence Mechanism of Crowd Intelligence

John Henry Holland, the main founder of the "emergence" theory, described the phenomenon of emergence in his book *Emergence: From Chaos to Order* (Holland, 1999). In complex adaptive systems, the phenomenon of 'emergence' can be found everywhere: ant communities, neural networks, immune systems, the Internet, and even the world economy. Whenever the overall behavior of a process is much more complex than the parts that make up it, it can be called emergence. Generally speaking, emergence refers to the unpredictable and complex phenomena created by the pre-set simple interactive behavior between individuals in a system.

As humans enter the Internet age, big data and artificial intelligence are constantly improving the intelligence of people, machines, and objects. The Internet, Internet of Things, Industry 4.0, etc. are constantly enhancing the connection capabilities between people, enterprises, government, and other institutions, intelligent robots, and intelligent objects. Cloud computing continues to strengthen the interaction capabilities between agents. The development of the above technologies results in the crowd intelligence network system having not only the characteristics of large scale and close connection but also the many intelligent agents in the system that are in a deep integration of physical space, information space, and consciousness space. In the three-dimensional superposition space, the interaction and influence of matter, information, and consciousness that conform to different motion laws make the behavioral results of the intelligent agents in the system show a broader unity of opposites. The schematic diagram of the crowd intelligence network system is shown in Figure 11.12, in which intelligent agents connect, influence, and interact with each other in various ways. The level of these agents determines the intelligence level of the entire crowd intelligence network system.

Yang *et al.* (2021) proposed a Visual health IoT framework V-HIoT, which is illustrated in Figure 11.13. This architecture realizes information generation,

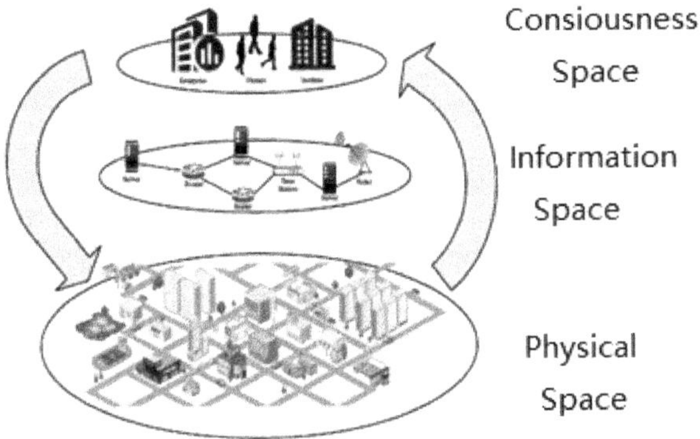

Fig. 11.12.   Schematic figure of crowd intelligence network system.

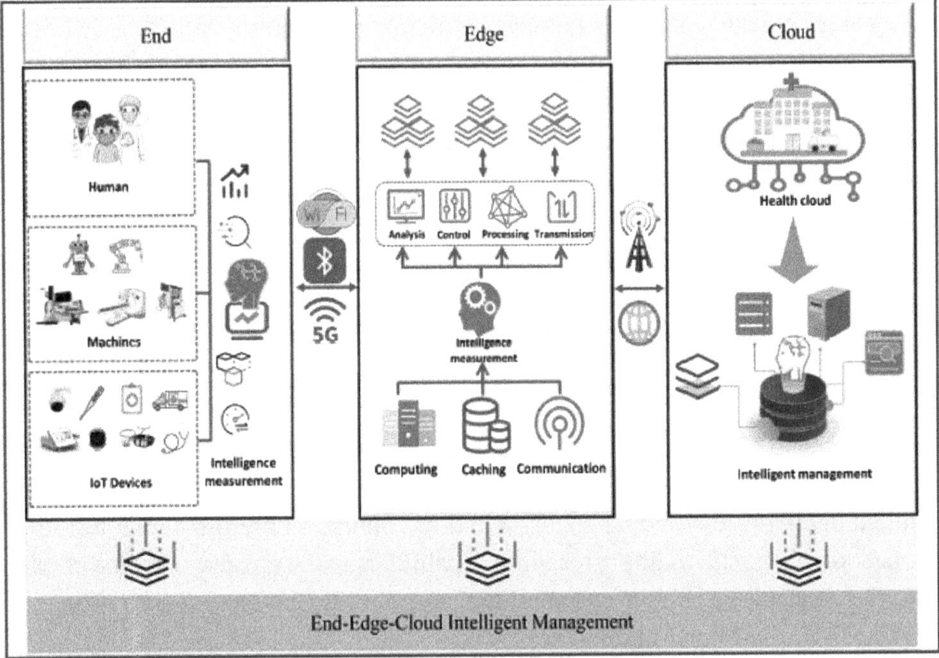

Fig. 11.13.   Framework of V-HIoT.

encoding transmission, and intelligent control of visual data through collaborative management of terminal, edge, and cloud systems. The entire intelligent V-HIoT architecture includes three modules: end processing module, edge control module, and cloud management module. The functions of each module are described as follows:

1. End processing module

In healthcare scenarios, there are many different HIoT devices at the terminal location, and the data they generate is heterogeneous. In addition, hospitals contain doctors, nurses, patients, and traditional machines that also generate large amounts of different types of information and data. Therefore, from people to machines to sensor devices, the processing of these heterogeneous data becomes more difficult due to their own characteristics and different environments. How to describe humans, machines, and things in a unified way has become increasingly important. Many factors need to be considered when uniformly describing heterogeneous devices, which leads to uncertainty in device interaction and data fusion. Therefore, we mainly focus on solving the problem of how to measure heterogeneous devices. This is also the primary problem to solve the effective use of various terminal devices. To address this problem, we proposed the concept of terminal intelligence. Terminal intelligence should at least include the characteristics of humans, machines, and things. The two core issues in ultimately intelligent measurement are how to include the characteristics of different types of agents and what factors to choose for measurement. Current solutions mainly include task-oriented evaluation and capability-oriented evaluation. The definition of ultimate intelligence is a very complex task.

End intelligence should contain many different types of agents. People, machines, and sensor devices are the most representative agents in smart medicine. Therefore, the definition of terminal intelligence should at least include the characteristics of people, machines, and sensor devices. Human intelligence is primarily the ability to solve problems. However, machines and sensor devices are mainly used for specific tasks. Previous research only focused on task-specific skills or only on general abilities, ignoring the mixed characteristics of heterogeneous subjects. Combining the characteristics of tasks and capabilities, a formal definition of terminal intelligence based on quality and time is proposed: terminal intelligence is a collection of agents' multiple response capabilities to the environment and emphasizes the concepts of response capabilities, response quality, and response time, as a description of terminal intelligence.

The key part of the intelligent system, the terminal intelligence diagram, is shown in Figure 11.13. Environment refers to all tasks in which humans, machines, and sensor devices are involved in smart healthcare. Responsiveness mainly includes response quality and response time. Response quality is the amount of data generated by the agent in the task, and response time is the time it takes the agent to complete the task. Using the task characteristics expressed by response quality and response time, we can describe the intelligent mixing properties of heterogeneous agents. According to this definition, the agent's response quality and response time in multiple new tasks constitute the final intelligence as a whole. According to the definition of end intelligence, its mathematical expression is as follows:

$$I_{end} = \sum_i^n \frac{Q_i}{T_i} \tag{11.1}$$

where Q_i represents the comprehensive evaluation of the agent's performance in the ith task and T_i represents the time it takes the agent to complete the ith task. The mathematical expression of agent reaction quality is as follows:

$$Q_i = \sum_i^N \mu_i \cdot D(i) \tag{11.2}$$

$$\sum_i^N \mu_i = 1 \tag{11.3}$$

where N represents the number of tasks completed by the agent, $i = 1, 2, 3 \ldots N$. $D(i)$ indicates the amount of data generated by the agent in the ith task. μ_i indicates the weight of each task among all tasks.

These heterogeneous HIoT devices are managed in a unified manner through end intelligence, and intelligent visual data is efficiently transmitted to the edge and the cloud. It provides a theoretical basis for collaborative management of end-edge-cloud systems.

2. Edge control module

The essence of edge computing is to transfer some functions of the cloud center to the edge of the network close to the end device to achieve efficient processing of data and related applications. In the context of HIoT generating large amounts of visual data, edge computing can not only reduce the pressure on cloud center data flows but also improve data processing efficiency. It has the advantages of low latency, low bandwidth, and high real-time performance. However, many resources in HIoT are dynamic, which brings challenges to the deployment of edge nodes. To address this challenge, we plan to implement an intelligent matching function between mission goals and available resources in the edge control module. First, edge intelligence is calculated on the edge side, and complex intelligent collaborative analysis is performed on the edge side and the terminal side. Then rationally design the calculation migration and corresponding collaborative optimization tasks to reduce the computational complexity during the task migration process. Finally, the latency of visual data processing tasks and the resource consumption of multi-collaboration processing are optimized. In order to realize this function, we first define edge intelligence based on end intelligence. Considering the heterogeneity of the internal structure

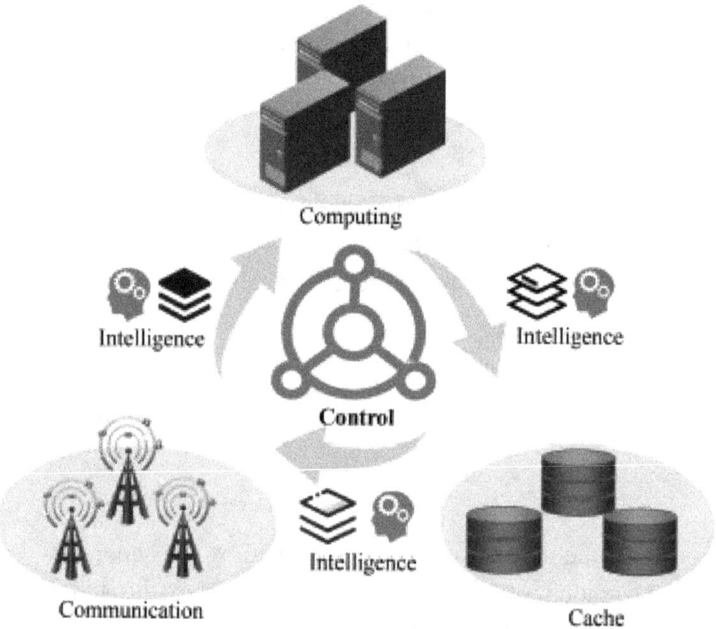

Fig. 11.14. Edge intelligence.

of edge servers, their common task is data processing. Therefore, we define edge intelligence based on data. Data processing typically includes computation, caching, and communication. Therefore, we regard edge intelligence as the comprehensive capabilities of edge servers for data computing, caching, and communication. The schematic diagram of edge intelligence is shown in Figure 11.14. We extend a three-efficiency model to computing for edge intelligence. Utility is expressed as data processing capacity per unit time. Therefore, the statistical functions can regulate the efficiency of computation, communication, and caching, where is the length of the file. The size of edge intelligence is then expressed using the multiplication of compute, communication, and cache utility. Therefore, the intelligence of the edge server can be calculated as

$$I_{edge} = U_{edge}^{comp}(r) \cdot U_{edge}^{coch}(r) \cdot U_{edge}^{comm}(r) \tag{11.4}$$

Through collaborative computing of edge intelligence and end intelligence, visual data processing tasks can be automatically assigned based on the intelligence level of the edge server. The edge control module can not only dynamically adjust resource allocation but also effectively improve the processing efficiency of visual data. With edge intelligence, intelligent control of visual HIoT is achieved through accurate prediction of available resources and rapid allocation of tasks.

3. Cloud management module

Cloud computing is an Internet-centric computing model that provides fast and secure data computing and storage services on websites. It gives everyone using the Internet access to the vast computing resources and data centers on the network. Cloud computing is mainly used for large-scale centralized data processing tasks. However, in actual HIoT systems, many devices are distributed in different geographical locations, which makes the performance of the devices vary greatly. It brings severe challenges to the transmission and management of visual data. Therefore, cloud centers are needed for smarter management of edge nodes and these geographically dispersed HIoT devices. In response to this challenge, we plan to implement intelligent management functions of terminal intelligence and edge intelligence in the cloud management module. In the cloud, the first step is to perform in-depth analysis, query, calculation, and storage of visual data to support flexible management of distributed devices. Then, through intelligent collaborative analysis with terminal devices and edge nodes, the processing efficiency and resource utilization of visual data are improved. To implement this function, cloud intelligence first needs to be defined. Similar to edge intelligence, the main task of the cloud center is to calculate, cache, and communicate visual data. Therefore, we believe that cloud intelligence is the comprehensive capabilities of cloud servers in data computing, caching, and communication, as shown in Figure 11.15. In terms of cloud intelligent

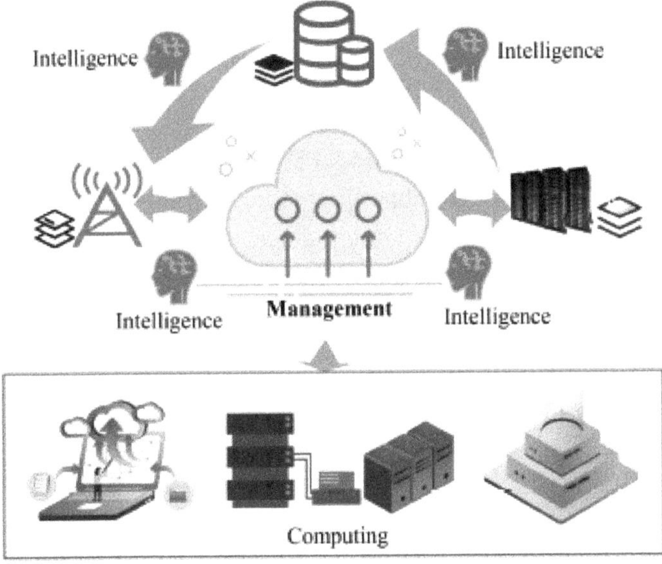

Fig. 11.15.   Cloud intelligence.

computing, the cloud center needs to manage edge and terminals at the same time. Therefore, in addition to the data connection between the cloud center and the end, data transmission between the cloud center and edge nodes is also required.

Through collaborative computing of terminal intelligence, edge intelligence, and cloud intelligence, edge nodes and end devices are reclassified and managed. Dynamically reorganize nodes with similar geographical location and performance. The cloud center builds a task-oriented HIoT group in real time by collecting all available edge nodes and terminal devices. The flexibility of group nodes is adjustable, and each group has a certain degree of independence and can complete specific tasks. Finally, the intelligent management of the collaborative system based on edge cloud is realized.

11.7 Multi-Agent Environment MAGE

Multi-Agent Environment (MAGE) is an agent-oriented software development, integration, and run environment developed by Intelligence Science Laboratory, Institute of Computing Technology, Chinese Academy of Sciences (Shi *et al.*, 2003). It provides users with a software deployment and system integration mode which includes requirement analysis, system design, and agent generation. It also provides multi-software reuse mode which is beneficial for agent software. It can also wrap other legacy systems to quickly integrate as a new software.

11.7.1 *The Architecture of MAGE*

MAGE system framework includes requirement analysis and modeling toolkit AUMP, visual agent studio, and multi-agent run-time environment MAGE. The structure of MAGE is shown in Figure 11.16. AUMP supports agent-oriented analysis and design. Vastudio can realize the model to programs, including different agent behaviors. Finally, agents can run in the agent-supporting environment.

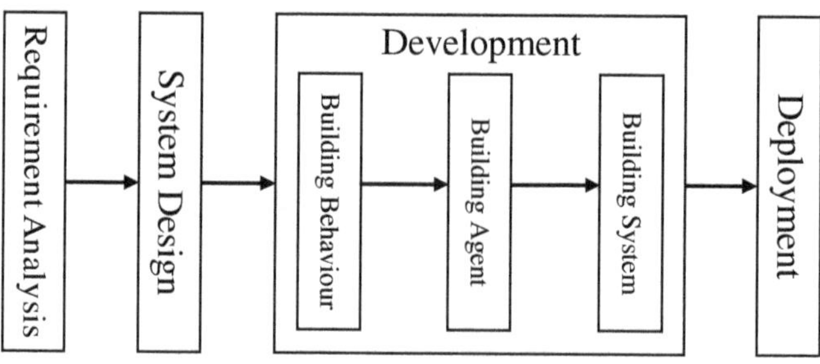

Fig. 11.16. MAGE architecture.

11.7.2 *Agent Unified Modeling Language*

Agent Unified Modeling Language (AUML) is an agent-oriented modeling language. Its main function is to help software designers and programmers model the software and describe the software development process from the beginning to the end. AUML unifies the agent-oriented and object-oriented methods.

AUMP is multi-window application software which runs on the Windows platform. It provides graphic user interface for users to edit and revise the models.

11.7.3 *Visual Agent Development Tool*

On the agent-oriented distributed platform, we combine the common object request broker agent and Internet to build the common agent request broker architecture (CARBA). As shown in Figure 11.17, CABAR consists of agent request broker (ARB), agent application facilities (AppFacilities), agent domain pattern (appPattern), and agent services.

CARBA is a distributed management mechanism which uses the ARB as its core component. It defines the way distributed agents send request and receive the response. AppFacilities provides the agent components from horizontal and vertical directions. AppPattern builds the templates according to the domain requirement, e.g., agent life circle, agent library, and name. CARBA can realize the following objectives:

(1) Decompose the system according to the function in heterogeneous and distributed environment.
(2) Integrate the components according to requirement.

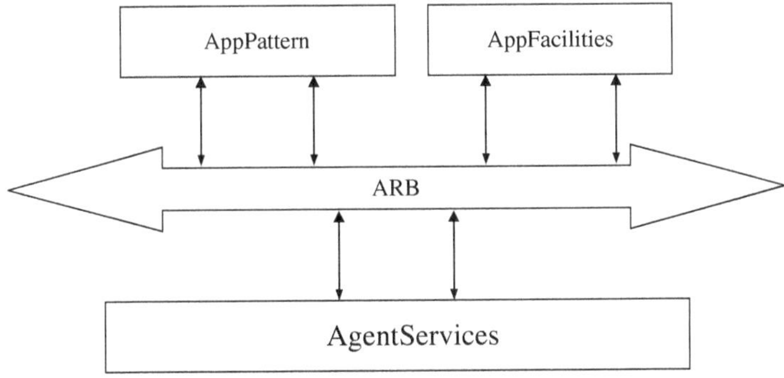

Fig. 11.17. CARBA architecture.

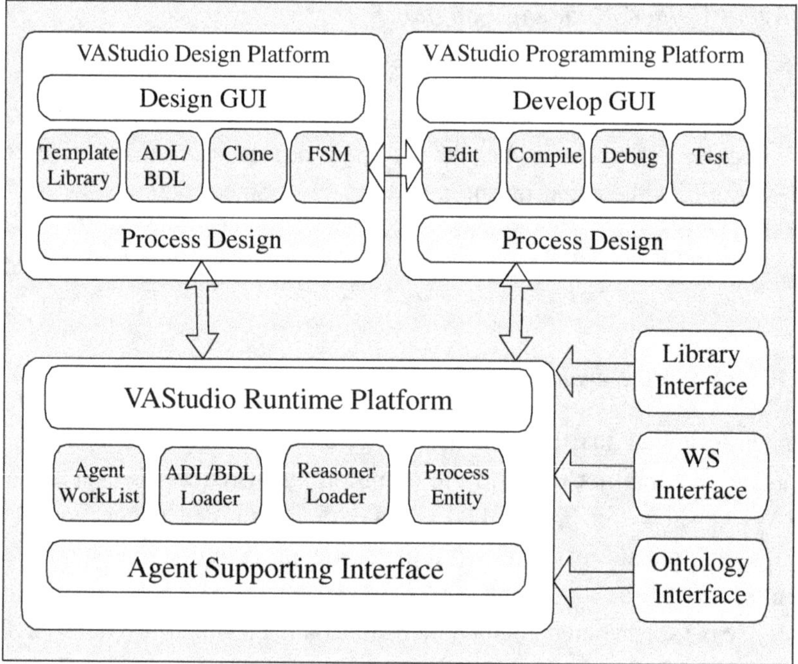

Fig. 11.18.   VAStudio structure.

Based on the above method, we developed a visual development toolkit named VAStudio (Visual Agent Studio). The objective of designing this toolkit is to provide a user-friendly GUI to support agent-oriented development. It is not just a programming editor but a design and coding environment. It can use GUI to step-by-step generate the agents. It also provides a series of basic tools, e.g., components management tool, behavior library, and agent library. (Figure 11.18).

11.7.4 *MAGE Running Platform*

MAGE platform follows the rule of FIPA. It is a realization of FIPA standard. It provides agent generation, localization, registration, communication services, and mobility and exit services (Figure 11.19):

(1) **Agent Management System (AMS):** It is an indispensable component in MAGE platform. AMS controls the agent access and use of the platform. One platform has only one AMS which records the agent directory, including agent name and address. AMS also provides white-page services for other agents. Every agent must register on AMS to get an effective Agent ID.

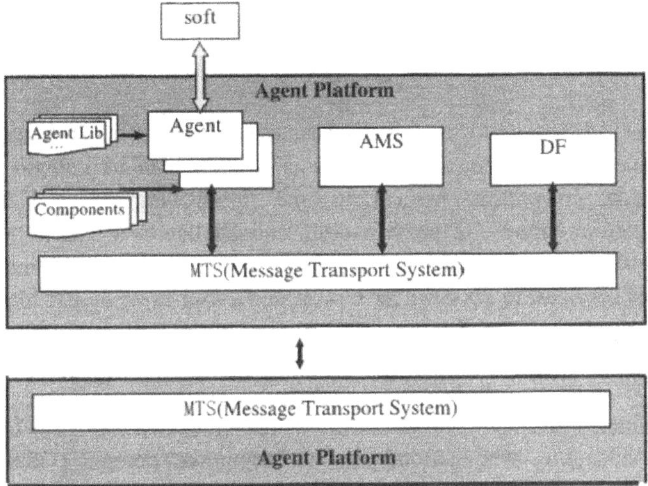

Fig. 11.19.   MAGE running platform.

(2) **Directory Facilitator (DF):** It is also indispensable in the platform. DF provides yellow-page service for other agents. Agents can register their services on DF and can search services provided by other agents. One platform can have several DFs.

(3) **Message Transport Service (MTS):** It provides communication service for agent from different platforms.

(4) **Agent:** It is a basic component of the platform. It is a unified and whole component to provide one or more services. It can access outer resources, user interface, and communication infrastructure. Every agent has a unique AID.

(5) **Agent platform (AP):** It provides physical infrastructure for agents which includes machine, operation system, agent support software, agent management components (DF, AMS, and MTS), and agents.

(6) **Software:** It includes all non-agent software. Agent can access them, e.g., by adding new service, acquiring new service protocol and new security protocol/ algorithm, new negotiation protocol, and accessing supporting toolkits.

(7) Agent library (Agent Lib).

11.8 Agent Grid Intelligence Platform

Web services provide a means of interoperability that is essential to achieve large-scale computation with a focus on middleware to support large-scale information processing. In a grid, the middleware is used to hide the heterogeneous nature and

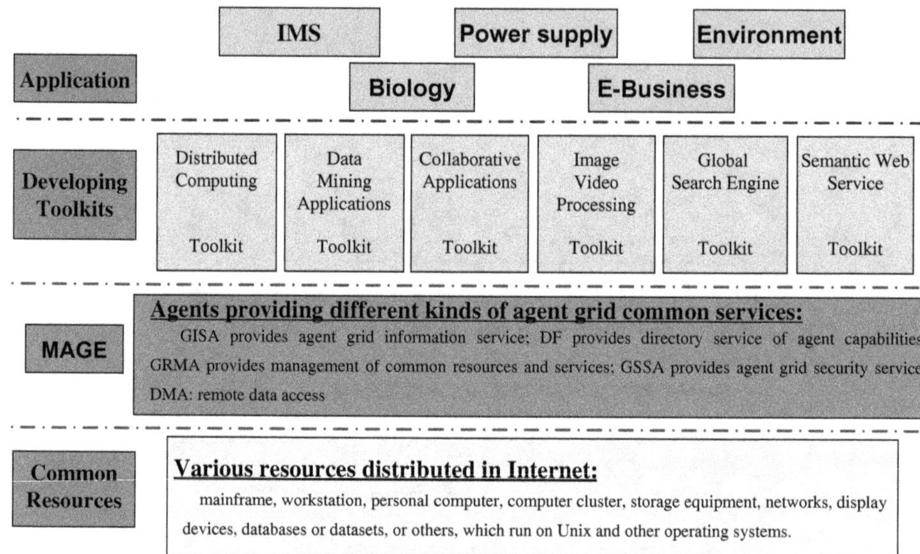

Fig. 11.20. Model of AGEGC.

provide users and applications with a homogeneous and seamless environment by providing a set of standardized interfaces to a variety of services. In 2002, we have proposed a four-layer model for agent-based grid computing to combine agent computing with grid computing through middleware concept shown in Figure 11.20 (Shi *et al.*, 2002):

(1) **Common resources:** They consist of various resources distributed on the Internet, such as mainframe, workstation, personal computer, computer cluster, storage equipment, databases or datasets, or others, which run on Unix and other operating systems.

(2) **MAGE environment:** It is the kernel of grid computing which is responsible for resources' location and allocation, authentication, unified information access, communication, task assignment, agent developing tools, and others.

(3) **Developing toolkit:** It provides developing toolkits, containing distributed computing, data mining applications, collaborative applications, image and video processing, global search engine, semantic Web services, and negotiation support, to let users effectively use middleware resources.

(4) **Application service:** It organizes certain agents automatically for specific purpose applications, such as power supply, oil supply e-business, distance education, and e-government.

In terms of the model of AGEGC, we have developed Agent Grid Intelligence Platform (AGrIP) which is a powerful developing environment for complex and large distributed intelligent systems. AGrIP has been applied to develop many group intelligent decision support systems, such as GEIS (Agent Grid-Based City Emergency Interactive Response System), RSMAS (Residential Power Demand Simulator Based on MAS), and so on.

Exercises

11.1 Describe briefly the solution process of distributed problem-solving.

11.2 Explain briefly the concept of agent and its basic structures.

11.3 What is deliberative agent? What is the reactive agent? Compare the difference between them.

11.4 What is the BDI model? Please explain the algorithm.

11.5 Describe the communication between the agents. Compare the KQML and FIPA's ACL.

11.6 What is the principle of the contract net protocol in distributed problem-solving?

11.7 How to realize the collaborative work of multi-agent systems?

11.8 How to construct a mobile crowd-sensing platform?

11.9 What is end-edge-cloud platform? The end-edge-cloud intelligent management should have which functions?

11.10 What components constitute the multi-agent environment MAGE? Please describe the main functions of each component.

11.11 Explain which aspect of AUML to extend the functions of UML.

11.12 Draw the architecture of visual agent development environment VAStudio and explain why it can realize software reuse.

Chapter 12

Artificial Intelligence for Science

Artificial intelligence has started to advance natural sciences by improving, accelerating, and enabling our understanding of natural phenomena at a wide range of spatial and temporal scales, giving rise to a new area of research known as artificial intelligence for science.

12.1 Introduction

After more than 60 years of evolution, especially driven by new theories and technologies such as mobile Internet, big data, supercomputing, sensor networks, and brain science, as well as the strong demand for economic and social development, artificial intelligence has accelerated its development, showing deep learning, cross-border integration, human–machine collaboration, open group intelligence, autonomous control, and other new features. Big data-driven knowledge learning, cross-media collaborative processing, human–computer collaboration to enhance intelligence, group-integrated intelligence, and autonomous intelligent systems have become the focus of the development of artificial intelligence. Brain-like intelligence inspired by the results of brain science research is ready to be developed into chip-based hardware. The trend of platformization has become more obvious, and the development of artificial intelligence has entered a new stage. At present, the overall advancement of the development of new-generation artificial intelligence-related disciplines, theoretical modeling, technological innovation, and software and hardware upgrades is triggering a chain of breakthroughs and promoting an accelerated leap from digitalization and networking to intelligence in all economic and social fields.

Artificial intelligence has become a new focus of international competition. Artificial intelligence is a strategic technology that will lead the future. The world's major developed countries regard the development of artificial intelligence as a major strategy to enhance national competitiveness and safeguard national security. They have stepped up the introduction of plans and policies and strengthened deployment around core technologies, top talents, standards, and regulations, striving to take the lead in the new round of international scientific and technological competition. At present, my country's national security and international competition situation are more complex. We must look at the world, put the development of artificial intelligence at the national strategic level, plan systematically and proactively, firmly grasp the strategic initiative of international competition in the new stage of artificial intelligence development, and create new competitive advantages, openning up new space for development and effectively ensure national security.

Artificial intelligence has become a new engine for economic development. As the core driving force of a new round of industrial transformation, artificial intelligence will further release the huge energy accumulated in previous scientific and technological revolutions and industrial changes, and create new powerful engines to reconstruct all aspects of economic activities, forming a New demands for intelligence in various fields from macro to micro have spawned new technologies, new products, new industries, new formats, and new models, triggered major changes in the economic structure, profoundly changed human production, lifestyles, and thinking patterns, and achieved an overall jump in social productivity. My country's economic development has entered a new normal, and the task of deepening supply-side structural reform is very arduous. It is necessary to accelerate the in-depth application of artificial intelligence, cultivate and expand the artificial intelligence industry, and inject new momentum into my country's economic development.

Artificial intelligence is widely used in education, medical care, elderly care, environmental protection, urban operations, judicial services, and other fields, which will greatly improve the accuracy of public services and comprehensively improve the quality of people's lives. Artificial intelligence technology can accurately perceive, predict, and provide early warning of major trends in infrastructure and social security operations, timely grasp group cognitive and psychological changes, and proactively make decisions and responses. It will significantly improve the ability and level of social governance, and is indispensable for effectively maintaining social stability.

In the last 10 years, AI has developed rapidly and has become comparable to humans in many tasks. And in recent years, the large models represented by ChatGPT are amazing. There are four basic paradigms in previous scientific research:

(1) Empirical paradigm

An observation based on experience is a summary of all things by a genius scientist. For example, the famous astronomer Kepler summed up the laws of the movement of celestial bodies through observation: "The orbit of all planets around the sun is elliptical, and the sun is in the common focus of all ellipses."

(2) Theoretical paradigm

Exponentists perform mathematical abstractions and deductions from experience, such as Newton's equations of motion used to describe classical mechanics, Maxwell's equations used to describe the relationship between electric and magnetic fields, etc.

(3) Computational paradigm

With the invention of computers, people began to have the ability to solve complex physical equations. For example, fluid equations can be solved by finite element or finite difference, which can help humans to accurately predict weather forecasts.

(4) Data-driven paradigm

Machine learning plays a very important role, using machine learning methods to analyze data, find patterns, and make predictions.

In recent years, everyone has begun to pay attention to a new paradigm called AI for Science, which is an organic combination of the first four paradigms, giving full play to the respective strengths of experience and theory, and integrating AI and computational science. AI for Science is a more comprehensive understanding of scientific discovery, so we call it the fifth paradigm of scientific discovery.

There are three levels of content in AI for Science which are as follows:

(1) At present, what most people in this field are doing is to use AI technology, or more precisely, deep learning technology for most people, for different scenarios, for scientific research, technological innovation, and achievement transformation in various disciplines. At present, this has a relatively large impact and a relatively fast effect. In this regard, the goals are generally clear and specific: such as solving quantum mechanical equations, solving complex fluid mechanical equations, suggesting pathways to synthesize chemical molecules, protein folding, designing small molecule drugs for targets, various image recognition, and so on. The progress and results produced in this area have indeed greatly accelerated the progress of science and technology in some respects.

(2) The second level is to use AI to discover new science. As an extreme example, training an AI with a large amount of planetary motion data should make it easy to predict the orbit of a planet and its position in orbit at any point in the future. Further, it is quite possible that it will also discover Kepler's three laws.

But at least for now, it's hard to imagine it going one step further and discovering Newton's laws. The discovery of Newton's laws established a completely new theoretical framework and essentially changed the scientific paradigm. In 2009, the journal *Science* published an article titled: "Distilling Free-Form Natural Laws from Experimental Data." Translated into Chinese is "extracting the laws of nature in free form from experimental data." The author used a computer to analyze simple harmonic vibration and the trajectory of the double pendulum (the motion of the double pendulum is chaotic), and the machine automatically found some meaningful analytical expressions for conserved quantities, such as Hamiltonian and momentum conservation. At first glance, it seems that machines can discover new science. But essentially, it's the man who asks the machine to find the conserved quantity of motion. Instead of machines, the concept of conserved quantities was proposed. Sure enough, two months later, Phil Anderson (a well-known physicist) published a review of the work in the journal *Science*, titled: "Machines Fall Short of Revolutionar."

(3) The third meaning, simply put, is Science of AI. AI is a powerful enabling technology, but there must be a science behind it, that is, there will be a corresponding Science. Computer scientists, mathematicians, and physicists have done a lot of work in this area. Half of last year's Nobel Prize in Physics was awarded to Giorgio Parisi for his work on spin glass, which mentions the impact of the statistical physics theory he developed through the spin glass model in the field of neuroscience and machine learning. I think there should be a broader and deeper scientific connotation in this area. It is not only the application of some existing concepts and theories but also the emergence of new concepts and theories. AI was originally intended to simulate human intelligence, and the network architecture of deep learning was also inspired by the neural network of the human brain. Intuitively, advances in neurobiology should further advance AI, and in turn, advances in AI should enlighten our understanding of the brain and human intelligence. Of course, it is also possible that in the end, AI and NI (natural intelligence) do not have much to do with it. But in any case, Science of AI should be there, and it is also the frontier of science.

On March 27, 2023, the Ministry of Science and Technology of China, together with the Natural Science Foundation of China, launched a special deployment of "Artificial Intelligence for Science," closely integrating key issues in basic disciplines such as mathematics, physics, chemistry, astronomy, scientific research needs are unfolding in key areas and focusing on drugs, genetic research, biological breeding, and new material research and development, and a cutting-edge technology research and development system for "artificial intelligence-driven scientific research" has been laid out.

12.2 Knowledge Discovery

Information is processed and transformed to form knowledge. Knowledge refers to the sum of all kinds of knowledge that people learn, accumulate, discover, and invent. It is a form of result in the process of human understanding. It is a reliable reflection of objective reality that is produced on the basis of human practice and has been tested by practice. Knowledge is the result of human brain innovation and the crystallization of human wisdom.

Knowledge discovery is the extraction and refinement of new patterns from data sets (Shi, 2001). The scope of knowledge discovery is very wide and can be economic, industrial, agricultural, military, social, commercial, scientific data, or data obtained from satellite observations. The form of data includes numbers, symbols, graphics, images, and sounds. Data organization methods also vary and can be structured, semi-structured, or unstructured. The results of knowledge discovery can be expressed in various forms, including rules, laws, scientific laws, equations, or concept networks.

The knowledge discovery process can be roughly understood as three steps: data preparation, data mining, and interpretation and evaluation of results (see Figure 12.1).

1. Data preparation
Data preparation can be divided into three sub-steps: data selection, data preprocessing, and data transformation. The purpose of data selection is to determine the

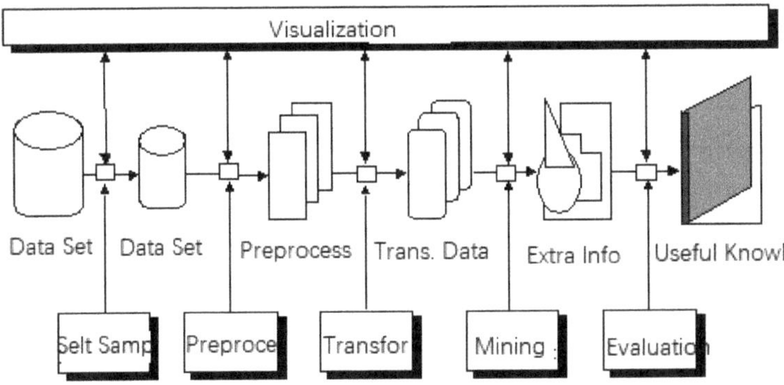

Fig. 12.1. Knowledge discovery procedure.

operation object of the discovery task, that is, target data, which is a set of data extracted from the original database according to the user's needs. Data preprocessing may generally include eliminating noise, deriving and calculating missing value data, eliminating duplicate records, and completing data type conversion (such as converting continuous value data into discrete data to facilitate symbolic induction or converting discrete data into continuous data) value type to facilitate neural network induction. When the object of data mining is a data warehouse, generally speaking, data preprocessing has been completed when generating the data warehouse. The main purpose of data transformation is to reduce data dimensionality or dimensionality reduction, that is, to find truly useful features from the initial features to reduce the number of features or variables to be considered during data mining.

2. Data mining stage

The data mining stage must first determine what the mining task or purpose is, such as data summary, classification, clustering, association rule discovery, or sequence pattern discovery. After determining the mining task, it is necessary to decide what mining algorithm to use. The same task can be implemented with different algorithms. There are two considerations in choosing the implementation algorithm: First, different data have different characteristics, so they need to be mined with algorithms related to them; second, the requirements of users or actual operating systems, some users may want to obtain descriptive, easy-to-understand knowledge (mining methods using rule representation are obviously better than methods such as neural networks), while some users or systems aim to obtain predictions as accurate as possible, possibly high predictive knowledge. After completing the above preparations, you can implement the data mining operation. Specific data mining methods are discussed in more detail in subsequent chapters. It should be pointed out that although data mining algorithms are the core of knowledge discovery and are currently the main direction of researchers' efforts, to obtain good mining results, one must fully understand the requirements or premise assumptions of various mining algorithms.

3. Interpretation and evaluation of results

The patterns discovered in the data mining stage may, after evaluation by users or machines, contain redundant or irrelevant patterns, in which case they need to be eliminated; it is also possible that the patterns do not meet user requirements, in which case the entire discovery process needs to return to the discovery stage. Previously, such as re-selecting data, using new data transformation methods, setting new data mining parameter values, or even changing to a mining algorithm (for example, when the discovery task is classification, there are multiple classification methods, and different methods are suitable for different data have different effects).

In addition, since KDD is ultimately oriented to human users, it may be necessary to visualize the discovered patterns or convert the results into another representation that is easy for users to understand, such as converting classification decision trees into "if…then…" rules.

Knowledge discovery tasks contain the following seven classifications:

(1) Data summary

The purpose of data summary is to condense the data and give it a compact description. The traditional and simplest method of data summary is to calculate the sum, average, variance, and other statistical values of each field in the database or to use histograms, pie charts, and other graphical representations. Data mining is mainly concerned with discussing data summary from the perspective of data generalization. Data generalization is a process of abstracting relevant data in a database from a low level to a high level. Since the information contained in the data or objects on the database is always the most original and basic information (this is to not miss any potentially useful data information). People sometimes hope to process or browse data from a higher-level view, so the data need to be generalized at different levels to adapt to various query requirements. There are currently two main techniques for data generalization: multi-dimensional data analysis methods and attribute-oriented induction methods. Multi-dimensional data analysis method is a data warehouse technology, also known as online analytical processing (OLAP). Data warehouse is a collection of historical data that is decision-support-oriented, integrated, stable, and different in time. The prerequisite for decision-making is data analysis. In data analysis, aggregation operations such as sum, total, average, maximum, and minimum are often used, and the calculation amount of such operations is particularly large. Therefore, a natural idea is to pre-compute and store the results of the aggregation operation to facilitate the use of the decision support system. The place where the results of aggregation operations are stored is called a multi-dimensional database. Multi-dimensional data analysis technology has been successfully applied in decision support systems, such as the famous SAS data analysis software package, Business Object's decision support system Business Object, and IBM's decision analysis tools, all use multi-dimensional data analysis technology. Multi-dimensional data analysis method is used for data summary, which is aimed at the data warehouse, which stores offline historical data. To process online data, researchers proposed an attribute-oriented induction method. Its idea is to directly generalize the data views that users are interested in (which can be obtained using general SQL query language) rather than storing generalized data in advance like multi-dimensional data analysis methods. The proposer of the method calls this data generalization technique the attribute-oriented induction method. After the original relationship is generalized, a generalized relationship is obtained, which summarizes the original relationship at a low level from a higher level. With the generalization relationship, various in-depth operations can be performed on it to generate knowledge that meets user needs, such as generating characteristic rules, discrimination rules, classification rules, and association rules based on the generalization relationship.

(2) Concept description

Users often require abstract and meaningful descriptions. An inductive abstract description can summarize a large amount of information about a class. There are two typical descriptions: feature description and discriminant description. Extract characteristic expressions about a set of data related to a learning task. These characteristic expressions express the overall characteristics of the data set. A discriminant description describes how two or more classes differ.

(3) Classification

Classification is a very important task in data mining and is currently the most commonly used in business. The purpose of classification is to learn a classification function or classification model (also often called a classifier) that can map data items in the database to a given category. Both classification and regression can be used for prediction. The purpose of prediction is to automatically derive a generalized description of given data from historical data records so that future data can be predicted. The difference from the regression method is that the output of classification is a discrete category value, while the output of regression is a continuous value. We do not discuss regression methods here. To construct a classifier, a training sample dataset is required as input. The training set consists of a set of database records or tuples. Each tuple is a feature vector composed of related field (also called attribute or feature) values. In addition, the training sample also has a category label. The form of a specific sample can be $(v1, v2, \ldots, vn; c)$, where vi represents the field value and c represents the category. Classifier construction methods include statistical methods, machine learning methods, and neural network methods. Statistical methods include Bayesian methods and non-parametric methods (nearest neighbor learning or example-based learning), and the corresponding knowledge representations are discriminant functions and prototype cases. Machine learning methods include decision tree methods and rule induction methods. The former is represented by a decision tree or a discriminant tree, while the latter is generally a production rule. The neural network method is mainly the BP algorithm, and its model representation is a forward feedback neural network model (an architecture composed of nodes representing neurons and edges representing connection weights). The BP algorithm is essentially a nonlinear discriminant function. In addition, a new method has emerged recently: rough set, whose knowledge representation is production rules. Different classifiers have different characteristics. There are three classifier evaluation or comparison scales: 1. prediction accuracy, 2. computational complexity, and 3. simplicity of model description. Prediction accuracy is the most commonly used comparison scale, especially for predictive classification tasks. The currently recognized method is the 10-fold stratified cross-validation method. The computational complexity depends on the specific implementation details and hardware environment. In data mining, since the operating object is a huge database, the space and time complexity issues will be a very important link. For descriptive classification tasks, the simpler the model description, the more popular it is; for example, the classifier construction method using rule representation is more useful, while the results produced by the neural network method are difficult to understand. In addition, it should be noted that the effect of classification is generally related to the characteristics of the data. Some data are noisy, some have missing values, some are sparsely distributed, some fields or attributes are highly correlated, some attributes are discrete, and some are continuous values or mixed. It is generally believed that there is no method suitable for data with various characteristics.

(4) Clustering

According to the different characteristics of the data, it is divided into different data categories. Its purpose is to make the distance between individuals belonging to the same category as small as possible, and the distance between individuals in different categories as large as possible. Clustering methods include statistical methods, machine learning methods, neural network methods, and database-oriented methods. Among statistical methods, clustering is called cluster analysis, which is one of the three major methods of multivariate data analysis (the other two are regression analysis and discriminant analysis). It mainly studies clustering based on geometric distance, such as Euclidean distance and Minkowski distance. Traditional statistical clustering analysis methods include systematic clustering, decomposition, joining, dynamic clustering, ordered sample clustering, overlapping clustering, and fuzzy clustering. This clustering method is a clustering based on global comparison. It needs to examine all individuals to determine the classification of classes; therefore, it requires that all data must be given in advance, and new data objects cannot be added dynamically. The cluster analysis method does not have linear computational complexity and is difficult to apply to situations where the database is very large. In machine learning, clustering is called unsupervised or teacher-less induction because compared with classification learning, the examples or data objects of classification learning have category labels, while the examples to be clustered do not have labels and need to be processed by a clustering learning algorithm, which is automatically determined. In many artificial intelligence literature, clustering is also called concept clustering because the distance here is no longer the geometric distance in statistical methods but is determined based on the description of the concept. When clustering objects can be added dynamically, concept clustering is called concept formation. In neural networks, there is a type of unsupervised learning method: self-organizing neural network method; such as Kohonen self-organizing feature mapping network and competitive learning network. In the field of data mining, the reported neural network clustering method is mainly the self-organizing feature mapping method. IBM specifically mentioned the use of this method for database clustering and segmentation in its data mining white paper.

(5) Correlation analysis

Discover interdependencies between features or data. Data dependency relationships represent an important type of discoverable knowledge. A dependency exists between two elements. If the value of one element A can be inferred from the value of another element B ($A \rightarrow B$), then B is said to depend on A. The so-called elements here can be fields or relationships between fields. Data dependencies have a wide range of applications. The results of dependency analysis can sometimes be provided directly to the end user. However, often strong dependencies reflect inherent domain structure rather than anything new or interesting. Finding dependencies automatically can be a useful approach. This type of knowledge can be used by other pattern extraction algorithms. Commonly used techniques include regression analysis, association rules, and belief networks.

(6) Deviation analysis

Abnormal instances in classification, exception patterns, deviations of observation results from expected values, and changes in magnitude over time are examples of deviation analysis. The basic idea is to look for meaningful differences between observations and reference quantities. By discovering anomalies, you can draw extra attention to special situations. Anomalies include the following patterns that may arouse people's interest: abnormal examples that do not satisfy regular classes, singular points that appear on the edge of other patterns, classes that are different from parent classes or sibling classes, certain classes that have undergone significant changes at different times, elements or sets, instances where there is a significant difference between the observed value and the expected value inferred by the model, etc. An important feature of bias analysis is that it can effectively filter out a large number of uninteresting patterns.

(7) Modeling

Through data mining, a mathematical model describing an activity or state is constructed. Knowledge discovery in machine learning is actually modeling some natural phenomena and rediscovering scientific laws, such as BACON (Shi, 1992) and SDS (Washio *et al.*, 1999).

The 21st century will be a century in which the knowledge economy dominates. A country with continuous innovation capabilities and a large number of high-quality human resources will have great potential to develop the knowledge economy. The national innovation system plays an important role in the era of knowledge economy. Building a national innovation system is an important measure and the right choice to seize the opportunities of the knowledge economy, meet the challenges of the knowledge economy, and implement the strategy of rejuvenating the country through science and education. National innovation capabilities are related to the future and destiny of the Chinese nation and are an important factor in determining my country's comprehensive national strength and international competitiveness in the overall world pattern.

The main functions of the national innovation system are knowledge innovation, knowledge dissemination, knowledge exchange, and knowledge utilization, specifically including the execution of innovation activities, the allocation of innovation resources (human, material, financial, information, knowledge resources, etc.), system, organization, management innovation, and the construction of related infrastructure. Vigorously promoting and improving the level, scale, and efficiency of knowledge production, dissemination, exchange and utilization are the basic tasks of the national innovation system.

12.3 Protein Structure Prediction

Protein structure prediction is the inference of the three-dimensional structure of a protein from its amino acid sequence, that is, the prediction of its secondary and tertiary structure from primary structure. Protein structure is the three-dimensional arrangement of atoms in an amino acid chain molecule. Protein structures range in size from tens to several thousand amino acids. By physical size, proteins are classified as nanoparticles, between 1 and 100 nm. Very large protein complexes can be formed from protein subunits. For example, many thousands of actin molecules assemble in a microfilament.

There are four distinct levels of protein structure:

1. Primary structure

The primary structure of a protein refers to the sequence of amino acids in the polypeptide chain. The primary structure is held together by peptide bonds that are made during the process of protein biosynthesis. The two ends of the polypeptide chain are referred to as the carboxyl terminus (C-terminus) and the amino terminus

(N-terminus) based on the nature of the free group on each extremity. Counting of residues always starts at the N-terminal end (NH_2-group), which is the end where the amino group is not involved in a peptide bond. The primary structure of a protein is determined by the gene corresponding to the protein. A specific sequence of nucleotides in DNA is transcribed into mRNA, which is read by the ribosome in a process called translation. The sequence of amino acids in insulin was discovered by Frederick Sanger, establishing that proteins have defining amino acid sequences. The sequence of a protein is unique to that protein and defines the structure and function of the protein. The sequence of a protein can be determined by methods such as Edman degradation or tandem mass spectrometry. Often, however, it is read directly from the sequence of the gene using the genetic code. It is strictly recommended to use the words "amino acid residues" when discussing proteins because when a peptide bond is formed, a water molecule is lost, and therefore proteins are made up of amino acid residues. Post-translational modifications such as phosphorylations and glycosylations are usually also considered a part of the primary structure and cannot be read from the gene. For example, insulin is composed of 51 amino acids in 2 chains. One chain has 31 amino acids, and the other has 20 amino acids.

2. Secondary structure

Secondary structure refers to highly regular local sub-structures on the actual polypeptide backbone chain. Two main types of secondary structures, the α-helix and the β-strand or β-sheets, were suggested in 1951 by Linus Pauling. These secondary structures are defined by patterns of hydrogen bonds between the main chain peptide groups. They have a regular geometry, being constrained to specific values of the dihedral angles ψ and φ on the Ramachandran plot. Both the α-helix and the β-sheet represent a way of saturating all the hydrogen bond donors and acceptors in the peptide backbone. Some parts of the protein are ordered but do not form any regular structures. They should not be confused with random coil, an unfolded polypeptide chain lacking any fixed three-dimensional structure. Several sequential secondary structures may form a "supersecondary unit."

3. Tertiary structure

Tertiary structure refers to the three-dimensional structure created by a single protein molecule (a single polypeptide chain). It may include one or several domains. The α-helices and β-pleated sheets are folded into a compact globular structure. The folding is driven by the *non-specific* hydrophobic interactions, the burial of hydrophobic residues from water, but the structure is stable only when the parts of a protein domain are locked into place by *specific* tertiary interactions, such as salt bridges, hydrogen bonds, and the tight packing of side chains and disulfide bonds.

The disulfide bonds are extremely rare in cytosolic proteins since the cytosol (intra-cellular fluid) is generally a reducing environment.

4. Quaternary structure
Quaternary structure is a three-dimensional structure consisting of the aggregation of two or more individual polypeptide chains (subunits) that operate as a single functional unit (multimer). The resulting multimer is stabilized by the same non-covalent interactions and disulfide bonds as in tertiary structure. There are many possible quaternary structure organizations. Complexes of two or more polypeptides (i.e., multiple subunits) are called multimers. Specifically, it would be called a dimer if it contains two subunits, a trimer if it contains three subunits, a tetramer if it contains four subunits, a pentamer if it contains five subunits, and so forth. The subunits are frequently related to one another by symmetry operations, such as a 2-fold axis in a dimer. Multimers made up of identical subunits are referred to with the prefix "homo-" and those made up of different subunits are referred to with the prefix "hetero-," for example, a heterotetramer, such as the two alpha and two beta chains of hemoglobin.

Protein structure prediction is one of the most important goals pursued by computational biology, and it is important in medicine (for example, in drug design) and biotechnology. Starting in 1994, the performance of current methods is assessed biannually in the CASP experiment (Critical Assessment of Techniques for Protein Structure Prediction). A continuous evaluation of protein structure prediction web servers is performed by the community project CAMEO3D.

12.4 Drug Development

AI drug research and development is the application of artificial intelligence technologies such as machine learning, natural language processing, and big data to all aspects of drug research and development, thereby promoting cost reduction and efficiency improvement of new drug research and development. Currently, it is mainly used in drug discovery and preclinical stages of drug research and development. With the continuous application of ChatGPT, the penetration of AI into the clinical development stage is expected to continue to accelerate.

From the perspective of online AI drug research and development, the layout of AI-assisted drug research and development is distributed in various stages of drug discovery, preclinical research, clinical trials, and drug sales. Among them, the most deployed link is the design, optimization, and synthesis of lead compounds, followed by compound screening, target discovery, and drug repositioning. Relatively few companies have laid out the crystal form prediction and dosage form design links.

At present, virtual screening of small molecule compounds, new target discovery, drug optimization design, and drug redirection are currently popular directions in the field of domestic AI drug research and development.

There are currently three main types of entrants in the domestic and foreign AI pharmaceutical market, namely large pharmaceutical companies, AI pharmaceutical start-ups, and leading Internet companies. Large pharmaceutical companies are further divided into traditional pharmaceutical companies and CRO companies. From the perspective of the AI drug research and development industry chain, the upstream is the AI model dataset supply and cloud computing platform. The medical data provided by the dataset is a key competitive barrier for the industry, and the cloud computing platform is used to ensure the computing power supply of the underlying architecture. The middle reaches of the industry chain are AI drug R&D companies and IT companies. Among them, AI drug R&D companies mainly cooperate with downstream companies in the form of pharmaceutical R&D outsourcing and rely on internal training tools and AI development tools to build models based on medical datasets and training; IT companies participate in AI drug research and development by building their own AI drug research and development platforms and providing computing power and computing framework services. The downstream is traditional pharmaceutical companies, and the midstream AI drug research and development companies will directly sell their services in the drug research and development stage to traditional pharmaceutical companies. Therefore, traditional pharmaceutical companies are the direct demanders of AI drug research and development.

AI drug research and development will enter the field of macromolecules, such as antibodies. In April 2022, Israeli pharmaceutical company Biolojic Design announced that its first-ever computationally designed antibody had entered clinical trials. In November, Canadian pharmaceutical company AbCellera and its partner Regeneron announced that they had advanced the first antibody drug candidate targeting an undisclosed G protein-coupled receptor (GPCR) into the preclinical development stage. In the same month, AI pharmaceutical company Exscientia announced that its AI technology platform will include human antibody design. Some media have made incomplete statistics and found that more than 20 companies around the world are discovering antibody drugs through AI technology. From a regional perspective, most of these companies are located in Europe and the United States. There are also companies in China, but they are still in a niche field. Xinhua Biotech announced that the first-in-class multi-functional antibody drug designed and developed using its self-developed AI platform has shown excellent safety and efficacy in preclinical animal experiments and has excellent performance in druggability. It will soon enter CMC and IND-enabling stage.

Automated laboratories have become a new attraction. In 2022, the quantity and quality of data will still be the core issues in the development of AI pharmaceuticals. The emergence of automated laboratories is precisely to solve this problem. In 2021, some AI pharmaceutical companies have begun to establish automated laboratories with the purpose of improving internal data generation capabilities to optimize AI models. According to incomplete statistics, Exscientia, Insilicon, Arctoris, Recursion, Insitro, etc. have all established automation laboratories. Insilicon Intelligent released the world's first fully automated robotic laboratory assisted by artificial intelligence in decision-making in December 2021. The intelligent robot laboratory focuses on target discovery, compound screening, personalized drug development, and translational medical research. Automation has become the next important module in the strategic layout of many AI pharmaceutical companies. At the beginning of 2021, the UK's Automata Labs raised US$50 million for automated laboratory research.

Drug development is the process of bringing a new pharmaceutical drug to the market once a lead compound has been identified through the process of drug discovery. It includes preclinical research on micro-organisms and animals, filing for regulatory status, such as via the United States Food and Drug Administration for an investigational new drug to initiate clinical trials on humans, and may include the step of obtaining regulatory approval with a new drug application to market the drug. The entire process — from concept through preclinical testing in the laboratory to clinical trial development, including Phase I–III trials — to approved vaccine or drug typically takes more than a decade.

12.5 Genetic Research

Genetic research is the study of human DNA to find out what genes and environmental factors contribute to diseases. If we find out what causes disease, we can better detect disease, better treat disease, and hopefully even prevent disease from happening in the first place!

In 1972, Paul Berg constructed the world's first recombinant DNA molecule. Since then, genetic engineering was born and soon set off an upsurge in genetic engineering research. In the past 27 years, genetic engineering technology has developed by leaps and bounds. It has made many remarkable achievements and has had a huge impact on many aspects, such as agricultural production, medical and health care, light industry, food, and environmental health. The first field trials of genetically modified plants were conducted in France and the United States in 1986. The experimental plant was a herbicide-resistant genetically modified tobacco. China was the first country to commercialize genetically modified plants, introducing virus-resistant tobacco in 1992.

In 2009, 11 genetically modified crops were commercially produced in 25 countries, mainly the United States, Brazil, Argentina, India, Canada, China, Paraguay, and South Africa. In 2010, scientists at the Craig Venter Institute created the first synthetic genome and inserted it into empty bacterial cells. The resulting bacteria, named "Cynthia," can replicate and produce proteins. In 2014, a bacterium was developed that replicates plasmids containing unique base pairs (non-adenine, cytosine, thymine, and guanine), the first organism to use an expanded genetic alphabet. Operations and steps if a certain genetic code fragment in the DNA of one organism is connected to the DNA chain of another organism, and the DNA is reorganized, new genetic material can be designed and new organisms can be created according to human wishes.

Genetic engineering generally includes the following four steps: Obtain DNA fragments that meet the requirements; construct gene expression vectors; introduce the target gene into the recipient cells; detection and identification of target genes:

1. Obtain DNA fragments

The first step is to select and isolate the gene that will be inserted into the genetically modified organism. Furthermore, genes to be inserted into genetically modified organisms must be combined with other genetic components in order for them to work properly. Genes can also be modified at this stage for better expression or effectiveness. In addition to the gene to be inserted, most constructs contain promoter and terminator regions and a selectable marker gene. The promoter region initiates the transcription of a gene and can be used to control the position and level of gene expression, while the terminator region terminates transcription. In most cases, a selectable marker that confers antibiotic resistance to the organism in which it is expressed requires determining which cells are transformed with the new gene.

2. Construct gene expression vectors

After the DNA molecule chains are cut, they must be sewn together to complete the gene splicing. In 1967, scientists in five laboratories discovered and extracted an enzyme almost simultaneously. This enzyme could connect two DNA fragments and repair the break in the DNA chain. After 1974, the scientific community officially recognized this discovery and called this enzyme DNA ligase. Since then, DNA ligase has become a veritable "molecular thread" that "sews" genes together. As long as a "molecular needle thread" is added to the two DNA fragments cut with the same "molecular scissors", the two DNA fragments will be reconnected. Plasmid vector DNA manipulation usually occurs within plasmids. Constructs are prepared using recombinant DNA techniques, such as restriction digestion, ligation, and molecular cloning.

3. Introduce the target gene into the recipient cells

Introduce the gene of interest into the recipient cells. Only about 1% of bacteria are naturally capable of taking up foreign DNA. However, this ability can be induced in other bacteria by external stimuli (such as heat or electric shock) to increase the permeability of their cell membranes to DNA; the absorbed DNA can integrate with the genome or exist as extrachromosomal DNA (such as plasmids). DNA is typically inserted into animal cells using microinjection, where it can be injected directly into the nucleus through the cell's nuclear membrane or through the use of viral vectors. In plants, DNA is often inserted using Agrobacterium-mediated recombination or gene gun technology. In Agrobacterium-mediated recombination, the plasmid construct contains T-DNA, which is responsible for inserting the DNA into the host plant genome. This plasmid is transformed into plasmid-free Agrobacterium before infecting plant cells. Agrobacterium will then naturally insert the genetic material into the plant cells. In biodynamics, particles of gold or tungsten are coated with DNA and injected into callus cells or plant embryos. Some genetic material will enter the cells and transform them. This method can be used on plants that are less susceptible to Agrobacterium infection and also allows transformation of plant plastids. Another transformation method used for plant and animal cells is electroporation. Electroporation involves subjecting plant or animal cells to an electrical shock, which renders the cell membrane permeable to plasmid DNA and, in some cases, allows the electroporated cells to incorporate DNA into their genomes. Due to their damage to cells and DNA, gene gun and electroporation are less efficient than Agrobacterium-mediated transformation and microinjection. Since there is usually only one cell used for transformation, this single cell must be grown into an organism. Bacteria consist of single cells and do not require clonal regeneration. In plants, this is achieved through the use of tissue culture. Each plant has different requirements for successful regeneration through tissue culture. If successful, an adult plant containing the transgene in every cell is produced. In animals, it is necessary to ensure that the inserted DNA is present in the embryonic stem cells.

4. Detection and identification of target genes

Selectable markers are used to differentiate between transformed and non-transformed cells. These markers are commonly present in transgenic organisms, although a variety of strategies have been developed to remove selectable markers from mature transgenic plants. When offspring are produced, the presence of the gene can be screened. All offspring from the first generation will be heterozygous for the inserted gene and must be mated together to produce homozygous animals.

Further testing uses PCR, Southern blotting, and DNA sequencing to confirm that the organism contains the new gene. That is, the genomic DNA of the genetically modified organism is extracted, and the DNA fragment containing the target gene is labeled with a radioactive isotope. This is used as a probe to hybridize the probe with the genomic DNA. If a hybridization band is displayed, it indicates that the target gene has been inserted in chromosomal DNA. This method is named the Southern Blot method after its discoverer. These tests can also confirm the chromosomal location and copy number of the inserted gene. The presence of a gene does not guarantee that it will be expressed at appropriate levels in the target tissue, so methods of finding and measuring the gene products (RNA and protein) are also used. These include Northern blotting, quantitative real-time polymerase chain reaction (RT-PCR), Western blotting, immunofluorescence, enzyme immunoassay (ELISA), and phenotypic analysis. For stable transformation, the gene should be passed on to offspring in a Mendelian mode of inheritance, so the offspring of the organism should also be studied.

It is becoming the most cutting-edge subject in biological research. Scientists predict that genetic engineering technology will be the mainstay. Biotechnology will become one of the leading new technologies in the 21st century.

12.6 Biological Breeding

Biological breeding is the process of cultivating new biological varieties using the principles of genetics, cell biology, modern bioengineering technology, and other methods. With the evolution of natural species and the advancement of human science and technology over thousands of years, world agricultural breeding has experienced three eras: primitive breeding, traditional breeding, and molecular breeding, forming various technical versions with typical characteristics of the times, that is, from the initial artificial domestication version 1.0 and hybrid breeding version 2.0 are gradually and iteratively upgraded to version 3.0 of transgenic breeding and version 4.0 of intelligent design breeding in the era of molecular breeding. Biological breeding is a molecular breeding technology that has moved from genetically modified breeding version 3.0 to intelligent design breeding version 4.0, integrating various cutting-edge technologies. From the end of the 20th century to the beginning of the 21st century, with the cross-integration of cutting-edge sciences such as omics, systems biology, synthetic biology, and computational biology, modern biological breeding technologies for cultivating revolutionary and subversive major varieties emerged at the historic moment, among which the most representative sexual technologies include whole-genome selection, gene editing, and synthetic biotechnology. The upstream of the biological breeding industry chain is breeding research and development institutions, the midstream is seed breeding (production), processing, and packaging institutions, and the downstream is promotion and sales institutions. Large-scale breeding companies rely on their strong scientific research and financial strength and are market-oriented to form a model that integrates scientific research, production, promotion, and sales, that is, the "breeding, breeding, and promotion" integrated model. This is the development trend of the breeding industry chain.

Development status:

1. High concentration of the international seed industry:

Since the 21st century, the global seed market has grown rapidly thanks to the development of biological breeding technology. According to statistics, the global seed market has grown from US$19.752 billion in 2005 to nearly US$60 billion in 2020, with a compound annual growth rate of nearly 9%. The main growth space comes from the rapid growth of vegetable seeds. Among them, the United States has long occupied the world's largest seed market, accounting for more than 35%. Since 2016, China has become the world's second largest seed market, accounting for approximately 23%. Currently, the global seed industry market presents a duopoly situation composed of China and the United States.

2. Leading companies are strong and strong:

The seed industry in developed countries has developed into an industrial system integrating scientific research, production, processing, sales, and technical services. A few large seed groups have monopolized most of the world's seed industry market. According to PhillipsMcDougall/HISMarkit data, from a global perspective, the seed industry is highly concentrated. There are only about 10 international seed company giants, occupying more than 50% of the market share, with annual sales revenue exceeding US$500 million and profit margins basically maintaining between 10% and 15%. There are a large number of second-tier seed companies, about 40, with sales revenue ranging from US$100 million to US$500 million and profit margins basically maintained at around 10%. The largest number of seed companies are in the third tier, with annual revenue of less than US$100 million and profit margins of about 5%. The industry as a whole shows the "Matthew Effect", a situation in which the strong will always be strong.

3. The target traits of genetically modified crops being developed continue to expand:

At present, the genetically modified crops grown commercially on a large scale at home and abroad are mainly first-generation genetically modified products, involving target traits, such as herbicide tolerance, insect resistance, virus resistance, and drought resistance. At the same time, in order to meet the diverse needs of planting, production, processing, or consumption, the target traits of genetically modified crops under development are constantly expanding, including herbicide resistance traits such as glufosinate resistance and dicamba resistance, disease resistance traits such as resistance to late blight, and resistance to cucumber mosaic disease, insect resistance traits such as resistance to potato beetles and rice brown planthopper, stress resistance traits such as salt-alkali tolerance and efficient nutrient utilization, etc., and quality improvement traits such as high lysine, high unsaturated fatty acids, delayed ripening, storage-resistant and antibrowning, etc. In recent years, the industrialization of directly edible genetically modified products cultivated using gene silencing technology has accelerated. Genetically modified potatoes that are resistant to browning and late blight, genetically modified apples that are resistant to browning, genetically modified lycopene pineapples, and fast-growing genetically modified salmon have been approved for marketing in the United States. In agriculture, the industrial application of genetically modified genes has expanded from initially non-edible cotton and feed crops to directly edible food crops, fruits, and farmed animals.

In China, Biological Breeding Industry Patent Navigation Research Results released by the Intellectual Property Development Research Center of the State Intellectual Property Office in 2021 show that my country currently ranks first in the world in the number of biological breeding patent applications. Among them, the number of molecular marker breeding patent applications has surpassed the United States, becoming the first application in the world, the country with the largest amount. Our country attaches great importance to biological breeding technology and its industrialization. The "Outline for Building a Powerful Country with Intellectual Property Rights" (2021–2035) proposes to accelerate the cultivation of a number of excellent new plant varieties with intellectual property rights and improve the quality of authorized varieties around cutting-edge technologies and key areas of biological breeding; the No. 1 Central Document of 2021 clearly states, strengthen the protection of intellectual property rights in the field of breeding.

In terms of domestic seed companies, Syngenta and Longping High-tech both rank among the top ten in the world, becoming China's first-tier seed companies in the world. There are many second-tier companies in China, and the gap in revenue scale is small. Beidahuang Kenfeng, Jiangsu Dahua, Tsuen Yin Hi-Tech, China Seed Group, and Denghai Seed Industry are shortlisted for the second tier. According to data from the Ministry of Agriculture, the number of seed industry companies in my country was 8,700 in 2010, which dropped to 4,316 in 2016. However, the number of seed companies has begun to rebound since 2017. As of the end of 2020, the number of certified seed companies in my country is around 6,000. In terms of crop types, domestic corn seed companies account for the largest number, accounting for about 28%; followed by wheat, accounting for about 21%. Rice seed companies include hybrid and conventional rice, accounting for about 18% in total.

In recent years, the Chinese government has launched a number of policies on the seed industry with the purpose of promoting the development of biotechnology in the seed industry, improving the supervision system of the seed industry market, and cracking down on illegal genetic modification and other illegal activities. Judging from the relevant policies that the government has successively introduced, and the government's attitude toward supporting and protecting genetically modified seed research and development enterprises, biological breeding technology is an important development trend in my country's seed industry in the future.

Development Trend

(1) Whole-genome selective breeding technology is widely used. Whole-genome selective breeding technology aggregates excellent genotypes at the whole-genome level and improves important agronomic traits through computational biology model prediction and high-throughput genotype analysis. Compared with traditional molecular marker-assisted selection, whole-genome selective breeding technology has two major advantages: First, the genomically mapped parent populations can be directly applied to breeding; second, it is more suitable for improving quantitative traits controlled by polygenes with smaller effects. Especially with the rapid advancement of high-throughput sequencing, omics big data, and gene chip technology, whole-genome selective breeding technology is increasingly being used in the practice of agricultural biological variety breeding. At present, whole-genome selection technology has brought revolutionary changes to animal and plant breeding, greatly improving the efficiency of animal and plant breeding, and has become a research hotspot in the field of international animal and plant breeding and the focus of competition among multinational companies. My country has initially established a rice whole-genome selective breeding technology system with breeding chips as the core, including using high-throughput SSR marker technology to identify and screen target genes, using OpenArray chip technology to identify and screen chromosome segment haplotypes, and using whole-genome breeding chips. Technical identification and screening of genetic background, etc.

(2) Gene editing breeding technology is advancing rapidly. Gene editing technology, especially the CRISPR/Cas9-mediated genome editing system, has become one of the most effective breeding tools in the agricultural field due to its precise orientation, simplicity, efficiency, and diversification. In recent years, its development

has been rapid and rapid, continuously upgrading. Single-base editing technology (base editing) developed based on the CRISPR-Cas system is a fast, efficient, and accurate new generation of gene editing technology. It uses cytosine deaminase or artificially evolved adenine deaminase to target a certain range of sites. Cytosine (C) or adenine (A) undergoes deamination reaction to achieve precise replacement of C-T or A-G. In 2020, Chinese scientists used cytosine and adenine double base editors to conduct directed or random mutagenesis of rice genes. The C>T single base mutagenesis efficiency was as high as 61.61%, and the C>T and A>G double base mutagenesis efficiency was as high as 61.61%. The conversion efficiency is also as high as 15.10%. Another brand-new guided editing system developed in recent years can achieve any type of base substitution and precise insertion and deletion of small fragments without relying on DNA templates and has been successfully used in rice and wheat. In addition, fusing the CRISPR-dCas9 system to target modifying enzymes can create a complete set of plant epigenetic editing tools. In 2021, an upgraded epigenetic editing system called CRISPRoff was reported to be able to cause target gene silencing through highly specific methylation without changing the DNA sequence and can be used for crop breeding and plant protection.

(3) Synthetic biology breeding technology leads the future. Synthetic biology technology adopts the modular concept and system design theory of engineering to transform and optimize existing natural biological systems or synthesize new artificial biological systems with predetermined functions from scratch, constantly breaking through the natural world of life. The law of inheritance marks that modern life science has moved from understanding life to a new stage of designing and transforming life. The application of synthetic biotechnology in the agricultural field provides photosynthesis (carbon fixation with high light efficiency), biological nitrogen fixation (fertilizer saving and efficiency increase), biological stress resistance (water-saving and drought tolerance), biological transformation (biomass resource utilization), and future synthetic foods (artificial meat and milk), and other global agricultural production problems provide revolutionary solutions.

12.7 New Materials

China aims to accelerate the industrialization of new materials like superconducting materials, graphene, and liquid metal that are now considered critical to the development of high-tech industries. These materials represent the direction of the development of the new materials industry, which is an important entry point for building new growth engines.

12.7.1 *Superconducting Materials*

Superconductivity is a set of physical properties observed in certain materials where electrical resistance vanishes and magnetic fields are expelled from the material. Any material exhibiting these properties is a superconductor. Unlike an ordinary metallic conductor, whose resistance decreases gradually as its temperature is lowered, even down to near absolute zero, a superconductor has a characteristic critical temperature below which the resistance drops abruptly to zero. (Combescot, 2022) An electric current through a loop of superconducting wire can persist indefinitely with no power source.

12.7.2 *Graphene*

Graphene is an allotrope of carbon consisting of a single layer of atoms arranged in a hexagonal lattice nanostructure (Geim & Novoselov, 2007). The name is derived from "graphite" and the suffix ene, reflecting the fact that the graphite allotrope of carbon contains numerous double bonds.

Each atom in a graphene sheet is connected to its three nearest neighbors by σ-bonds and a delocalized π-bond, which contributes to a valence band that extends over the whole sheet. This is the same type of bonding seen in carbon nanotubes and polycyclic aromatic hydrocarbons, and (partially) in fullerenes and glassy carbon. The valence band is touched by a conduction band, making graphene a semimetal with unusual electronic properties that are best described by theories for massless relativistic particles. Charge carriers in graphene show linear, rather than quadratic, dependence of energy on momentum, and field-effect transistors with graphene can be made that show bipolar conduction. Charge transport is ballistic over long distances; the material exhibits large quantum oscillations and large and nonlinear diamagnetism. Graphene conducts heat and electricity very efficiently along its plane. The material strongly absorbs light of all visible wavelengths, which accounts for the black color of graphite; yet a single graphene sheet is nearly transparent because of its extreme thinness. Microscopically, graphene is the strongest material ever measured.

Scientists theorized the potential existence and production of graphene for decades. It has likely been unknowingly produced in small quantities for centuries, through the use of pencils and other similar applications of graphite. It was possibly observed in electron microscopes in 1962 but studied only while supported on metal surfaces.

In 2004, the material was rediscovered, isolated, and investigated at the University of Manchester, by Andre Geim and Konstantin Novoselov. In 2010, Geim and Novoselov were awarded the Nobel Prize in Physics for their "groundbreaking experiments regarding the two-dimensional material graphene." High-quality graphene proved to be surprisingly easy to isolate.

Graphene has become a valuable and useful nanomaterial due to its exceptionally high tensile strength, electrical conductivity, transparency, and being the thinnest two-dimensional material in the world. The global market for graphene was $9 million in 2012, with most of the demand from research and development in semiconductors, electronics, electric batteries, and composites.

The IUPAC (International Union for Pure and Applied Chemistry) recommends use of the name "graphite" for the three-dimensional material and "graphene" only when the reactions, structural relations, or other properties of individual layers are discussed A narrower definition, of "isolated or free-standing graphene," requires that the layer be sufficiently isolated from its environment but would include layers suspended or transferred to silicon dioxide or silicon carbide.

12.7.3 *Liquid Metal*

A liquid metal is a metal or a metal alloy which is liquid at or near room temperature. The only stable liquid elemental metal at room temperature is mercury, which is molten above $-38.8\,°C$. Three more stable elemental metals melt just above room temperature: caesium (Cs), which has a melting point of $28.5\,°C$; gallium (Ga) ($30\,°C$); and rubidium (Rb) ($39\,°C$). The radioactive metal francium (Fr) is probably liquid close to room temperature as well. Calculations predict that the radioactive metals copernicium (Cn) and flerovium (Fl) should also be liquid at room temperature.

Alloys can be liquid if they form a eutectic, meaning that the alloy's melting point is lower than any of the alloy's constituent metals. The standard metal for creating liquid alloys used to be mercury, but gallium-based alloys, which are lower both in their vapor pressure at room temperature and toxicity, are being used as a replacement in various applications.

Due to their excellent characteristics and manufacturing methods, liquid metals are often used in wearable devices, medical devices, interconnected devices, and so on. Typical uses of liquid metals include thermostats, switches, barometers, heat transfer systems, and thermal cooling and heating designs. Uniquely, they can be used to conduct heat and/or electricity between non-metallic and metallic surfaces.

Liquid metal is sometimes used as a thermal interface material between coolers and processors because of its high thermal conductivity. The PlayStation 5 video game console uses liquid metal to help cool high temperatures inside the console. Liquid metal-cooled reactors also use them.

Liquid metal can be used for wearable devices and for spare parts.

Liquid metal can sometimes be used for biological applications, ie making interconnects that flex without fatigue. As Galinstan is not particularly toxic, wires made from silicone with a core of liquid metal would be ideal for intracardiac pacemakers and neural implants where delicate brain tissue cannot tolerate a conventional solid implant. In fact, a wire constructed of this material can be stretched to 3 or even 5 times its length and still conduct electricity, returning to its original size and shape with no loss.

12.8 Climate Change

Climate change describes global warming and its effects on Earth's climate system. The current rise in global average temperature is more rapid than previous changes and is primarily caused by humans burning fossil fuels. Fossil fuel use, deforestation, and some agricultural and industrial practices add to greenhouse gases, notably carbon dioxide and methane. Greenhouse gases absorb some of the heat that the Earth radiates after it warms from sunlight. Larger amounts of these gases trap more heat in Earth's lower atmosphere, causing global warming.

Climate change came to international public attention in the late 1980s. Due to media coverage in the early 1990s, people often confused climate change with other environmental issues like ozone depletion. In popular culture, the climate fiction movie *The Day After Tomorrow* (2004) and the Al Gore documentary *An Inconvenient Truth* (2006) focused on climate change.

Climate change is causing a range of increasing impacts on the environment. Deserts are expanding, while heat waves and wildfires are becoming more common. Amplified warming in the Arctic has contributed to melting permafrost, glacial retreat, and sea ice loss. Higher temperatures are also causing more intense storms, droughts, and other weather extremes. Rapid environmental change in mountains, coral reefs, and the Arctic is forcing many species to relocate or become extinct. Even if efforts to minimize future warming are successful, some effects will continue for centuries. These include ocean heating, ocean acidification, and sea level rise.

Climate change threatens people with increased flooding, extreme heat, increased food and water scarcity, more disease, and economic loss. Human migration and conflict can also be a result. The World Health Organization (WHO) calls climate change the greatest threat to global health in the 21st century. Societies and ecosystems will experience more severe risks without action to limit warming.

Adapting to climate change through efforts like flood control measures or drought-resistant crops partially reduces climate change risks, although some limits to adaptation have already been reached.

Reducing emissions requires reducing energy use and generating electricity from low-carbon sources rather than burning fossil fuels. This change includes phasing out coal and natural gas-fired power plants while vastly increasing electricity generated from wind, solar, and nuclear power. This electricity will need to replace fossil fuels for powering transportation, heating buildings, and operating industrial facilities. Carbon can also be removed from the atmosphere, for instance by increasing forest cover and farming with methods that capture carbon in soil.

12.8.1 *Climate Model*

A climate model is a representation of the physical, chemical, and biological processes that affect the climate system. Models also include natural processes like changes in the Earth's orbit, historical changes in the Sun's activity, and volcanic forcing. Models are used to estimate the degree of warming future emissions will cause when accounting for the strength of climate feedback or reproduce and predict the circulation of the oceans, the annual cycle of the seasons, and the flows of carbon between the land surface and the atmosphere.

The physical realism of models is tested by examining their ability to simulate contemporary or past climates. Past models have underestimated the rate of Arctic shrinkage and underestimated the rate of precipitation increase. Sea level rise since 1990 was underestimated in older models, but more recent models agree well with observations. The 2017 United States published National Climate Assessment notes that "climate models may still be underestimating or missing relevant feedback processes." Additionally, climate models may be unable to adequately predict short-term regional climatic shifts.

A subset of climate models adds societal factors to a simple physical climate model. These models simulate how population, economic growth, and energy use affect — and interact with — the physical climate. With this information, these models can produce scenarios of future greenhouse gas emissions. This is then used as input for physical climate models and carbon cycle models to predict how atmospheric concentrations of greenhouse gases might change. Depending on the socioeconomic scenario and the mitigation scenario, models produce atmospheric CO_2 concentrations that range widely between 380 and 1400 ppm.

12.8.2 *Long-Term Impacts*

The long-term effects of climate change on oceans include further ice melt, ocean warming, sea level rise, and ocean acidification. On the timescale of centuries to millennia, the magnitude of climate change will be determined primarily by anthropogenic CO_2 emissions. This is due to CO_2's long atmospheric lifetime. Oceanic CO_2 uptake is slow enough that ocean acidification will continue for hundreds to thousands of years. These emissions are estimated to have prolonged the current interglacial period by at least 100,000 years. Sea level rise will continue over many centuries, with an estimated rise of 2.3 metres per degree Celsius (4.2 ft/°F) after 2000 years.

Extreme weather leads to injury and loss of life, and crop failures to malnutrition. Various infectious diseases are more easily transmitted in a warmer climate, such as dengue fever and malaria. Young children are the most vulnerable to food shortages. Both children and older people are vulnerable to extreme heat. The World Health Organization (WHO) has estimated that between 2030 and 2050, climate change would cause around 250,000 additional deaths per year. They assessed deaths from heat exposure in elderly people, increases in diarrhea, malaria, dengue, coastal flooding, and childhood malnutrition. Over 500,000 more adult deaths are projected yearly by 2050 due to reductions in food availability and quality. By 2100, 50%–75% of the global population may face climate conditions that are life-threatening due to combined effects of extreme heat and humidity.

12.8.3 *Solution Strategy*

Accelerate the green and low-carbon transformation of development methods. Adhere to green and low-carbon development as a solution to solve the fundamental problems of ecological environment and accelerate the promotion of industrial structure and energy structure. The transportation structure will be adjusted and optimized. Implement a comprehensive conservation strategy and promote various resources. We should save and use intensively and speed up the construction of waste recycling system. Complete support for green Develop fiscal, taxation, financial, investment, price policy, and standard systems, and develop green and low-cost carbon industry, improve the market-oriented allocation system of resource and environmental factors, and accelerate energy conservation and carbon reduction. Promote technology research and development, promotion and application, advocate green consumption, and promote the formation of a green and low-carbon mode of production and lifestyle. Actively and steadily promote carbon peak and carbon neutrality. Adhere to national coordination, save priority, and the principles of two-wheel drive, smooth internal and external circulation, and risk prevention will be implemented to achieve carbon peak and carbon neutrality. The "1+N" policy system gradually shifts to "dual control" of the total amount and intensity of carbon emissions. Build a clean, low-carbon, safe and efficient energy system, accelerate the construction of new power systems, and improve national oil and gas security capabilities. Improve the carbon emission statistical accounting system and improve the carbon Emissions rights market trading system. Consolidate and improve the carbon sink capacity of the ecosystem.

Climate change can be mitigated by reducing the rate at which greenhouse gases are emitted into the atmosphere, and by increasing the rate at which carbon dioxide is removed from the atmosphere. In order to limit global warming to less than 1.5°C, global greenhouse gas emissions need to be net-zero by 2050, or by 2070 with a 2°C target. This requires far-reaching, systemic changes on an unprecedented scale in energy, land, cities, transport, buildings, and industry. The United Nations Environment Programme estimates that countries need to triple their pledges under the Paris Agreement within the next decade to limit global warming to 2°C. An even greater level of reduction is required to meet the 1.5°C goal. With pledges made under the Paris Agreement as of October 2021, global warming would still have a 66% chance of reaching about 2.7°C (range: 2.2–3.2°C) by the end of the century. Globally, limiting warming to 2°C may result in higher economic benefits than economic costs.

There is no single pathway to limit global warming to 1.5 or 2°C, most scenarios and strategies see a major increase in the use of renewable energy in combination with increased energy efficiency measures to generate the needed greenhouse gas reductions. To reduce pressures on ecosystems and enhance their carbon sequestration capabilities, changes would also be necessary in agriculture and forestry, such as preventing deforestation and restoring natural ecosystems by reforestation.

12.9 Platform of Artificial Intelligence for Science

Here, we take MindSpore as an example to introduce the driving forces of AI frameworks which are mainly five elements "ABCDE," as shown in Figure 12.2 (Jin, 2023):

(1) **Application+Bigdata:** The application of the AI framework is the model and algorithm of AI.
(2) **Big data:** Big data is the data required by the AI model.

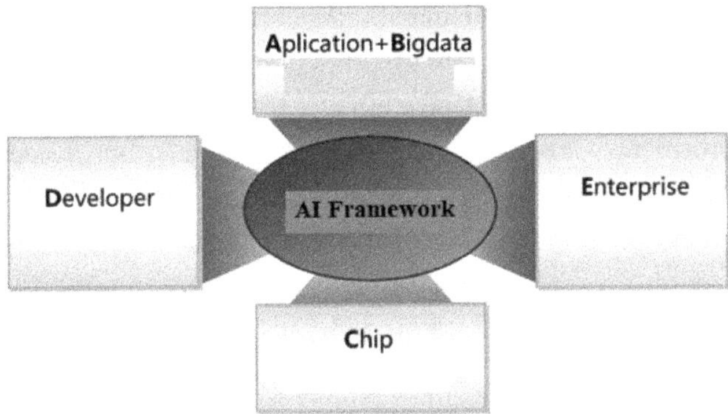

Fig. 12.2.   Artificial intelligence platform for science.

(3) **Chip:** AI chip represents the algorithm development of AI;

(4) **Developer:** Algorithm researchers and engineers for AI.

(5) **Enterprise:** AI deployment and AI responsibility.

We hope MindSpore can lead the evolution of AI frameworks in five directions:

(1) **Beyond AI:** Evolving from deep learning frameworks to general tensor differentiable computing frameworks.

   MindSpore will provide a more universal AI compiler, providing possibilities for supporting a wider range of application types.

(2) **Scale out:** Evolving from manual parallelism to automatic parallelism.

   MindSpore not only has good performance and scalability in supporting large-scale training, but more importantly, it aims to lower the threshold in this area and make distributed training easier.

(3) **Scale up:** Evolving from graph computation separation optimization to graph computation joint optimization.

   MindSpore provides key technologies, such as graph computation joint optimization, automatic operator generation, and depth map optimization, fully leveraging the computing power of AI chips.

(4) **Full scenario AI:** Evolving from end-to-end cloud separation architecture to end-to-end cloud unified architecture.

   MindSpore's cloud-side framework and end-to-end framework achieve a unified architecture, facilitating cloud-based training/end-to-end fine-tuning or federated learning for end-to-end cloud collaboration.

(5) **Enterprise-level trustworthiness:** Evolving from consumer-level AI to enterprise-level AI.

MindSpore will have built-in capabilities, such as adversarial training, differential privacy, dense AI, federated learning, and interpretable AI.

The Shengsi Big Model Platform aims to provide AI learners and developers with a platform for online learning projects, models, big model experiences, and datasets. The platform has also added classic datasets from various fields to help learners solve a series of difficulties in the AI learning process, such as the difficulty of obtaining high-quality datasets and the difficulty of using large-scale datasets for model training locally, providing users with support for various business scenarios.

Exercises

12.1 Illustrate the knowledge discovery procedure and take an example to explain how to mine from big data set.

12.2 How to divide the protein structure prediction levels?

12.3 How to develop a new drug? Take an example to explain the procedure and benefit of new drug development.

12.4 Try to list the steps for genetic engineering.

12.5 What is biological breeding? Why it is important to people's lives and social development?

12.6 What is superconducting material? How to prepare superconducting materials?

12.7 What kinds of factors cause climate change?

12.8 What can artificial intelligence do against climate change?

Bibliography

Abualsaud, K., Elfouly, T.M., Khattab, T., *et al.* (2018). A Survey on Mobile Crowd-Sensing and its Applications in the IOT Era. *IEEE*. Doi: 10.1109/ACCESS.2018.2885918.

Alghamdi, R. and Alfalqi, K. (2015). A survey of topic modeling in text mining. *International Journal of Advanced Computer Science and Applications*, 6(1): 147–153.

Amari, S. and Wu, S. (1999). Improving support vector machine classifier by modifying kernel functions. *Neural Networks*, 12: 783–789.

Andrieu, C., De Freitas, N., Doucet, A., *et al.* (2003). An introduction to MCMC for machine learning. *Machine Learning*, 50(1): 5–43.

Arnold, A., Nallapati, R. and Cohen, W.W. (2007). A comparative study of methods for transductive transfer learning. *Proc. Seventh IEEE Int'l Conf. Data Mining Workshops*, pp. 77–82.

Argyriou, A., Evgeniou, T. and Pontil, M. (2008). Convex multi-task feature learning. *Machine Learning*, 73.

Baader, F., Calvanese, D., McGuinness, D., Nardi, D. and Patel-Schneider, P. (eds.), (2003). *The Description Logic Handbook: Theory, Implementation and Applications*. Cambridge: Cambridge University Press.

Baars, B.J. and Franklin S. (2009). Consciousness is computational: The LIDA model of global workspace theory. *International Journal of Machine Consciousness*, 1(1): 23–32.

Bach, J. (2009). *Principles of Synthetic Intelligence*: Psi: An Architecture of Motivated Cognition.

Bahdanau, D., Cho, K. and Bengio, Y. (2015). Neural machine translation by jointly learning to align and translate. *In International Conference on Learning Representations*.

Bayer, J., Wierstra, D., Togelius, J., *et al.* (2009). *Evolving Memory Cell Structures for Sequence Learning. Artificial Neural Networks – ICANN 2009*. Lecture Notes in Computer Science. Springer, Berlin, Heidelberg. 5769: 755–764.

Bellman, R. (1957). *Dynamic Programming*. Princeton Press.

Ben-David, S., Lu, T., Luu, T. and Pál, D. (2010). Impossibility theorems for domain adaptation. In *Proceedings of the International Conference on Artificial Intelligence and Statistics (AISTATS)*, volume 9, pages 129–136.

Berners-Lee, T., Hendler, J. and Lassila, O. (2001). The semantic Web. *Scientific American*, 281(5): 29–37.

Berners-Lee, T., Hall, W., Hendler, J.A., O'Hara, K., Shadbolt, N. and Weitzner, D.J. (2006). A Framework for Web Science. *Foundations and Trends in Web Science*, 1(1): 1–130.

Blei, D.M., NG, A.Y. and Jordan, M.I. (2003). Latent dirichlet allocation. *Machine Learning Research*, 3: 993–1022.

Bonawitz, K., Ivanov, V., Kreuter, B., *et al.* (2016). Practical secure aggregation for privacy-preserving machine learning. In *Conference on Computer and Communications Security*. arXiv: 1611.04482V1[cs.CR].

Brooks, R.A. (1991a). Intelligence without representation. *Artificial Intelligence*, 47: 139–159.

Brooks, R.A. (1991b). Intelligence without reasoning. In *Proceedings of IJCAI'91*, Sydney.

Brooks, T., Peebles, B., Holmes, C., *et al.* (2024). Sora Technical Report. OpenAI, February 16.

Burges, J. C. J. (1998). A tutorial on support vector machines for pattern recognition. *Data Mining and Knowledge Discovery*, 2(2): 167.

Castrignanò, A., Quarto, R., Venezia, A., G. Buttafuoco, G. (2017). A geostatistical approach for modelling and combining spatial data with different support. *Advances in Animal Biosciences Journal*, 8, 594–599.

Chickering, D., and Heckerman, D. (1996). Efficient approximations for the marginal likelihood of incomplete data given a Bayesian network. Technical Report MSR-TR-96-08, Microsoft Research, Redmond, WA.

Christian, S., Youssef, W. L., Tomaso, Y. J., *et al.* (2015). Going deeper with convolutions. *2015 IEEE* Conference on Computer Vision and Pattern Recognition (CVPR) 1–9.

Christiano, P.F., Leike, J., Brown, T. B., *et al.* (2017). Deep Reinforcement Learning from Human Preferences. arXiv:1706.03741 [stat.ML].

Ciresan, D.C., Meier, U., Gambardella, L.M., *et al.*, (2010). Deep Big Simple Neural Nets Excel On Handwritten Digit Recognition. Neural Computation – NECO. arXiv:1003.0358 [cs.NE].

Clark, P. and Niblett, T. (1989). The CN2 induction algorithm. *Machine Learning*, 3(4): 26.

Combescot, R. (2022). Superconductivity. Cambridge University Press.

Cooper, G. F. and Herskovits, E. (1992). A bayesian method for the induction of probabilistic networks from data. *Machine Learning*, (9): 309–347.

Dai, J.F., Qi, H.Z., Qiong, Y.W. etc. (2017). Deformable Convolutional Networks. arXiv:1703.06211v3[cs.CV].

Day, O. and Khoshgoftaar, T.M. (2017). A survey on heterogeneous transfer learning. *Journal of Big Data*, 4: 29.

Decker, K. (1995). Environment Centered Analysis and Design of Coordination Mechanisms. PhD thesis, University of Massachusetts.

Doran, J.E., Franklin, S., Jennings, N.R. and Norman, T.J. (1997). On cooperation in multi-agent systems. http://eprints.ecs.soton.ac.uk/2193/1/FOMAS-PANEL-KER.pdf.

Doyle, J. (1979). A truth maintenance system. *Artificial Intelligence*, 12(3): 231–272.

Durfee, E.H. and Lesser, V.R. (1991). Partial Global Planning: A Coordination Framework for Distributed Hypothesis Formation. IEEE *Transactions on Systems, Man, and Cybernetics*, Special Issue on Distributed Sensor Networks, SMC-21(5): 1167–1183.

Dong, Chunling and Zhang Qin. (2020). The cubic dynamic uncertain causality graph: A methodology for temporal process modeling and diagnostic logic inference. *IEEE Transactions on Neural Networks and Learning Systems*, 31(10): 4239–4253.

de Kleer, J. (1986). An assumption-based TMS. *Artificial Intelligence*, 28: 127–162.

Doyle, J. (1979). A truth maintenance system. *Artificial Intelligence*, 12(3): 231–72.

Dutta, P. (1999). *Strategies and Games: Theory and Practice*. The MIT Press.

Eliasmith C., and Anderson C.H. (2003). *Neural Engineering: Computation, Representation, and Dynamics in Neurobiological Systems*. TLFeBOOK.

Eliasmith C., Stewart T.C., Choo X., *et al.* (2012). Large-scale model of the functioning brain. *Science*, 338, 1202–1205.

Endsley, M.R. (1995). Toward a Theory of Situation Awareness in Dynamic Systems. *Human Factors Journal* 37(1): 32–64.

Fudenberg, D. and Jean T. (1991). *Game Theory*. MIT Press.

Fernández, S., Graves, A. and Schmidhuber, J. (2007). Sequence labelling in structured domains with hierarchical recurrent neural networks. Proc. *20th Int. Joint Conf. on Artificial Intelligence, Ijcai.* 774–779.

Franklin, S., Madl,T., D'Mello, S. and Javier Snaider, J. (2014). LIDA: A systems-level architecture for cognition, emotion, and learning. *IEEE Transactions on Autonomous Mental Development,* 6(1): 19–41.

Freund, Y. and Schapire, R.E. (1995). A decision-theoretic generalization of on-line learning and an application to boosting. Proc. *of the Second European Conference on Computational Learning.*

Fu, K.S. (1974). Syntactic Method in Pattern *Recognition.* Academic Press, New York.

Gasser, L., Bragaza, C. and Herman, N. (1987). MACE: A flexible testbed for distributed AI research. In *Distributed Artificial Intelligence,* Pitman Publishing.

Geim, A.K. and Novoselov, K.S. (2007). The rise of graphene. *Nature Materials,* 6(3): 183–191.

Gers, F.A. and Schmidhuber, E. (2001). LSTM recurrent networks learn simple context-free and context-sensitive languages. *IEEE Transactions on Neural Networks,* 12(6): 1333–1340.

Gers, F.A., Schraudolph, N.N. and Schmidhuber, J. (2017). Learning Precise Timing with LSTM Recurrent Networks. Retrieved 2017-06-13, pp. 115–143.

Goertzel, B. (2009). OpenCogPrime: A Cognitive Synergy Based Architecture for General Intelligence. *International Conference on Cognitive Informatics,* Hong Kong.

Goertzel, B. and Cassio, P., (Eds.), (2006). *Artificial General Intelligence.* Springer-Verlag.

Goodfellow, I., Bengio, Y. and Courville, A. (2016). *Deep Learning.* MA: The MIT Press.

Graves, Alex; Fernández, Santiago; Gomez, Faustino (2006). Connectionist temporal classification: Labelling unsegmented sequence data with recurrent neural networks. In *Proceedings of the International Conference on Machine Learning,* ICML: 369–376.

Grosz, B. and Sarit K. (1996). Collaborative plans for complex group actions. *Artificial Intelligence,* 86(2): 269–357.

Gruber, T.R. (1993). A Translation Approach to Portable Ontology Specifications. *Knowledge Acquisition,* 5(2): 199–220.

Hammerstein, P. and Selten, R. (1994). Game theory and evolutionary biology. In *Handbook of Game Theory with Economic Applications,* 2, 929–993.

Hardy, S., Henecka, W., Ivey-Law, H., *et al.* (2017). Private federated learning on vertically partitioned data via entity resolution and additively homomorphic encryption. arXiv preprint arXiv:1711.10677.

Harsanyi, J.C. (1967/1968). Games with Incomplete Information Played by Bayesian Players, I-III. *Management Science,* 14 (3): 159–183 (Part I), 14 (5): 320–334 (Part II), 14 (7): 486–502 (Part III).

Hebb, D.O. (1949). *The Organization of Behavior.* Wiley.

Heckerman, D. (1997). Bayesian networks for data mining. *Data Mining and Knowledge Discovery,* 1: 79–119.

Hewitt, C. (1991). Open systems semantics for distributed artificial intelligence. *Artificial Intelligence,* 47: 79–106.

Hewitt, C. and DeJong, P. (1983). Analyzing the roles of descriptions and actions in open systems, In *AAAI.*

Hinton, G.E. (2002). Training products of experts by minimizing contrastive divergence. *Neural Computation,* 14, 1771–1800.

Hinton, G.E. (2010). *A Practical Guide to Training Restricted Boltzmann Machines.* UTML TR 2010–003.

Hinton, G.E. and Salakhutdinov, R. (2006). Reducing the dimensionality of data with neural networks. *Science,* 313(5786): 504–507.

Hochreiter, S. and Schmidhuber, J. (1997). Long short-term memory. *Neural Computation,* 9(8): 1735–1780.

Hoffman, T. (2001). Unsupervised learning by probabilistic latent semantic analysis. *Machine Learning*, 42(1–2): 177–196.

Holland, J.H. (1999). *Emergence From Chaos To Order*. Basic Books.

Huberman, B.A. and Tad H. (1987). Phase Transitions in Artificial Intelligence Systems. *Artificial Intelligence*, 33(2): 155–171.

Hunicke, R., Leblanc, M. and Zubek, R. (2004). MDA: A Formal Approach to Game Design and Game Research. In *Proceedings of the Challenges in Games AI Workshop, Nineteenth National Conference of Artificial Intelligence*, San Jose, CA: AAAI Press.

ImageNet (2014). http//www.image-net.org/, Stanford Vision Lab, Stanford University.

Jin, X.F. (2023). The Evolution Trend of AI Framework and MindSpore's Concept. https:www.mindspore.cn/, 2023.3.31.

Kraus S. (1997). Negotiation and Cooperation in Multi-Agent Environments. *Artificial Intelligence Journal*, Special Issue on Economic Principles of Multi-Agent System.

Krizhevsky, A., Sutskever, I. and Hinton, G.E. (2012). ImageNet Classification with Deep Convolutional Neural Networks. In: *Annual Conference on Neural Information Processing Systems Lake Tahoe*: 1106–1114.

LeCun, Y., Boser, B., Denker, J.S., *et al.* (1989). Handwritten digit recognition with a back-propagation network. In: *Advances in Neural Information Processing Systems*. Denver: Morgan Kaufmann, 396–404.

LeCun, Y., Bottou, L., Bengio, Y., *et al.* (1998). Gradient-based learning applied to document recognition [J]. Proc of the IEEE, 86(11): 2278–2324.

Lesser, V.R. (1991). A retrospective view of fa/c distributed problem solving. *IEEE Transactions on Systems, Man, and Cybernetics*, Special Issue on Distributed Artificia.

Lesser, V.R. and Erman, L.D. (1980). Distributed interpretation: a model and experiment. *IEEE Transaction on Computer*, C–29: 1144–1163.

Lesser, V.R. and Corkill, D.D (1983). The Distributed Vehicle Monitoring Testbed: A Tool for Investigating Distributed Problem Solving Networks. *AI Magazine*, 4(3): 15–33.

Levesque, H.J. (1990). All I know: A study in autoepistemic logic. *Artificial Intelligence*, 42: 263–309.

Li, T., Sahu, A.K., Talwalkar, A. and Smith, V. (2019). Federated Learning: Challenges, Methods, and Future Directions. arXiv: 1908.07873V1.

Li, Zhixin, Liu, Xi, Shi, Zhiping, Shi, Zhongzhi. (2009). Learning image semantics with latent aspect model. In *Proceedings of the 2009 IEEE International Conference on Multimedia and Expo (ICME'09)*. 366–369.

Li, Zhixin, Shi, Zhiping, Liu, Xi and Shi, Zhongzhi. (2010). Automatic image annotation with continuous PLSA. In *Proceedings of the 35th IEEE International Conference on Acoustics, Speech and Signal Processing (ICASSP'10)*, 806–809.

Lin, F. and Reiter, R. (1994). How to progress a database (and why) I: logic foundations. *4th International Conference on Principles of Knowledge Representation and Reasoning*.

Markram, H. (2006). The blue brain project. *Nat Rev Neurosci. 7,* 153–160.

McCarthy, J. (1980). Circumscription — A form of non-monotonic reasoning. *Artificial Intelligence*, 13(1–2): 27–39.

McCarthy, J. (2005). The future of AI-A manifesto. *AI Magazine*, 26(4): 39.

McCulloch, W. S. and Pitts. (1943). A logic calculus of the ideas immanent in nervous activity. *Bulletin of Mathematical Biophysics*, 5: 115–133.

McDermott, D. and Doyle, J. (1980). Non-monotonic logic I. *Artificial Intelligence*, 13(1–2): 41–72.

Minsky, M (1961). Steps toward artificial intelligence. *Proceedings of the IRE* 49: 8–30.

Minsky, M. (1985). *Society of Mind*. Simon and Schuster Inc.

Moore, R.C. (1985). Semantical considerations on nonmonotonic logic. *Artificial Intelligence*, 25(1): 75–94.

Moore, R.C. (1993). Autoepistemic logic revisited. *Artificial Intelligence*, 59(1–2): 27–30.

Nair R. and Milind, T. (2005). Hybrid BDI-POMDP Framework for Multiagent Teaming. *J. Artif. Intell. Res. (JAIR)*, 23: 367–420.

Nash, J. (1950). Equilibrium points in n-person games. *Proceedings of the National Academy of Sciences*, 36(1): 48–49.

Nash, J. (1951). Non-Cooperative Games. *The Annals of Mathematics*, 54(2): 286–295.

Newell, A. and Simon, H.A. (1976). Computer science as empirical inquiry: Symbols and search. *Communications of the Association for Computing Machinery*, 19(3): 113–126.

Newton, J. (2018). Evolutionary Game Theory: A Renaissance. *Games*, 9(2): 31.

Nigam, K., McCallum, A., Thrun, S. and Mitchell, T. (1998). Learning to classify text from labeled and unlabeled documents. *In Proceedings of the Fifteenth National.*

Nisan, N., Roughgarden, T., Tardos, E., Vazirani, V.V., (Eds.). (2007). *Algorithmic Game Theory.* Cambridge University Press.

Osawa, E.I. (1993). A Schema for Agent Collaboration in Open Multiagent Environments. *IJCAI-93*, 352–359.

Pan, S.J. and Yang, Q. (2010). A Survey on Transfer Learning. *IEEE Transaction on TKDE*, 22(10): 1345–1359.

Pan, S.J., Tsang, I.W., Kwok, J.T. and Yang, Q. (2011). Domain adaptation via transfer component analysis. *IEEE Transaction on Neural Networks*, 22(2): 199–210.

Pan, Y. (2016). Heading toward artificial intelligence 2.0. *Engineering*, 2: 409–413.

Pearl, J. (2000). *Causality: Models, Reasoning and Inference.* Cambridge University Press.

Pearl, J. and Mackenzie, D. (2018). The Book of Why: The New Science of Causal and Effect. Basic Books.

Pednault, E.P.D., *et al.* (1989). ADL: Exploring the middle ground between STRIPS and the situation calculus. *Proceedings of the First International Conference on Principles of Knowledge Representation and Reasoning*, 324–332.

Phong, L.T., Aono, Y., Hayashi, T., *et al.* (2018). Privacy-Preserving Deep Learning via Additively Homomorphic Encryption. *IEEE Transaction on Information Forensics and Security*, 13(5): 1333–1345.

Phong, L.T. and Phuong, T.T. (2019). *Privacy-Preserving Deep Learning via Weight Transmission.* arXiv.org/abs/1904.03272.

Poo, M.M., Du, J.L., Nancy, Y., *et al.* (2016). China brain project: Basic neuroscience, brain diseases and brain-inspired computing. *Neuron*, 92(3): 591–596.

Pratt, L.Y. and Thrun, S. (1997). *Machine Learning — Special Issue on Inductive Transfer.* Springer.

Qi, G.J., Aggarwal C. and Huang T. (2011). Towards semantic knowledge propagation from text corpus to web images. In *Proceedings of the 20th International Conference on World Wide Web.* New York: ACM.

Qi, J., Zhou, Q., Lei, L. and Zheng, K. (2021). *Federated Reinforcement Learning: Techniques, Applications, and Open Challenges.* arXiv: 2108.11887v2 [cs.LG].

Radford, A., Karthik Narasimhan, T.S. and Ilya S. (2018). Improving language understanding with unsupervised learning. OpenAI Technical Report.

Rao, A.S. and Georgeff, M.P. (1992). An Abstract Architecture for Rational Agents. In *Proc. of the 3rd International Conference on Principles of Knowledge Representation and Reasoning*, Morgan Kaufmann.

Redko, I., Morvant, E., Habrard, A., *et al.* (2020). A survey on domain adaptation theory: learning bounds and theoretical guarantees. arXiv:2004.11829 [cs.LG].

Reiter, R. (1978). On closed world data bases. In Gallaire, H. and Minker, J. (Eds.), *Logic and Data Bases*, Plenum Press.

Reiter, R. (1980). A logic for default reasoning. *Artificial Intelligence*, 13(1–2): 81–132.

Rosenblatt, F. (1958). The Perceptron: A probabilistic model for information storage and organization in the brain. *Phychological Review*, 65: 386–408.

Rumelhart, D.E. and McClelland, J.L. (Eds.), (1986). *Parallel Distributed Processing: Explorations in the Microstructure of Cognition*, vol. I, Cambridge, MA: MIT Press.

Russell, S.J. and Norvig, P. (1995, 2003). *Artificial Intelligence: A Modern Approach.* Prentice Hall.

Saha, S. and Ahmad, T. (2010). *Federated Transfer Learning: Concept and Applications.* arXiv: 2010.15561v3.

Schmidhuber, J. (2014). *Deep Learning in Neural Networks: An Overview.* http://www.idsia.ch/~juergen/deep-learning-overview.html.

Schmidhuber, J. (2015). Deep Learning in Neural Networks: An Overview. Neural Networks. 61: 85–117.

Shi, Z.Z. (1988). *Knowledge Engineering* (In Chinese). Tsinghua University Press.

Shi, Z.Z. (1990a). Hierarchical model of mind. Invited Speaker. *Chinese Joint Conference on Artificial Intelligence.*

Shi, Z.Z. (1990b). Logic — object based knowledge model. *Chinese Journal of Computer*, (10).

Shi, Z.Z. (1992). *Principles of Machine Learning.* International Academic Publishers.

Shi, Z.Z. (2000). *Intelligent Agent and Application* (In Chinese). Science Press.

Shi, Z.Z. (2001). *Knowledge Discovery.* Tsinghua University Press.

Shi, Z.Z. (2003). Perspectives on intelligence science. *Scientific Chinese*, 8: 47–49.

Shi, Z.Z. (2021). *Intelligence Science: Leading the Age of Intelligence.* Elsevier and Tsinghua University Press.

Shi, Z.Z., Dong, M.K., Zhang, H.J. and Sheng, Q. (2002). Agent-based Grid Computing. Keynote Speech, International Symposium on Distributed Computing and Applications to Business, Engineering and Science, Wuxi, December 16–20.

Shi, Z.Z., Dong, Mingkai, Jiang, Yuncheng, Zhang, Haijun. (2005). A Logic Foundation for the Semantic Web. Science in China, Series F Information Sciences, 48(2): 161–178.

Shi, Z. Z. and Huang, Z. Q. (2019). Cognitive model of brain-machine integration. *AGI-19 Proceedings*, Springer, pp. 168–177.

Shi, Z.Z., Zhang, H.J. and Dong, M.K. (2003). *MAGE: Multi-Agent Environment. ICCNMC-03.* IEEE CS Press, pp.181-188.

Silver, D. (2016). Google's DeepMind creating artificial agents to achieve human-level performance. https://futuristech.info/posts/, E Futuretech Tech Info, 06/22/2016

Smith, J.M. and Price, G.R. (1973). The Logic of Animal Conflict. *Nature.* 246: 15–18.

Smith, R.G. (1980). The contract-net protocol: high-level communication and control in a distributed problem solver. *IEEE Transaction on Computers*, C–29(12): 1104–1113.

Smolensky, P. (1986). Information processing in dynamical systems: Foundations of harmony theory. Rumelhart, D.E. and McClelland J.L. (Eds.), *Parallel distributed processing*, I: 194–281.

Snaider, J., McCall, R. and Franklin, S. (2012). Time production and representation in a conceptual and computational cognitive model. *Cognitive Systems Research*, 13(1): 59–71.

State Council. (2017). *New Generation of Artificial Intelligence Development Plan.* State Council Document. No. 35.

Stevo B. and Ante F. (1976). The influence of pattern similarity and transfer learning upon the training of a base perceptron B2. *Proceedings of Symposium Informatica* 3-121-5, Bled.

Sutskever, I. and Tielemant, T. (2010). On the convergence properties of contrastive divergence. *Journal of Machine Learning Research - Proceedings Track*, 9: 789–795.

Sutskever, I., Vinyals, O. and Le, Q.V. (2014). Sequence to sequence learning with neural networks. In *Advances in Neural Information Processing Systems*, pp. 3104–3112.

Sycara, K., Decker, K., Pannu, A., Williamson, M., Zeng. (1996). Distributed Artificial Agents. URL: http://www.cs.cmu.edu/~softagents/.

Tesauro, G.J. (1992). Practical issues in temporal difference learning. *Machine Learning*, 8: 257–277.

Tian, Q. and Zhongzhi S. (1997). Model-theoretical foundation of action and progression. *Science Press, Series E-Technological Sciences*, 40(4): 430–438.

Tieleman, T. (2008). Training restricted Boltzmann machines using approximations to the likelihood gradient. *Proceedings of the 25th international conference on Machine learning*, 1064–1071.

Tieleman, T. and Hinton, G. (2009). Using fast weights to improve persistent contrastive divergence. *Proceedings of the 26th Annual International Conference on Machine Learning.* New York, NY, USA.

Thorpe, T.L. (1997). Vehicle traffic light control using sarsa. https://citeseer.ist.psu.edu/thorpe97vehicle.html.

Valiant, L.G. (1984). A theory of the learnable. *Communications of the ACM*, 27(11): 1134–1142.

Vapnik, V. N. (1995). *The Nature of Statistical Learning Theory.* Springer-Verlag, New York.

Vapnik, V.N. (1998). *Statistical Learning Theory.* Wiley-Interscience Publication, John Wiley&Sons, Inc.

Vapnik, V.N. and Chervonenkis, A.J. (1991). A necessary and sufficient conditions for consistency in the empirical risk minimization method. *Pattern Recognition and Image Analysis*, 1(3): 284–305.

Vapnik, V., Golowich, S. and Smola A. (1997). Support vector method for function approximation, regression estimation, and signal processing. In *Advances in Neural Information Processing Systems 9.*

Vaswani, A., Noam S., Niki, P., *et al.* (2017). Attention is all you need. In *Advances in Neural Information Processing Systems*, P5998–6008.

von Neumann, J. (1928). Zur Theorie der Gesellschaftsspiele (On the Theory of Games of Strategy). Mathematische Annalen (*Mathematical Annals*) (in German), 100(1): 295–320.

von Neumann, J. and Oskar, M. (1944). *Theory of games and economic behavior.* Princeton University Press.

von Neumann, J. (1959). On the Theory of Games of Strategy. In Tucker, A.W. and Luce, R.D. (Eds.), *Contributions to the Theory of Games.* Vol. 4. Translated by Bargmann, Sonya. Princeton, New Jersey: Princeton University Press. pp. 13–42.

Wang, C.K. and Feng, J. (2024). Fedpv-Fs: A feature selection method for federated learning in insurance precision marketing. *Proceedings of IIP.*

Wang, P. (2000). *Non-Axiomatic Reasoning System.* AAAI-00 Proceedings.

Wang, P. (2007). *The Logic of Intelligence.* Springer.

Wang, P. (2022). *Outline of a General Theory of Intelligence.* Shanghai Science and Technology Education Press.

Wang, P. and Goertzel, B. (2008). Introduction: Aspects of Artificial General Intelligence. In Shi, P., Goertzel, B. and Stan F.S. (Eds.), *AGI-08 Proceedings of the 1st AGI conference*, IOS Press.

Wang, P. and Goertzel, B. (2012). *Theoretical Foundations of Artificial General Intelligence.* Springer Science & Business Media.

Washio, T., Motoda, H. and Niwa, Y. (1999). Discovering admissible model equations from observed data based on scale-types and identity constraints, *IJCAI-99.*

Watkins, C. and Peter, D. (1989). Q-learning. *Machine Learning*, 8, 279–292.

Werbos, P.J. (1974). Beyond regression: New tools for prediction and analysis in the behavioral sciences. Ph.D. Thesis, Harvard University, Cambridge, MA.

Wooldridge, M. (2002). *An Introduction to MultiAgent Systems*, John Wiley & Sons Ltd.

Wooldridge, M. and Jennings, N.R. (1995). Intelligent agents: Theory and practice. *Knowledge Engineering*, 10(2).

Wu, Yonghui, Schuster, M., Zhifeng Chen, *et al.* (2016). Google's Neural Machine Translation System: Bridging the Gap. arXiv:1609.08144v2.

Xu, Y., Scerri, P., Yu, B., Okamoto, S., Lewis, M. and Sycara, K. (2005). An integrated token-based algorithm for scalable coordination. In *Proceedings of the Fourth International Joint Conference on Autonomous Agents and Multiagent Systems (AAMAS)*, Utrecht, NL 407–414.

Yang, Q., Liu Y., Chen, T., *et al.* (2019). Federated machine learning: Concept and applications. *ACM Transaction on Intelligent System Technology*, 10(2): 1–19.

Yang, Q., Liu Y., Chen, Y., *et al.* (2020). *Federated Learning*. Morgan & Claypool Publishers.

Yang, Q., Zhang, Y., Dai, W. and Pan, S.J. (2020). *Transfer Learning*. Cambridge University Press.

Yang, Z., Liang, B. and Ji, W. (2021). An intelligent end-edge-cloud architecture for visual IoT assisted healthcare systems. *IEEE Internet of Things Journal*, 823: 16779–16786.

Yao, A. (1977). Probabilistic computations: Toward a unified measure of complexity. *Proceedings of the 18th IEEE Symposium on Foundations of Computer Science (FOCS)*, pp. 222–227.

Zhang, Q. (2012). Dynamic uncertain causality graph for knowledge representation and reasoning: Discrete DAG cases. *J. Comput. Sci. Technol*, 27(1): 1–23.

Zhang, Q. (2015). Dynamic uncertain causality graph for knowledge representation and reasoning: Continuous variable, uncertain evidence, and failure forecast. *IEEE Trans. Syst. Man. Cybern. Syst.*, 45(7): 990–1003.

Zhong, Y.X. (2023). *Unified Intelligence Theory*. China Science Press.

Index

C

categorized by domain data, 218
categorized by migration method, 218
causal
    diagram, 108–109
    discovery framework, 103
    inference, 107
    model, 104
    reasoning, 71
    structure, 103
characteristic function form, 122
ChatGPT, 4, 260, 326
China brain project, 260
Chomsky, Noam, 2
circumscription logic, 46
classification, 217, 332
client–server architecture, 238
climate change, 349
climate model, 350
closed world assumption, 7, 37, 70
closeness method, 217
cloud management module, 317
clustering, 333
CNN architecture, 168
cognitive modeling, 5
cognitive process, 274
collaborative crowd action, 291
common resources, 322
common sense reasoning, 77
commonsense knowledge, 25
computational complexity, 102
computational learning, 91
computational paradigm, 327
computer science, 119
concept description, 332
concept of agent, 281
connectionism, 9
connectionist temporal classification (CTC),
  179
considering that neural network can extract
  feature information well, 246
construct gene expression vectors, 340
contract net, 297
contract net protocol, 295
contrastive divergence (CD), 163, 165
convolutional layer, 170, 172–173
convolutional neural network (CNN), 167,
  169–170
cooperative plan, 300
correlation analysis, 334

correlation coefficient method, 215
cross-media analysis reasoning technology, 14
cross-media perceptual computing theory, 12
crowd
    computing, 277–280
    algorithms, 280
    model, 280
    operating system, 280
cyber system, 277
    intelligence, 280, 312
    intelligence computing, 279
    perception, 309
CUT, 33, 36
CUT component, 34
cyclic transfer, 239

D

data
    intelligence period, 3
    mining, 16
    stage, 330
    preparation, 329
    summary, 331
    data-driven paradigm, 327
    deductive reasoning, 6, 73
deep
    belief networks, 166
    learning, 3, 157–159, 185
    reinforcement learning, 3, 246
    transfer learning, 226
DeepMind, 243
default logic, 42, 77
definitions of logic programming, 28
deliberate-agent algorithm, 287
deliberative agent, 286
DENDRAL, 2
dependency-directed backtracking, 57
description logic, 62, 70
detection and identification of target genes,
  341
developing toolkit, 322
development trend, 345
deviation analysis, 334
directory facilitator (DF), 321
distributed artificial intelligence, 19, 23
distributed partial order constraint, 302
distribution density, 88
domain adaptation transfer learning, 231
dominant strategy, 120
drug development, 337